Featured Papers in Inorganic Materials 2025

Featured Papers in Inorganic Materials 2025

Guest Editors

Roberto Nisticò
Eleonora Aneggi
Hicham Idriss
Torben R. Jensen

MDPI

Basel • Beijing • Wuhan • Barcelona • Belgrade • Novi Sad • Cluj • Manchester

Guest Editors
Roberto Nisticò
Department of Materials Science
University of Milano-Bicocca
Milano
Italy

Eleonora Aneggi
Department of Agricultural, Food, Environmental and Animal Sciences (Di4A)
University of Udine
Udine
Italy

Hicham Idriss
Department of Chemistry
University College London
London
UK

Torben R. Jensen
Interdisciplinary Nanoscience Center and Department of Chemistry
Aarhus University
Aarhus
Denmark

Editorial Office
MDPI AG
Grosspeteranlage 5
4052 Basel, Switzerland

This is a reprint of the Special Issue, published open access by the journal *Inorganics* (ISSN 2304-6740), freely accessible at: https://www.mdpi.com/journal/inorganics/special_issues/0ZMH6814BQ.

For citation purposes, cite each article independently as indicated on the article page online and as indicated below:

Lastname, A.A.; Lastname, B.B. Article Title. *Journal Name* **Year**, *Volume Number*, Page Range.

ISBN 978-3-7258-7126-1 (Hbk)
ISBN 978-3-7258-7127-8 (PDF)
https://doi.org/10.3390/books978-3-7258-7127-8

Contents

About the Editors

Roberto Nisticò

Roberto Nisticò (R.N.) is Associate Professor in General and Inorganic Chemistry at the University of Milano-Bicocca (Department of Materials Science, Italy). His research activity mainly focuses on several aspects at the interface between the inorganic chemistry, and the materials science, always looking for novel and appealing technological solutions for a sustainable future. His research area is mainly focused on the development of metal oxides and other inorganic systems for the environmental remediation of contaminated (waste)water, catalysis, energy production, and the design of new generations of nanocomposites. A significant portion of his research is dedicated to the surface engineering of magnetic nanomaterials. R.N. has published approximately 100 papers in international peer-reviewed journals (h-index: 30, Scopus database).

Eleonora Aneggi

Eleonora Aneggi (E.A.) is Associate Professor in General and Inorganic Chemistry at the University of Udine (Department of Agricultural, Food, Environmental and Animal Sciences—DI4A, Italy). Her research activity focuses on the development of advanced catalytic and functional materials for environmental applications, with particular emphasis on sustainable solutions for the removal and abatement of atmospheric and water pollutants. Her work focuses on catalytic processes for environmental remediation, the design of materials with tailored properties, and green technologies aimed at reducing environmental impact. E.A. has authored/co-authored 80 scientific publications in international peer-reviewed journals (h-index: 26, Scopus database).

Hicham Idriss

Hicham Idriss (H.I.) received his BSc, MSc, PhD, and Habilitation from the U. of Strasbourg (France). After postdoctoral positions at the Departments of Chem. Eng., U. of Delaware and U. of Illinois (Urbana), USA, he started his academic career at the Department of Chem., U. of Auckland, New Zealand (1995 to 2008) where he was coordinator of the Bachelor of Technology (Materials). H.I. then moved to the U. Aberdeen (UK) to hold the Aberdeen Energy Futures Professor and Chair position. Wanting to focus his activity on hydrogen production from renewables he moved to corporate research at SABIC/KAUST (Saudi Arabia) in 2011 where he was appointed as Chief Scientist then Fellow. After retiring from SABIC in 2021, H.I. moved to the Institute of Functional Interfaces at KIT (Germany) until Sep. 2025. His main research area is in catalysis and surface reactions of oxides. H.I. is Professor (Hon.) at the Depart. Chem., UCL, UK, since 2013.

H.I. has about 350 scientific publications and 15 granted patents (h-index: 66 Scopus database).

Torben R. Jensen

Torben R. Jensen (T.R.J.) is professor of inorganic and materials chemistry at The Interdisciplinary Nanoscience Center (iNANO) and Department of Chemistry at Arhus University where he also received a doctor of science degree (DSc). The discovery and detailed characterization of new inorganic compounds have always been a driving force for my research. During the past decade we developed many synthesis strategies and prepared and characterized multitudes of new metal hydrides, derivatives and nano-composites with fascinating structures and properties. The initial focus was new materials for hydrogen storage, which then move towards solid state battery materials. The efforts also led to several versatile silver closo-borates, which displayed fast cationic conductivity, semiconductor band gap, and can form silver nano-filaments. We also discovered a novel mechanism for cationic conductivity owing to weak interactions, structural flexibility and the presence of neutral molecules in the solid state. Recently, we discovered new nano-interface effects, which significantly increase the cationic conductivity. T.R.J. has published around 390 scientific publications, which received ca 20000 citations (h-index: 71, Scopus database).

inorganics

MDPI

Editorial

Featured Papers in Inorganic Materials 2025

Eleonora Aneggi [1,*], Roberto Nisticò [2,*], Hicham Idriss [3] and Torben R. Jensen [4]

1 Department of Agricultural, Food, Environmental and Animal Sciences (Di4A), University of Udine, 33100 Udine, Italy
2 Department of Materials Science, University of Milano-Bicocca, U5, INSTM, Via R. Cozzi 55, 20125 Milano, Italy
3 Department of Chemistry, University College London, London WC1E 6BT, UK; h.idriss@ucl.ac.uk
4 Interdisciplinary Nanoscience Center and Department of Chemistry, Aarhus University, DK-8000 Aarhus C, Denmark; trj@chem.au.dk
* Correspondence: eleonora.aneggi@uniud.it (E.A.); roberto.nistico@unimib.it (R.N.)

1. Introduction and Overview of the Special Issue

Following the success of the previous edition [1], the Inorganic Materials section of Inorganics presents the Special Issue Featured Papers in Inorganic Materials 2025. This Special Issue aims to highlight recent advances and emerging trends in inorganic materials research, with particular emphasis on materials designed to address key societal challenges related to energy transition, environmental sustainability, and biomedical applications [2–7].

In recent years, inorganic materials science has undergone a significant evolution driven by the demand for efficient, sustainable, and multifunctional material systems. Advances in synthesis, characterization [8], and modeling now enable precise control over composition, structure [9,10], and defects across multiple length scales [11,12]. As a result, contemporary inorganic chemistry increasingly focuses on the rational design of materials systems, in which functionality arises from the synergistic interplay between nanoscale architecture, interfaces, and hierarchical organization rather than from composition alone.

This Special Issue collects 10 peer-reviewed contributions, including 7 original research articles and 3 review papers, which together provide a representative overview of the state of the art in inorganic materials research. The selected works demonstrate how controlled structural manipulation at the nanoscale can induce enhanced or entirely new macroscopic properties, enabling progress in both fundamental understanding and technological applications. For clarity, the contributions are grouped into three thematic subsections reflecting major research directions in the field: Energy Production and Storage, Nanostructures and Surfaces, and Environmental and Biomedical Applications. An overview of the published articles is provided in Table 1.

The first subsection, The Challenge of Clean Energy: Production and Storage, comprises four contributions dedicated to inorganic materials for sustainable energy technologies [13–16]. These studies address photocatalytic hydrogen production, electrocatalytic water splitting, solid-state hydrogen storage, and ion-conducting materials for advanced battery systems. Together, they illustrate how compositional tuning, defect engineering, and morphology control can substantially improve energy conversion and storage performance.

The second subsection, Nanostructures and Surfaces, includes three contributions focused on controlled synthesis strategies and surface engineering [17–19]. These works demonstrate that relatively simple and scalable approaches, such as solvothermal synthesis, capacitive coupling, and low-cost chemical deposition, can be effectively used to tailor particle morphology, surface charge distribution, and defect chemistry, with direct implications for electronic, sensing, and catalytic applications.

Table 1. Overview of the contributions included in the Special Issue Featured Papers in Inorganic Materials 2025 [13–22].

Subsection	Contribution n.	Authors and Title
The Challenge of Clean Energy: Production and Storage	1	Bitsos, D.R.; Salepis, A.; Orfanos, E.; Coutsolelos, A.G.; Kosheleva, R.I.; Mitropoulos, A.C.; Ladomenou, K. Exploring Metal- and Porphyrin-Modified TiO_2-Based Photocatalysts for Efficient and Sustainable Hydrogen Production. Inorganics 2025, 13, 121.
	2	Mane, S.M.; Wagh, K.S.; Lee, S.; Teli, A.M.; Chavan, G.T.; Shin, J.C.; Lee, J. Solvent-Driven Structural Modulation of Co-Ni_3S_2 and Impact on Electrochemical Water Splitting. Inorganics 2025, 13, 359.
	3	Grigorova, E.; Çakmak, G.; Yüce, H.; Markov, P. Electrochemical and Gas-Solid Hydrogen Storage Properties of a Multi-Metal Magnesium-Based Alloy Obtained by Ball Milling. Inorganics 2025, 13, 299.
	4	Milanova, M.; Yang, X.; Vargas, P.; Rosero-Navarro, N.C.; Harizanova, R.; Jivov, B.; Aleksandrov, L.; Iordanova, R.; Shopska, M.; Koleva, S. Structure and Electrochemical Performance of Glasses in the $Li_2O–B_2O_3–V_2O_5–MoO_3$ System. Inorganics 2025, 13, 285.
Nanostructures and Surfaces	5	Tuccio, C.; Armetta, F.; Chillura Martino, D.F.; Skaudžius, R.; Saladino, M.L. Controlled Synthesis of Alkali Metal Hydroxide Particles via Solvothermal Processing. Inorganics 2025, 13, 373.
	6	Chini, E.; Gentili, D.; Liscio, A.; Cavallini, M. SiO_2 Electret Formation via Stamp-Assisted Capacitive Coupling: A Chemophysical Surface Functionalisation. Inorganics 2026, 14, 21.
	7	Ruvalcaba-Manzo, S.G.; Ramírez-Bon, R.; Ochoa-Landín, R.; Castillo, S.J. A Comprehensive Study of the Optical, Structural, and Morphological Properties of Chemically Deposited ZnO Thin Films. Inorganics 2025, 13, 331.
Environmental and Biomedical Applications	8	Li, M.; Yang, L.; Song, Y.; Hou, H.; Fang, Y.; Liu, Y.; Xie, L.; Lu, D. Research Progress on the Preparation, Modification, and Applications of g-C_3N_4 in Photocatalysis and Piezoelectric Photocatalysis. Inorganics 2025, 13, 300.
	9	Sezanova, K.; Tuparova, Y.; Shestakova, P.; Markov, P.; Kovacheva, D.; Rabadjieva, D. Calcium Phosphate Ceramic Powders Prepared from Mechanochemically Activated Precursors. Inorganics 2025, 13, 313.
	10	Jiang, Z.; Xiang, H.; Tang, X. Smart Inorganic Nanomaterials for Tumor Microenvironment Modulation. Inorganics 2025, 13, 337.

The third subsection, Environmental and Biomedical Applications, highlights the expanding role of inorganic materials in addressing environmental and health-related challenges [20–22]. The included contributions cover piezo-photocatalytic systems for environmental remediation and energy conversion, nanostructured bioceramics for bone regeneration, and inorganic nanomaterials as active modulators of the tumor microenvironment. Collectively, these studies exemplify the transition of inorganic materials from

https://doi.org/10.3390/inorganics14020059

passive components to active, multifunctional systems capable of interacting dynamically with complex biological and environmental environments.

Overall, this Special Issue provides a coherent snapshot of current advances in inorganic materials research, emphasizing the central role of nanoscale design, interfacial control, and functional integration. The collected contributions not only reflect the diversity and vitality of the field but also point toward future directions in which inorganic chemistry will continue to play a key role in the development of sustainable and high-performance technologies.

2. The Challenge of Clean Energy: Production and Storage

The transition toward sustainable energy systems requires inorganic materials capable of improving both energy production and storage efficiency. In this context, four contributions in this Special Issue address key challenges associated with photocatalysis, electrocatalysis, hydrogen storage, and solid-state batteries.

Bitsos et al. (Contribution 1) [13] present a comprehensive review on the integration of porphyrins onto TiO_2 matrices for photocatalytic hydrogen evolution. This hybrid approach mimics natural photosynthetic systems by extending light absorption into the visible region, thereby overcoming intrinsic limitations of TiO_2. Recent studies confirm that organic photosensitizers [23] represent an effective strategy to exploit a broader portion of the solar spectrum.

Mane et al. (Contribution 2) [14] demonstrate the importance of precursor chemistry in electrocatalytic materials. By varying the solvent during the synthesis of Co-doped Ni_3S_2, the authors achieve controlled petal-like morphologies that significantly reduce overpotentials in water-splitting reactions. These results are consistent with current understanding of electrode–electrolyte interfacial dynamics, where surface morphology plays a decisive role in electrochemical performance [24].

Grigorova et al. (Contribution 3) [15] investigate multi-metal Mg–Ni–Al–V–Fe alloys synthesized via high-energy ball milling for hydrogen storage applications. The materials exhibit hydrogen absorption capacities of up to 1.6 wt.% at room temperature, highlighting the effectiveness of ball milling in introducing structural defects that lower kinetic barriers for hydrogenation [25].

Milanova et al. (Contribution 4) [16] focus on glassy materials in the Li_2O–B_2O_3–V_2O_5–MoO_3 system for solid-state battery applications. Partial substitution of boron with molybdenum increases the fraction of non-bridging oxygen, enhancing ionic conductivity. The presence of mixed-valence vanadium and molybdenum oxides facilitates both electronic and ionic transport, making these glasses promising candidates for electrodes or solid electrolytes in next-generation energy-storage devices, in agreement with recent works [26,27] on mixed-conduction glasses.

3. Development of Nanostructures and Surfaces

Surface engineering and controlled synthesis strategies play a central role in the development of advanced inorganic materials [28–30]. The contributions in this subsection demonstrate how morphology and surface properties can be tailored without relying on complex additives or high cost processes

Tuccio et al. (Contribution 5) [17] report a one-pot solvothermal method for the synthesis of alkaline and alkaline-earth metal hydroxides (LiOH, NaOH, KOH, $Ba(OH)_2$, $Sr(OH)_2$, and $Mg(OH)_2$). By controlling nucleation in non-aqueous solvent systems, the authors achieve precise tuning of particle size and morphology, yielding acicular, fibrous, or lamellar nanostructures. The proposed approach is scalable, environmentally compatible, and

https://doi.org/10.3390/inorganics14020059

suitable for applications ranging from cultural heritage conservation to tissue engineering and advanced construction materials.

Chini et al. (Contribution 6) [18] combine concepts from classical carbonyl cluster chemistry with modern nanotechnology. Metal clusters are investigated as model systems bridging molecular and solid-state chemistry, providing insight into electron-transfer processes relevant to heterogeneous and single-atom catalysis [31]. In addition, the authors introduce the Stamp-Assisted Capacitive Coupling (SACC) technique for fabricating SiO_2 electrets with sub-micrometric charge patterns, achieving surface charge densities up to 300 nC cm^{-2}. This method offers new opportunities for scalable sensor and energy-harvesting devices.

Ruvalcaba-Manzo et al. (Contribution 7) [19] explore low-cost chemical bath deposition as a route to fabricate ZnO thin films. The study highlights the critical role of intrinsic defects, particularly oxygen vacancies, in determining structural and optoelectronic properties [32]. Thermal annealing at 600 °C induces a transformation toward flower-like morphologies, improving crystallinity and phase purity and enhancing the performance of ZnO-based transparent and flexible electronic devices.

4. Environmental and Biomedical Applications

Inorganic materials are increasingly central to environmental protection and biomedical technologies. The contributions in this subsection illustrate how inorganic systems can actively interact with complex chemical and biological environments [7,33,34].

Li et al. (Contribution 8) [20] provide a critical review of graphitic carbon nitride (g-C_3N_4) for piezo-photocatalytic applications. By coupling piezoelectric effects with photocatalysis, mechanical energy from environmental vibrations can generate internal electric fields that enhance charge separation [35]. This synergistic mechanism leads to significant improvements in catalytic efficiency for pollutant degradation, as well as for sustainable hydrogen and hydrogen peroxide (H_2O_2) production.

Sezanova et al. (Contribution 9) [21] report advances in the synthesis of nanostructured tetracalcium phosphate (TTCP) ceramics for bone regeneration. Through mechanochemical activation of precursors, the authors reduce phase-transition temperatures and obtain highly reactive nanostructured materials. These ceramics exhibit enhanced bioactivity, promoting hydroxyapatite formation, cellular adhesion, and accelerated bone mineralization, in line with current trends in regenerative medicine [36].

Jiang et al. (Contribution 10) [22] review the emerging role of inorganic nanomaterials [6] as active modulators of the tumor microenvironment (TME). Metal oxides such as MnO_2, Fe_3O_4, and CeO_2 are shown to regulate local oxygen levels and pH, supporting catalytic nanomedicine approaches based on chemodynamic therapy. By generating oxygen in situ and alleviating tumor hypoxia, these nanomaterials enhance the effectiveness of radiotherapy and immunotherapy, paving the way for multifunctional and personalized cancer treatments [37,38].

5. Conclusions

This Special Issue highlights the vitality and diversity of contemporary inorganic materials research. The collected contributions demonstrate that progress in the field increasingly relies on the integration of controlled synthesis, advanced characterization, often performed in situ or under operating conditions, and application-driven design.

A recurring theme across the contributions is the central role of precision in materials engineering. Control over defects, interfaces, and hierarchical organization emerges as a key driver for transforming fundamental insights into functional technologies. The convergence

https://doi.org/10.3390/inorganics14020059

of structural control, targeted functionality, and application-oriented performance positions inorganic chemistry as a cornerstone for next-generation materials.

The editors hope that this Special Issue will not only contribute to the advancement of academic knowledge but also stimulate future research toward innovative, sustainable, and translational solutions. The combined focus on synthesis, characterization, and application remains essential for the continued progress of inorganic materials science.

Acknowledgments: The Editors sincerely thank all the Authors for their valuable contributions and for choosing to share their significant and original research results in this Special Issue, the Reviewers for their time and efforts in improving the quality of the articles, and the Editorial Staff of Inorganics for their support in the preparation of this volume.

Conflicts of Interest: The authors declare no conflicts of interest.

References

1. Nisticò, R.; Jensen, T.R. Featured Papers in Inorganic Materials 2024. *Inorganics* **2025**, *13*, 110. [CrossRef]
2. Kwasi-Effaha, C.C.; Okpako, O. Comprehensive review of emerging trends in thermal energy storage mechanisms, materials and applications. *Front. Energy Res.* **2025**, *13*, 1651471. [CrossRef]
3. Liang, Z.; Zhang, Y.; Liu, L.; Jiao, M.; Ye, C. Single-Atom Catalysts for Hydrogen Evolution Reaction: The Role of Supports, Coordination Environments, and Synergistic Effects. *Nanomaterials* **2025**, *15*, 1175. [CrossRef] [PubMed]
4. Poudel, T.P.; Oyekunle, I.P.; Deck, M.J.; Chen, Y.; Hou, D.; Ojha, P.K.; Ogbolu, B.O.; Huang, C.; Xiong, H.; Hu, Y.-Y. $Li_{1.6}AlCl_{3.4}S_{0.6}$: A low-cost and high-performance solid electrolyte for solid-state batteries. *Chem. Sci.* **2025**, *16*, 2391–2401. [CrossRef] [PubMed]
5. Zhang, S.; Gao, F.; Qiu, X.; Fang, M.; Tan, X. Research progress of piezocatalysis in energy and environment fields. *Chem. Commun.* **2026**. [CrossRef]
6. Lee, S.; Lee, J.; Park, M.; Kim, J.; Byun, H.; Kim, J.; Kim, W.J. Inorganic nanoparticle-based cancer immunotheranostics. *Coord. Chem. Rev.* **2025**, *542*, 216848. [CrossRef]
7. Zhan, S.; Fu, Y.; Yang, H.Y.; Lee, D.S. Recent developments and prospects of inorganic nanozymes for biomedical applications. *Biomater. Sci.* **2025**, *13*, 6755–6774. [CrossRef]
8. Bonn, M.; Domke, K.; Sulpizi, M.; McCrory, C.; Peng, Z. In situ and operando characterization: Advancing our understanding of functional materials. *J. Chem. Phys.* **2025**, *162*, 140402. [CrossRef] [PubMed]
9. Han, G.; Vasylenko, A.; Daniels, L.M.; Collins, C.M.; Corti, L.; Chen, R.; Niu, H.; Manning, T.D.; Antypov, D.; Dyer, M.S.; et al. Superionic lithium transport via multiple coordination environments defined by two-anion packing. *Science* **2024**, *383*, 739–745. [CrossRef]
10. Wang, Y.; Yu, H.; Wang, H.; Chen, T. Hierarchically Porous Titanosilicate Hollow Spheres Containing TS-1 Zeolite Precursors for Oxidative Desulfurization. *Inorganics* **2025**, *13*, 37. [CrossRef]
11. Mishra, R.K.; Verma, K.; Singh, D.S. Defect engineering in nanomaterials: Impact, challenges, and applications. *Smart Mater. Manuf.* **2024**, *2*, 100052. [CrossRef]
12. Chen, W.; Bao, M.; Meng, F.; Ma, B.; Feng, L.; Zhang, X.; Qiu, Z.; Gao, S.; Zhong, R.; Xi, S.; et al. Designer topological-single-atom catalysts with site-specific selectivity. *Nat. Commun.* **2025**, *16*, 574. [CrossRef]
13. Bitsos, D.R.; Salepis, A.; Orfanos, E.; Coutsolelos, A.G.; Kosheleva, R.I.; Mitropoulos, A.C.; Ladomenou, K. Exploring Metal- and Porphyrin-Modified TiO_2-Based Photocatalysts for Efficient and Sustainable Hydrogen Production. *Inorganics* **2025**, *13*, 121. [CrossRef]
14. Mane, S.M.; Wagh, K.S.; Lee, S.; Teli, A.M.; Chavan, G.T.; Shin, J.C.; Lee, J. Solvent-Driven Structural Modulation of Co-Ni_3S_2 and Impact on Electrochemical Water Splitting. *Inorganics* **2025**, *13*, 359. [CrossRef]
15. Grigorova, E.; Çakmak, G.; Yüce, H.; Markov, P. Electrochemical and Gas-Solid Hydrogen Storage Properties of a Multi-Metal Magnesium-Based Alloy Obtained by Ball Milling. *Inorganics* **2025**, *13*, 299. [CrossRef]
16. Milanova, M.; Yang, X.; Vargas, P.; Rosero-Navarro, N.C.; Harizanova, R.; Jivov, B.; Aleksandrov, L.; Iordanova, R.; Shopska, M.; Koleva, S. Structure and Electrochemical Performance of Glasses in the Li_2O-B_2O_3-V_2O_5-MoO_3 System. *Inorganics* **2025**, *13*, 285. [CrossRef]
17. Tuccio, C.; Armetta, F.; Martino, D.F.C.; Skaudžius, R.; Saladino, M.L. Controlled Synthesis of Alkali Metal Hydroxide Particles via Solvothermal Processing. *Inorganics* **2025**, *13*, 373. [CrossRef]
18. Chini, E.; Gentili, D.; Liscio, A.; Cavallini, M. SiO_2 Electret Formation via Stamp-Assisted Capacitive Coupling: A Chemophysical Surface Functionalisation. *Inorganics* **2026**, *14*, 21. [CrossRef]
19. Ruvalcaba-Manzo, S.G.; Ramírez-Bon, R.; Ochoa-Landín, R.; Castillo, S.J. A Comprehensive Study of the Optical, Structural, and Morphological Properties of Chemically Deposited ZnO Thin Films. *Inorganics* **2025**, *13*, 331. [CrossRef]

https://doi.org/10.3390/inorganics14020059

20. Li, M.; Yang, L.; Song, Y.; Hou, H.; Fang, Y.; Liu, Y.; Xie, L.; Lu, D. Research Progress on the Preparation, Modification, and Applications of g-C_3N_4 in Photocatalysis and Piezoelectric Photocatalysis. *Inorganics* **2025**, *13*, 300. [CrossRef]
21. Sezanova, K.; Tuparova, Y.; Shestakova, P.; Markov, P.; Kovacheva, D.; Rabadjieva, D. Calcium Phosphate Ceramic Powders Prepared from Mechanochemically Activated Precursors. *Inorganics* **2025**, *13*, 313. [CrossRef]
22. Jiang, Z.; Xiang, H.; Tang, X. Smart Inorganic Nanomaterials for Tumor Microenvironment Modulation. *Inorganics* **2025**, *13*, 337. [CrossRef]
23. Gonuguntla, S.; Kamesh, R.; Pal, U.; Chatterjee, D. Dye sensitization of TiO_2 relevant to photocatalytic hydrogen generation: Current research trends and prospects. *J. Photochem. Photobiol. C Photochem. Rev.* **2023**, *57*, 100621. [CrossRef]
24. Prior, T.; Figueira, J.; Freitas, Â.; Carvalho, D.; Gomes, B.M.; Baptista, M.C.; Lebre, H.; Martins, R.; Pereira, L.; Pinto, J.V.; et al. Surface Morphology and Electrochemical Behavior of Microstructured Cu Electrodes in All-Solid-State Sodium Batteries. *Molecules* **2025**, *30*, 3493. [CrossRef]
25. Hosseinigourajoubi, S.; Schade, C.; Huot, J. Effect of Mechanical Processing on First Hydrogenation of Gas-Atomized $Ti_{0.488}Fe_{0.46}Mn_{0.052}$ Alloy. *Hydrogen* **2025**, *6*, 114. [CrossRef]
26. Tariq, F.; Abdullah, A.; Kulkarni, M.A.; Thaalbi, H.; Ha, J.S.; Lee, J.K.; Ryu, S.-W. Highly efficient visible-light-driven CdS-loaded ZnO-GaN nanowire photoanode fabricated on Si for H_2 evolution. *J. Alloys Compd.* **2024**, *973*, 172901. [CrossRef]
27. Yadav, A.K. Superionic Glasses: Potential and Challenges in Solid-State Batteries and Fuel Cells. *ChemistrySelect* **2025**, *10*, e03258. [CrossRef]
28. Salehi, M.M.; Mohammadi, M.; Akbarzadeh, A.R.; Babamoradi, M.; Maleki, A.; Zare, E.N. Advancements in preceramic inorganic polymers for environmental applications: Properties, synthesis, and potential uses. *RSC Adv.* **2025**, *15*, 25514–25541. [CrossRef]
29. Singh, L.; Balakrishnan, S.; William, J.J.; Parveen, S.; Periasamy, P.A.; Kumar, R.; Srivastava, A.; Kumar, S.; Shringi, A.K.; Kumari, D.; et al. Mechanistic Insights in Surface Engineering of Micro-/Nanocomposite Phase Change Materials for Thermal Energy Storage: A Review. *Adv. Energy Sustain. Res.* **2025**, *6*, 2500067. [CrossRef]
30. Xu, W.; Bai, Y.; Yin, Y. Surface Engineering of Nanostructured Energy Materials. *Adv. Mater.* **2018**, *30*, 1802091. [CrossRef] [PubMed]
31. Fang, J.; Chen, Q.; Li, Z.; Mao, J.; Li, Y. The synthesis of single-atom catalysts for heterogeneous catalysis. *Chem. Commun.* **2023**, *59*, 2854–2868. [CrossRef] [PubMed]
32. Maarisetty, D.; Baral, S.S. Defect engineering in photocatalysis: Formation, chemistry, optoelectronics, and interface studies. *J. Mater. Chem. A* **2020**, *8*, 18560–18604. [CrossRef]
33. Gheisizadeh, A.; Darghiasi, S.F.; Farazin, A. Sustainable chemistry approaches for biomedical applications of energy materials: A converging frontier. *Sustain. Chem. Energy Mater.* **2025**, *2*, 100021. [CrossRef]
34. Zhang, H.; Zhao, Z.; Wu, C. Bioactive Inorganic Materials for Innervated Multi-Tissue Regeneration. *Adv. Sci.* **2025**, *12*, 2415344. [CrossRef]
35. Jiang, Z.; Tan, X.; Huang, Y. Piezoelectric effect enhanced photocatalysis in environmental remediation: State-of-the-art techniques and future scenarios. *Sci. Total Environ.* **2022**, *806*, 150924. [CrossRef]
36. Vallet-Regí, M.; Ruiz-Hernández, E. Bioceramics: From Bone Regeneration to Cancer Nanomedicine. *Adv. Mater.* **2011**, *23*, 5177–5218. [CrossRef]
37. Ghassemifar, V.; Zahedi, A.; Meigoli, M.S.S.; Mokhtari, T.; Zadeh, S.S.M.; Parsafar, M.; Zenjanab, M.K.; Esfahlan, R.J. Harnessing nanotechnology to modulate hypoxic tumor microenvironments: Enhanced strategies for oncological innovations. *Biomed. Pharmacother.* **2025**, *191*, 118543. [CrossRef]
38. Akpe, V.; Cock, I.E. Advances in Cancer Treatment Through Nanotheranostics and Emerging Therapies. *J. Nanotheranostics* **2025**, *6*, 29. [CrossRef]

https://doi.org/10.3390/inorganics14020059

inorganics

MDPI

Review

Exploring Metal- and Porphyrin-Modified TiO_2-Based Photocatalysts for Efficient and Sustainable Hydrogen Production

Dimitrios Rafail Bitsos [1], Apostolos Salepis [1], Emmanouil Orfanos [1], Athanassios G. Coutsolelos [2,3], Ramonna I. Kosheleva [1], Athanassios C. Mitropoulos [1] and Kalliopi Ladomenou [1,*]

[1] Hephaestus Laboratory, School of Chemistry, Faculty of Science, Democritus University of Thrace, Kavala University Campus, St. Lucas, 65404 Kavala, Greece; dibitso@chem.duth.gr (D.R.B.); amitrop@chem.duth.gr (A.C.M.)

[2] Laboratory of Bioinorganic Chemistry, Chemistry Department, University of Crete, 70013 Heraklion, Greece; acoutsol@uoc.gr

[3] Foundation for Research and Technology (FORTH), Institute of Electronic Structure and Laser (IESL), 70013 Heraklion, Greece

* Correspondence: kladomenou@chem.duth.gr

Abstract: Photocatalytic H_2 production is one of the most promising approaches for sustainable energy. The literature presents a plethora of carefully designed systems aimed at harnessing solar energy and converting it into chemical energy. However, the main drawback of the reported photocatalysts is their stability. Thus, the development of a cost-effective and stable photocatalyst, suitable for real-world applications remains a challenge. An ideal photocatalyst for H_2 production must possess appropriate band-edge energy positions, an effective sacrificial agent, and a suitable cocatalyst. Among the various photocatalysts studied, TiO_2 stands out due to its stability, abundance, and non-toxicity. However, its efficiency in the visible spectrum is limited by its wide bandgap. Metal doping is an effective strategy to enhance electron–hole separation and improve light absorption efficiency, thereby boosting H_2 synthesis. Common metal cocatalysts used as TiO_2 dopants include platinum (Pt), gold (Au), copper (Cu), nickel (Ni), cobalt (Co), ruthenium (Ru), iron (Fe), and silver (Ag), as well as bimetallic combinations such as Ni-Fe, Ni-Cu, Nb-Ta, and Ni-Pt. In all cases, doped TiO_2 exhibits higher H_2 production performance compared to undoped TiO_2, as metals provide additional reaction sites and enhance charge separation. The use of bimetallic dopants further optimizes the hydrogen evolution reaction. Additionally, porphyrins, with their strong visible light absorption and efficient electron transfer properties, have demonstrated potential in TiO_2 photocatalysis. Their incorporation expands the photocatalyst's light absorption range into the visible spectrum, enhancing H_2 production efficiency. This review paper explores the principles and advancements in metal- and porphyrin-doped TiO_2 photocatalysts, highlighting their potential for sustainable hydrogen production.

Keywords: hydrogen production; water splitting; photocatalysis; TiO_2; metal doping; porphyrins

1. Introduction

Since the beginning of the industrial revolution, there has been a significant challenge in transitioning from coal to liquid fuels [1]. This issue stems from the fact that energy production and consumption do not occur in the same location, leading to geopolitical confrontations and energy losses due to fuel transportation. The first oil crises of the

https://doi.org/10.3390/inorganics13040121

1970s resulted in high production costs that industrialized countries such as Germany and the United States could not sustain. Consequently, these nations faced embargoes from oil-producing countries due to political and economic disputes. Unlike previous energy crises driven solely by traditional supply and demand forces, the current crisis, which began unofficially in 2020, is unique and complex, influenced by a set of factors specific to our time [2]. One major factor in the recent crisis is the difficult transition to green energy, exacerbated by poorly planned decarbonization efforts. The shift from fossil fuels to renewable energy sources has been hindered by a lack of investment in infrastructure, inadequate statistical forecasting, and inefficient organization of renewable energy resources. As a result, the stable balance between energy production and consumption has been disrupted. Many countries struggle to maintain energy self-sufficiency, while electricity imports contribute to blackouts and rising fossil fuel prices. For instance, the price of natural gas increased by 400% in the first quarter of 2022, and the price of a barrel of oil reached EUR 93—the highest in the past three years [3]. The energy crisis has triggered high inflation, economic slowdowns, and long-term instability, potentially leading to prolonged economic crises and widespread poverty. To address this challenge, energy production technologies must leverage existing distribution systems, be accessible to all countries, and remain environmentally sustainable. One fuel that meets all these criteria is hydrogen [4].

Hydrogen is the lightest and most abundant element in the universe, but it is rarely found on Earth in its pure form. Instead, it is typically found bonded to other elements, such as in water or hydrocarbons. Despite its limited availability in nature, hydrogen is a crucial fuel source with the potential to revolutionize the energy industry. One of hydrogen's key advantages is its clean combustion, producing only water vapor with no greenhouse gas emissions or air pollution. Additionally, hydrogen fuel cells can convert the chemical energy of hydrogen into electricity, with water as the only byproduct [5]. However, hydrogen fuel cells remain relatively expensive, and challenges persist in its production, storage, and transportation [6]. Despite these obstacles, many companies and governments are investing in hydrogen technology as a means to reduce greenhouse gas emissions and transition to cleaner, renewable energy sources [7]. Exploring and developing new hydrogen production methods could pave the way for a cleaner and more sustainable energy future for our planet [8,9].

The advantages of TiO_2, as a photocatalyst, stem from its high chemical stability, abundance, and ability to generate electron–hole pairs under light irradiation. These properties enable TiO_2 to facilitate redox reactions on its surface, making it widely studied for hydrogen production via photocatalysis. However, TiO_2 has a wide bandgap (~3.2 eV for anatase), meaning it primarily absorbs UV light, which accounts for only 5% of the solar spectrum. To improve its efficiency, doping TiO_2 with metals like Pt, Au, Ni, Cu or incorporating metal-based porphyrins, extends its absorption range into the visible spectrum, improves charge separation, and suppresses the recombination of photo-generated electron–hole pairs [10,11]. Porphyrins, organic compounds with a conjugated ring structure, are being explored as potential dopants for TiO_2 due to their ability to absorb light in the visible solar spectrum. Their unique optical and electronic properties, particularly in the Soret and Q bands, make them ideal for enhancing TiO_2's photocatalytic activity beyond the UV range. By transferring energy to TiO_2 semiconductors, porphyrins facilitate efficient solar-driven hydrogen production [12,13]. This review highlights the development of metal- and porphyrin-doped TiO_2 catalysts, which provide additional reaction sites, enhance light absorption and improve charge carrier separation, ultimately boosting photocatalytic hydrogen production.

2. Semiconductor Photocatalysis

Photocatalysis utilizes light energy to accelerate chemical reactions between a catalyst and a substrate. In the context of hydrogen production, a photocatalyst enables the conversion of water into hydrogen and oxygen by generating electron–hole pairs under solar irradiation [14]. The main appeal of this process lies in its potential to produce hydrogen fuel in a clean, sustainable, and cost-effective manner, without releasing harmful emissions [15]. Consequently, the development of efficient and stable photocatalytic materials is a key area of research in renewable energy. Semiconductors play a fundamental role in in this process, acting as the primary photocatalysts. Semiconductors play a pivotal role in this process, acting as the primary photocatalysts. When exposed to light, these materials absorb photon energy, which excites electrons from the valence band to the conduction band, leaving behind positively charged holes in the valence band (Scheme 1).

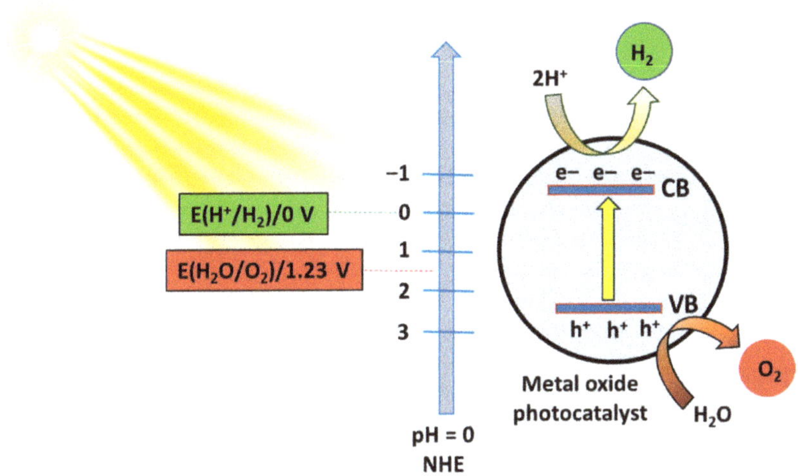

Scheme 1. General scheme of photocatalytic hydrogen production using a metal oxide photocatalyst and a sacrificial electron donor (SED).

The absorption of light is a prerequisite for photocatalytic reactions to proceed [16,17]. For a photon to induce an electronic transition, its energy must match or exceed the energy difference between the ground state and the excited state—a requirement de-fined by the material's bandgap [18]. Upon absorption, an electron is promoted from the ground singlet state (S_0) to an excited singlet state (S_1), where electron spins remain paired. In some cases, intersystem crossing can occur, allowing the molecule to transition into a triplet state (T_1), where the spins are unpaired. Although the triplet state is lower in energy and longer-lived, transitions to it are spin-forbidden and occur less readily. According to selection rules, direct excitation from S_0 to T_1 is not allowed, while excitation from S_0 to S_1 is spin-allowed [19]. The absorption of a photon radiation occurs very quickly. The molecule's energy can be lost by emitting a photon of the same energy (resonance fluorescence) or it can lose its vibrational energy through collisions with neighboring molecules and be degraded to its lowest vibrational level (vibrational relaxation). In semiconductor physics, when a semiconductor is irradiated with photons of equal to or greater than the bandgap energy ($hv \geq EBG$), it is excited, generating an electron–hole pair ($e-/h^+$) [20]. These charge carriers can then migrate to the surface of the semiconductor, where they participate in redox reactions. An internal electric field at the semiconductor–electrolyte interface helps separate the charges, overcoming their mutual Coulombic attraction [21].

Once at the surface, electrons and holes may become trapped near or on the surface of the semiconductor, which can prolong their lifetimes and enhance their reactivity. Once at the surface, electrons and holes may become trapped near or on the surface of the semiconductor, which can prolong their lifetimes and enhance their reactivity [22]. These traps are characterized as follows:

$$e-_{CB} \rightarrow e-_{tr}$$

$$h^+_{VB} \rightarrow h^+_{tr}$$

These charge carriers drive the water-splitting reactions, which can occur under both acidic and alkaline conditions. In alkaline media, the reduction half-reaction is as follows:

$$2H_2O + 2e- \rightarrow H_2 + 2OH-$$

In acidic media, often catalyzed by noble metals like platinum, the oxidation half-reaction proceeds as follows:

$$2H_2O \rightarrow O_2 + 4H^+ + 4e-$$

Together, these reactions facilitate the generation of hydrogen and oxygen gases, providing a pathway to produce hydrogen fuel for energy and industrial uses [12,23,24]. To improve photocatalytic efficiency, semiconductors can be doped with specific elements and engineered to increase their surface area. Such modifications enhance light absorption, charge separation, and surface reactivity. Overall, semiconductors are indispensable in the development of advanced photocatalytic systems for sustainable hydrogen production.

3. Titanium Dioxide as Semiconductor in Photocatalysis

Various semiconductors have been explored for photocatalytic applications. While many demonstrate promising characteristics, their long-term viability is often hindered by high charge recombination rates, poor stability in aqueous environments, or toxicity [25–29]. Among the most studied are n-type semiconductors with wide bandgaps, such as CdS, WO_3, and ZnO (Figure 1) [30,31]. These materials possess favorable conduction and valence band positions for photocatalysis; however, their large energy bandgaps (EBG) necessitate ultraviolet (UV) irradiation for activation. This is a significant limitation, as the majority of solar radiation falls within the visible spectrum (~480 nm), with only a small fraction in the UV range. On the other hand, semiconductors like CdS, CdSe, ZnS, and PbS exhibit narrower bandgaps (~2 eV), enabling excitation by visible light [16,28,32,33]. Despite this advantage, their inherent toxicity and instability in aqueous media severely restrict their practical application in photocatalytic systems.

Titanium dioxide (TiO_2) stands out as one of the most widely utilized photocatalysts due to its non-toxic nature, chemical stability, low cost, and ease of synthesis in various nanostructured forms such as nanoparticles, nanotubes, and nanowires. These morphologies play a significant role in modulating its photocatalytic performance TiO_2 primarily exists in two crystalline phases: anatase and rutile [34,35]. Anatase, with a bandgap of approximately 3.2 eV, features a higher surface area, superior electron mobility, and lower recombination rates, making it more efficient for photocatalysis under UV light. Rutile, although less reactive, exhibits greater thermal stability. Combining anatase and rutile phases, or doping them with metal or non-metal elements, has been shown to enhance charge separation and extend light absorption, thereby improving overall photocatalytic efficiency [36,37]. TiO_2 has been extensively studied for hydrogen generation via water splitting under both UV and visible light conditions [33,38–40]. However, a major limitation is the high recombination rate of photogenerated charge carriers, which reduces the efficiency of the overall process [41,42].

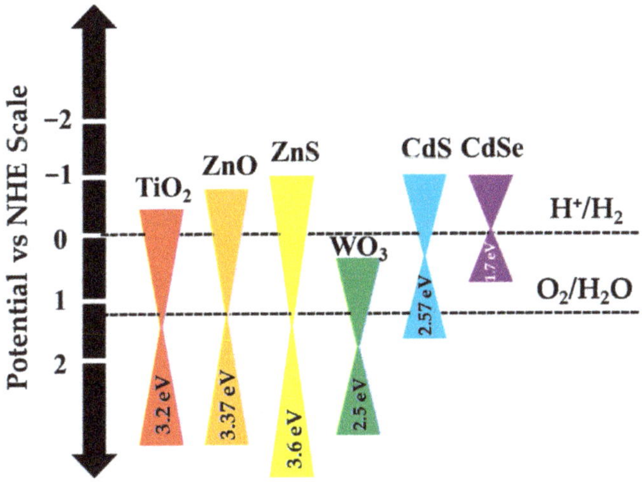

Figure 1. Energy gaps of most well-known semiconductors.

A key mechanism in TiO_2 photocatalysis involves the partial reduction of titanium cations. Under light excitation, Ti^{4+} ions—either on the surface or within the bulk of the material—can trap photogenerated electrons, forming Ti^{3+} species located just below the conduction band [38,40,43]. This process is represented by the following equilibrium:

$$e-_{CB} + Ti^{+4}\,OH \rightleftharpoons Ti^{+3}\,OH$$

This equation describes the trapping of conduction band electrons by surface-bound Ti^{4+} sites, resulting in the formation of Ti^{3+} species slightly below the conduction band edge. These electron traps contribute to extending charge carrier lifetimes, thereby enhancing photocatalytic efficiency, especially when the trapping occurs near the surface, where redox reactions take place.

3.1. The Importance of TiO_2 Modification

TiO_2 possesses several advantageous characteristics as a semiconductor and photocatalyst, making it superior to other available catalysts for photocatalytic reactions [29,39]. However, its relatively large bandgap limits photon absorption efficiency, while the recombination of photo-induced electron–hole pairs further limits its photocatalytic efficiency [10,44,45]. These limitations reduce the overall performance of TiO_2-based systems by contributing to energy losses.

To improve photocatalytic hydrogen production, it is essential to employ strategies that suppress charge carrier recombination and facilitate efficient charge separation and transfer from the catalyst surface to the reactants. Another key aspect is expanding the light absorption range of the material to harness a broader portion of the solar spectrum For enhanced efficiency and performance, TiO_2-based photocatalysts should have the following abilities:

i. Extend light absorption into the visible region.
ii. Efficiently separate charge carriers and promote their transfer to surface reactants.
iii. Incorporate suitable cocatalysts to boost photocatalytic activity.

Sacrificial agents also play a vital role in photocatalysis by donating electrons to prevent charge carrier recombination, thereby sustaining the photocatalytic process. Organic compounds such as methanol and ethanol are commonly employed as sacrificial agents in hydrogen production. They act as hole scavengers, helping to prolong the active life of the photocatalyst [46]. Moreover, the addition of cocatalysts such as platinum (Pt) or nickel (Ni) on the TiO_2 surface significantly enhances hydrogen evolution. These cocatalysts improve electron transfer, reduce the overpotential required for water splitting, and provide additional active sites for the reaction [47].

3.2. Metal Modification of TiO_2

Transition metals such as Pt, Au, Ag, Ni, and Pd are commonly used in order to modify TiO_2, as their electronic structure allows them to participate in catalytic reactions. These metals have their last electron entering an inner orbital, leaving the outer orbital with two electrons, which enables them to act as additional catalysts for electron trapping [11].

The deposition of metals on TiO_2 is one of the most widely used surface modification techniques to suppress the recombination of photogenerated electron–hole pairs. In practice, metals serve as electron traps, promoting efficient charge separation due to their high work function, which enhances electron acceptance. Metal enrichment can be achieved through various methods without compromising the photocatalyst's performance.

Doping TiO_2 with transition metal nanoparticles has gained significant interest in improving photocatalysis through the surface plasmon resonance (SPR) phenomenon [48,49]. In this process, noble metals such as Au, Pt, and Ag exhibit strong SPR effects in the visible range. At elevated temperatures, the SPR effect is further enhanced, leading to increased photocatalytic activity (Figure 2).

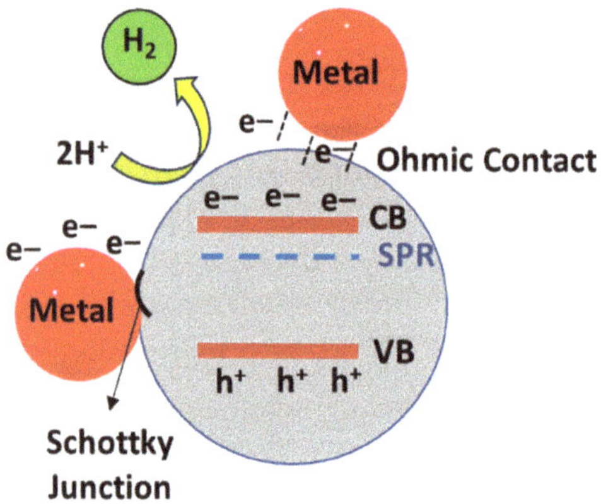

Figure 2. The graph shows the surface plasmon resonance when TiO_2 contacts a metal, forming Schottky junctions and Ohmic contacts (CB: conduction band; VB: valence band; SPR: surface plasmon resonance).

Upon modifying the bandgap of TiO_2 and enhancing its electronic properties, various surface modifications have been implemented, including dye sensitization [50–52], the use of nanocomposites [53–55], and doping with non-metals [56,57] and metals [58]. Among these approaches, metal doping stands out as a superior choice, as it can alter the electronic properties of TiO_2, act as an electron trap, facilitate electron–hole transfer, shift bandgap absorption toward the visible spectrum, and provide additional active sites for photocatalytic hydrogen evolution. Table 1 summarizes the key metal-doped TiO_2 photocatalysts reported in the literature. Composites of noble metals with TiO_2 have been proven to enhance the photocatalytic activity of titania in H_2 production. In one study, Pt/TiO_2 nanocrystals were synthesized to investigate the impact of Pt loading on photocatalytic performance. The results demonstrated that incorporating Pt significantly improved the hydrogen evolution rate, with 1.0 wt% Pt-loaded Pt/TiO_2 nanocrystals, featuring (001) exposed facets, achieving the highest photocatalytic activity (11.2 mmol $h^{-1}g^{-1}$) (entry 1, Table 1) [59]. Additionally, various platinum composites were deposited on TiO_2 P25 and tested for hydrogen production using methanol as a sacrificial agent.

Table 1. Summary of metal-doped TiO_2 photocatalysts and their hydrogen production efficiencies.

Entry	Photocatalyst	Metal Concentration	Light Source	H_2 Production Efficiency	Reference
1	Pt/TiO_2	1.0 wt% Pt	AM 1.5 G sunlight	11.2 mmol h^{-1} g^{-1}	[59]
2	Pt/TiO_2	5.0 wt% Pt	UV (280–400 nm)	27.6 mmol h^{-1} g^{-1}	[60]
3	Au/TiO_2	2.0 wt% Au	Vis (500 ± 20 nm)	0.05702 mmol h^{-1} g^{-1}	[61]
4	Ag/TiO_2	1.5 wt% Ag	UV (365 nm)	23.5 mmol h^{-1} g^{-1}	[62]
5	Ru/TiO_2	3.0 wt% Ru	UV (280–400 nm)	4.7 mmol h^{-1} g^{-1}	[63]
6	Cu/TiO_2	2.0 wt% Cu	UV (365 nm)	5.0 mmol h^{-1} g^{-1}	[64]
7	Ni/TiO_2	2.0 wt% Ni	UV (365 nm)	2.3 mmol h^{-1} g^{-1}	[64]
8	Ni/TiO_2	5.0 wt% Ni	UV (365 nm)	1.608 mmol $h^{-1}g^{-1}$	[65]
9	Co/TiO_2	2.0 wt% Co	UV (365 nm)	2.25 mmol h^{-1} g^{-1}	[64]
10	Co/TiO_2	1.0 wt% Co	Sunlight	0.22 mmol h^{-1} g^{-1}	[66]
11	Fe/TiO_2	1.1 wt% Fe	UV (250 nm) to visible (750 nm)	0.0155 mmol h^{-1} g^{-1}	[67]
12	$Ni\text{-}Fe/TiO_2$	1 wt% Ni–Fe (1:3)	Sunlight	8.27 mmol h^{-1} g^{-1}	[68]
13	$Ni\text{-}Cu/TiO_2$	1 wt% Ni–Cu (1:1)	UV (365 nm)	186 μL h^{-1} cm^{-2}	[69]
14	$Ag\text{-}Ni/TiO_2$	Ag-Ni (1.5:1.5)	UV (365 nm)	2.9339 mmol h^{-1} g^{-1}	[70]
15	$Nb\text{-}Ta/TiO_2$ (NTTO)	2 wt% Nb and 4 wt% Ta	Simulated solar illumination	1.168 mmol h^{-1} g^{-1}	[71]
16	$Ni\text{-}Pt/TiO_2$	0.01 wt% Ni and 1.0 wt% Pt	15 W black lamp (352 nm)	3.983 mmol g^{-1} h^{-1}	[72]

The structural and electronic properties of the catalysts were analyzed using multiple techniques, revealing that optimizing the platinum–titania interface enhanced photocatalytic performance by approximately 250%, reaching 27.6 mmol $h^{-1}g^{-1}$ (entry 2, Table 1) [60]. The study also identified a synergistic effect between the noble metal and the oxide support at interface sites, which promoted the conversion of carboxylate-type surface species into carbon dioxide and maximized hydrogen production (Figure 3).

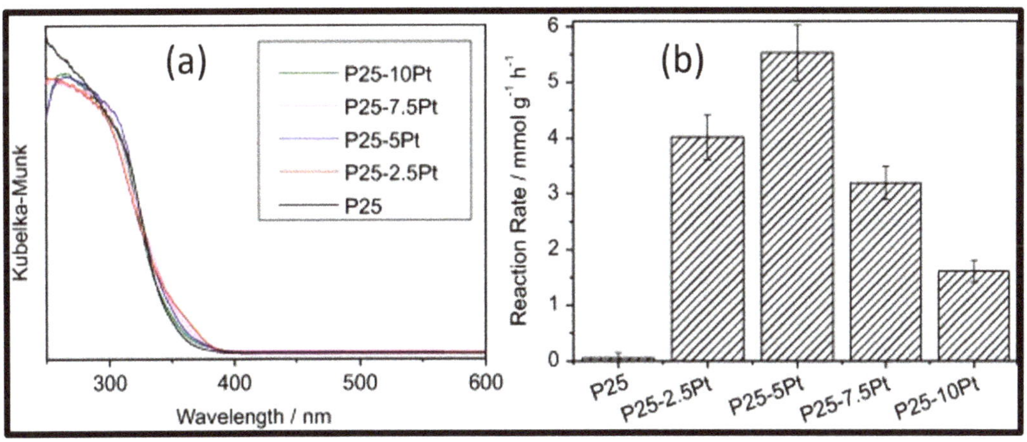

Figure 3. (**a**) UV–visible spectra of Pt and P25 samples (**b**) Hydrogen evolution rate, per gram of catalyst, in the 1:9 CH_3OH: H_2O mixture [60]. Reproduced from ref. [60] with permission from Elsevier, License number 5600041458742.

Gold (Au) has been widely used in photocatalytic hydrogen production due to its plasmonic properties. Au nanoparticles enhance visible light absorption and facilitate electron transfer from the TiO_2 semiconductor to the gold nanoparticles. The charged Au sites provide large surface areas for light stimulation and hydrogen production, significantly improving photocatalytic efficiency when combined with TiO_2 and advancing research toward more efficient photocatalytic systems [73]. A study reported the synthesis of mesoporous plasmonic Au–TiO_2 photocatalysts with varying Au concentrations using a copolymer-assisted sol–gel method [58]. These nanocomposites demonstrated excellent hydrogen production from photocatalytic water reduction under visible light, with 2 wt% Au–TiO_2 exhibiting the highest activity. The proposed mechanism was attributed to three key factors: (i) copolymer-induced defect/impurity states in the TiO_2 matrix, (ii) plasmon excitation of gold nanoparticles, and (iii) plasmon-excited electron transfer from Au nanoparticles to TiO_2, leading to hydrogen generation. The photocatalytic experiments showed hydrogen production rates of 0.05702 mmol $h^{-1}g^{-1}$ under UV-Vis light and 7 µmol $h^{-1}g^{-1}$ under visible light (500 ± 20 nm) (entry 3, Table 1) [61]. Another study explored the development of efficient metal-doped semiconductors for hydrogen production, focusing on Ag-doped TiO_2 photocatalysts synthesized through a cost-effective chemical reduction method. These catalysts exhibited a high hydrogen production rate of 23.5 mmol g^{-1} h^{-1} (entry 4, Table 1) with a quantum yield of 19%, attributed to the presence of oxygen vacancies and the Ti-Ag-O phase. The study also investigated the effects of reaction time, photocatalyst dosage, and sacrificial agents on the water-splitting process, providing valuable insights into the strategic design of metal–support hybrid configurations for photocatalytic hydrogen production [62]. The role of ruthenium (Ru) in TiO_2-based photocatalysts was also examined for hydrogen evolution. A series of Ru/TiO_2 catalysts were tested under UV and visible light irradiation in methanol–water mixtures (entry 5, Table 1) [63]. To further enhance TiO_2's photocatalytic performance for hydrogen production, researchers developed low-cost M/TiO_2 semiconductor catalysts using the impregnation method (Figure 4c). This involved depositing five different first-row transition metals (Fe, Co, Ni, Cu, or Zn) onto a commercially available TiO_2 support. The photocatalytic activity of these M/TiO_2 catalysts was investigated under UV (Figure 4b) and visible light irradiation (Figure 4a), with a metal loading of 2 wt%. Among them, Cu/TiO_2 (Figure 4d) demonstrated the highest hydrogen production efficiency, with rates of approximately 5.0 mmol $h^{-1}g^{-1}$ under UV irradiation and 0.22 mmol $h^{-1}g^{-1}$ under visible light irradiation (entry 6, Table 1) [64]. These values were significantly higher than those obtained with undoped TiO_2 (Figure 4d), by factors of 16 and 3 for UV and visible light, respectively. Copper-doped TiO_2 (Cu-TiO_2) is commonly used for photocatalytic hydrogen production due to the following: (i) increased electron–hole separation: Copper doping enhances charge separation, reducing recombination and leading to more efficient hydrogen production [64]; (ii) enhanced light absorption: Cu doping extends TiO_2's light absorption into the visible region, improving its ability to utilize sunlight for photocatalysis; iii) stability: Cu-TiO_2 has been found to be stable under photocatalytic conditions, making it a viable material for long-term hydrogen production [64].

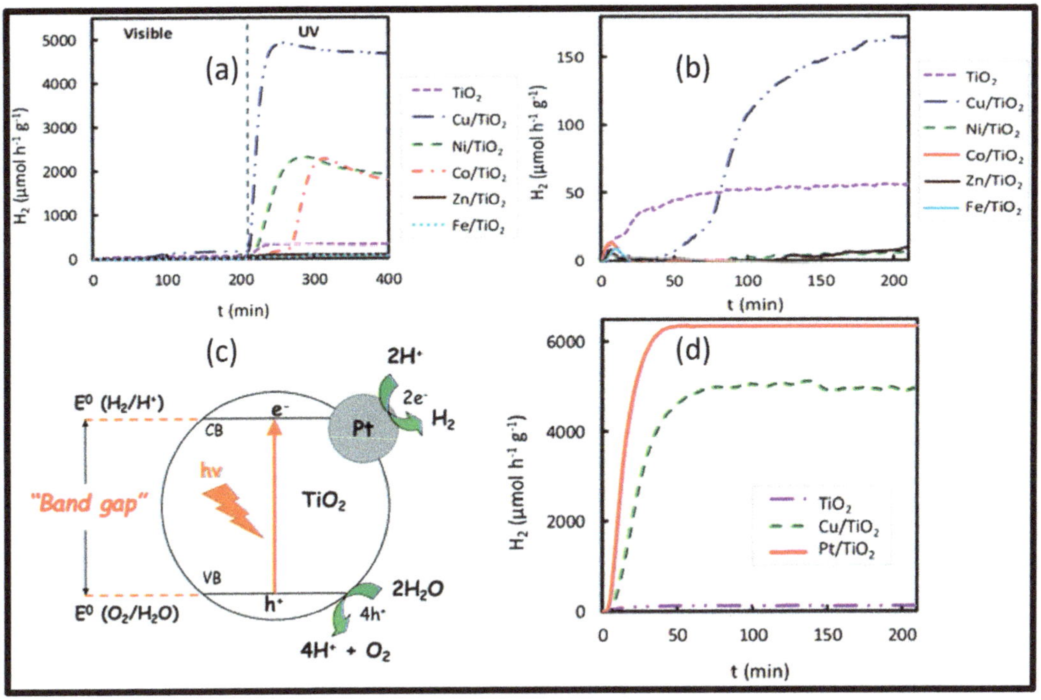

Figure 4. Hydrogen production for impregnated M/TiO_2 photocatalysts (M = Fe, Co, Ni, Cu, Zn), 2 wt% M; (**a**) under both visible and UV light irradiation, (**b**) under visible lamp irradiation, (**c**) schematic diagram of the water-splitting reaction on Pt/TiO_2 and (**d**) comparison of the catalytic activity of bare commercial TiO_2, 2 wt% Cu/TiO_2 and 2 wt% Pt/TiO_2 in the hydrogen photoproduction under UV irradiation. Reproduced from ref. [64] with permission from the Royal Society of Chemistry.

In addition, Ni/TiO_2 and Co/TiO_2 photocatalysts demonstrated high rates of H_2 production under UV light, with values of 2.3 mmol h^{-1} g^{-1} (entry 7, Table 1) and 2.25 mmol h^{-1} g^{-1}, (entry 9, Table 1), respectively (Figure 4a) [64]. Compared to other noble metal-based photocatalysts, such as Pt or Au, Ni is a more cost-effective material, making Ni-TiO_2 an attractive option for large-scale hydrogen production. A study investigated the use of titania nanotubes (NTs) prepared by rapid breakdown anodization for hydrogen generation via water splitting. The NTs were sensitized with platinum (Pt), palladium (Pd), and nickel (Ni) nanoparticles (NPs) to assess their efficiency for water splitting. The metal NPs were deposited on the TiO_2 NTs via chemical reduction. The TiO_2 NTs had a bundled morphology, while the metal NPs appeared as spherical deposits. The addition of metal NPs increased H_2 generation, with lower NP deposits producing better results. Ni NPs had the highest H_2 generation rate among the tested metals. The 5% Ni/TiO_2 NTs had a H_2 value rate of 1.608 mmol h^{-1} g^{-1} under 365 nm UV light (entry 8, Table 1) [65]. In another report, cobalt-doped TiO_2 photocatalysts were synthesized and studied towards hydrogen production. The cobalt-doped TiO_2 showed a maximum hydrogen evolution of 0.22 mmol h^{-1} g^{-1} (entry 10, Table 1) in pure water under solar irradiation. The study systematically investigated various reaction parameters such as the amount of cobalt on TiO_2, glycerol concentration, substrate effect (alcohols), and pH of the solution. Under UV irradiation, a 3- to 4-fold increase in activity was observed initially, but the activity diminished over time due to the loss of cobalt ions. However, under solar irradiation, the cobalt-doped TiO_2 photocatalysts showed stable and continuous activity [66]. In a following report Fe-ion-doped TiO_2 thin films were studied towards photocatalytic water-splitting under visible light irradiation. The study showed that low concentration metal ion doping near the conducting indium tin oxide (ITO)–TiO_2 interface was important to avoid the formation of recombination centers. The H_2 production rate was higher for Fe-doped TiO_2 compared to undoped TiO_2 due to the ability of Fe ions to trap both electrons and holes [67]. The study also found that TiO_2 films produced a constant H_2 generation rate 0.0155 mmol h^{-1} g^{-1} (entry 11, Table 1) for long periods of time, eliminating the back-reaction effect [67]. Even though there are some earth abundant metals doped on TiO_2 that achieve great H_2 evolution, on average the performances of the noble metals are higher. In order to improve the catalytic performance of non-noble metals, bimetallic systems such as Ni-Fe [68] (entry 12, Table 1) and Ni-Cu [69] were prepared and studied as photocatalysts. The TiO_2 doped with bimetallic catalysts showed enhanced photocatalytic activity compared to the single cocatalysts. For example, the TiO_2 nanotubes coated with Ni-Cu (1:1 atomic ratio) reached 186 μL h^{-1} g^{-1} (entry 13, Table 1) hydrogen evolution rate, which was 4.6 and 3 times higher compared to Ni/TiO_2 and Cu/TiO_2, respectively. Therefore, the bimetallic layers are able to provide synergistic effect and enhance H_2 evolution compared to monometallic-doped TiO_2. In a following example, Ni is deposited on Ag surface to form a core–shell Ag-Ni nanostructure which was used to synthesize Ag-Ni/TiO_2 photocatalyst via one-step photo-induced deposition method [70]. The Ag-Ni/TiO_2 (1.5:1.5) photocatalyst displayed a photocatalytic activity of 2.9339 mmol $h^{-1}g^{-1}$, about 79 times more than TiO_2 (entry 14, Table 1). Continuing with bimetallic catalysts, Nb-Ta co-doped TiO_2 (NTTO) (entry 15, Table 1) nanoparticles achieved a hydrogen evolution rate of 1.168 mmol h^{-1} g^{-1} under simulated solar illumination. This performance, while lower than noble metal-doped systems (e.g., Pt/TiO_2 at 27.6 mmol h^{-1} g^{-1} in entry 2, Table 1), surpasses several non-noble metal-doped catalysts (e.g., Fe/TiO_2 at 0.0155 mmol h^{-1} g^{-1} in entry 11, Table 1). The NTTO nanoparticles leverage dual doping with Nb and Ta to reduce the bandgap (2.90 eV vs. 3.27 eV for undoped TiO_2) and suppress electron–hole recombination by 30%, as demonstrated by photoluminescence studies. This bandgap engineering enhances visible light absorption and charge carrier separation, though the absence of a noble metal cocatalyst limits its efficiency compared to Pt- or Au-based systems. Nevertheless, NTTO's stability and scalable synthesis under mild conditions highlight its potential for sustainable applications. The Nb-Ta/TiO_2 system, while innovative in bandgap tuning, highlights the trade-off between noble metal free design and catalytic efficacy, emphasizing that dual-metal strategies involving noble metals remain pivotal for maximizing H_2 evolution rates [71]. In contrast, Ni-Pt/TiO_2 (entry 16, Table 1) achieves a significantly higher hydrogen evolution rate of 3.983 mmol h^{-1} g^{-1} under UV light (352 nm), outperforming most non-noble metal-doped catalysts (e.g., Cu/TiO_2 at 5.0 mmol h^{-1} g^{-1} in entry 6, Table 1) and even some noble metal systems (e.g., Ru/TiO_2 at 4.7 mmol h^{-1} g^{-1} in entry 5, Table 1). The study investigated various metals (Ag, Cu, Pd) co-deposited with Pt on TiO_2 for photocatalytic hydrogen production. **Ag/Pt** underperformed due to Ag's low work function (4.26 eV), which caused electron backflow from TiO_2 to Ag, reducing charge separation efficiency. Higher Ag loading further hindered performance by blocking light and active sites. **Cu/Pt** showed moderate improvement (1.25× higher H_2 than Pt/TiO_2) at 0.01 wt% Cu, as Cu's work function (4.65 eV) aligned with TiO_2, enabling a weak Schottky junction. However, excess Cu obscured active sites and filtered light. **Pd/Pt** had a high work function (5.12 eV) but lower H_2 yield than Ni/Pt, likely due to less effective synergy in charge transfer or unfavorable hydrogen evolution potential. **Ni/Pt** outperformed others due to Ni's high work function (5.25 eV) and redox synergy with Pt. Pt deposits first (higher redox potential: 1.44 V vs. Ni's −0.257 V), forming a Schottky junction that traps electrons, while Ni prolongs charge carrier lifetime. Optimal Ni loading (0.01 wt%) avoided site blockage and light obstruction, achieving 341× higher H_2 than pure TiO_2. The Ni-Pt interface facilitated efficient electron transfer and suppressed recombination, surpassing the limitations of Ag, Cu, and Pd in balancing Schottky effects, redox dynamics, and catalyst dispersion. The superior performance of Ni-Pt/TiO_2 arises from the synergistic interplay between Ni and Pt. Platinum, with its high work function (5.65 eV), forms a Schottky junction with TiO_2 (4.2–4.5 eV), effectively trapping photogenerated electrons and reducing recombination. Simultaneously, Ni enhances electron migration from Pt to Ni, prolonging charge carrier lifetimes and providing additional active sites. The optimal loading (0.01 wt% Ni and 1.0 wt% Pt) balances light absorption and catalytic site availability while preventing $Ni(OH)_2$ formation—a common issue in Ni-doped systems. The dominance of Ni-Pt/TiO_2 over other bimetallic systems (e.g., Ni-Cu/TiO_2 at 186 μL h^{-1} cm^{-2} in entry 13 or Ag-Ni/TiO_2 at 2.9339 mmol h^{-1} g^{-1} in entry 14, Table 1) underscores the critical role of Pt's low overpotential for H^+ reduction and its ability to stabilize Ni in metallic form. While non-noble metals like Cu or Ni improve cost-effectiveness, their lower work functions and higher recombination rates limit efficiency. Noble metals like Au (entry 3, Table 1: 0.057 mmol h^{-1} g^{-1}) or Ag (entry 4, Table 1: 23.5 mmol h^{-1} g^{-1}) exhibit strong plasmonic effects but face scalability challenges. Ni-Pt/TiO_2 strikes a balance by combining Pt's electron-trapping efficiency with Ni's affordability and structural benefits, making it a superior choice for high-performance, scalable photocatalytic hydrogen generation [72].

4. Porphyrin-Doped TiO_2 for Photocatalytic Hydrogen Production

Porphyrin-doped TiO_2 has emerged as a promising material for photocatalytic hydrogen production due to its ability to enhance light absorption and charge separation. Porphyrins, with their strong absorption in the visible spectrum, extend TiO_2's photocatalytic activity beyond the UV range, making better use of solar energy. Their interaction with TiO_2 facilitates efficient charge transfer, reducing recombination losses and improving hydrogen generation efficiency [74]. This modification offers a sustainable approach to solar-driven water splitting, advancing the development of clean energy technologies.

Porphyrins are organic molecules consisting of four pyrrole rings interconnected by methine bridges (-CH=), forming a highly conjugated system with 22 π-electrons, 18 of which contribute to its aromatic stability and planar geometry through sp^2 hybridization. This extensive conjugation allows porphyrins to strongly absorb visible light, making them highly effective photosensitizers. Their absorption spectrum includes the Soret band (400–450 nm) and Q bands (500–700 nm), which enhance visible light capture. Additionally, porphyrins can be functionalized with various substituents, allowing the fine-tuning of their electronic and optical properties. When incorporated into a TiO_2 photocatalytic system, porphyrins improve spectral response, enhance electron transfer efficiency, and provide stability under reaction conditions [75–79]

The Soret and Q bands are key features of porphyrins' electronic spectra, making them highly effective as photosensitizers in photocatalytic applications [80,81]. The Soret band, also known as the B band, appears in the 400–450 nm range, typically in the blue region of the visible spectrum. This strong absorption results from an electronic transition in the porphyrin molecule, often from the ground state to an excited singlet state, and is attributed to the $\pi \rightarrow \pi^*$ transition within the highly conjugated porphyrin ring. This enables porphyrins to capture high-energy visible light photons, which are then transferred to TiO_2, enhancing photocatalytic activity.

The Q bands, occurring in the 500–700 nm range (red region of the visible spectrum), are weaker than the Soret band and result from lower-energy $\pi \rightarrow \pi^*$ transitions. These bands are split into multiple peaks due to asymmetry in electron distribution, particularly in metal-free porphyrins. Despite their lower intensity, the Q bands significantly contribute to visible light absorption, extending TiO_2's light capture into the red region of the spectrum. Together, these absorption bands enable porphyrins to harness a broad range of solar energy, facilitating efficient charge transfer and enhancing hydrogen production [13,82,83].

4.1. Mechanism of Hydrogen Production in Porphyrin-Doped TiO_2 Systems

In porphyrin-doped TiO_2 systems, photocatalytic hydrogen production relies on the efficient separation of charge carriers (electrons and holes) generated under visible light. The mechanism typically involves three main steps: light absorption, electron transfer and hole scavenging and redox reactions. Porphyrins, with their extensive π-conjugated systems, strongly absorb visible light (400–700 nm). The excited electron is transferred to the conduction band of TiO_2 (Scheme 2), enabling TiO_2 to utilize visible light, which it normally cannot absorb. These conduction band electrons participate in water reduction to generate hydrogen (H_2), while the holes remain in the porphyrin molecule. To prevent electron–hole recombination, a sacrificial agent—typically an electron donor such as methanol or ethanol—quenches the hole in the porphyrin (Scheme 2). This maintains charge separation, ensuring continuous hydrogen production. Additionally, cocatalysts like platinum (Pt) can enhance electron transfer and improve hydrogen evolution efficiency. Porphyrins play a crucial role in extending TiO_2's light absorption into the visible spectrum, which accounts for approximately 43% of solar radiation. This significantly enhances the photocatalytic efficiency of hydrogen production [14,77–79,81].

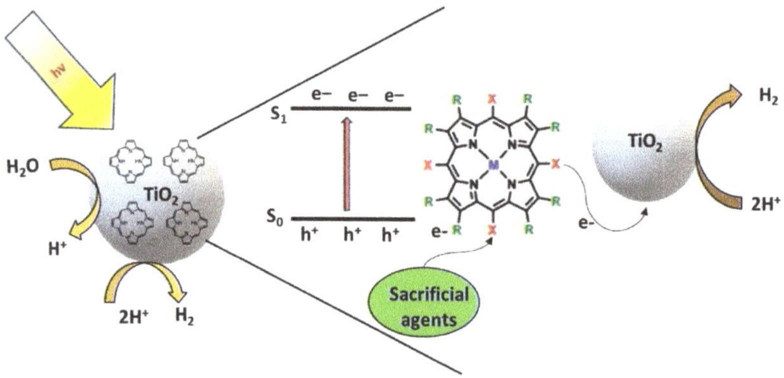

Scheme 2. Photocatalytic hydrogen production and water splitting capabilities of a porphyrin-doped TiO_2 photocatalyst using a sacrificial agent.

4.2. Types of Porphyrins and Their Suitability

Porphyrin-based systems offer extensive tunability through structural modifications. By incorporating different transition metal centers (e.g., Zn, Cu, Co) or functional groups (e.g., carboxylate, phosphonate), their photophysical properties can be tailored for specific applications. Free-base porphyrins, such as tetraphenylporphyrin (H_2TPP), are widely used due to their simple synthesis and tunability; however, they typically exhibit lower stability and electron transfer efficiency compared to metal porphyrins. Among metal porphyrins, zinc porphyrins (ZnTPP) are among the most studied for TiO_2 doping, as they offer strong visible light absorption and efficient electron transfer, making them highly effective for hydrogen production. Copper and cobalt porphyrins have also been explored for their redox properties, providing a balance between cost and performance [84]. Functionalized porphyrins, which incorporate groups such as carboxyl or phosphonic acids, exhibit enhanced binding affinity to the TiO_2 surface. These functional groups facilitate strong anchoring, ensuring efficient charge transfer between the photosensitizer and the semiconductor while preventing porphyrin aggregation, thereby preserving their optical and electronic properties [80,81]. Furthermore, porphyrins can act synergistically with metallic cocatalysts such as platinum, enhancing hydrogen evolution by improving electron transfer kinetics at the TiO_2 interface. These advantages establish porphyrin-doped TiO_2 as a promising material for sustainable solar-to-hydrogen conversion, positioning it as an efficient and versatile candidate for future energy applications [12,51,54].

4.3. Recent Examples of Porphyrin-Doped TiO_2 Systems for Hydrogen Production

Zinc and platinum porphyrins are among the most studied due to their strong visible light absorption and efficient electron transfer capabilities [52,78,79,82,85]. Functionalized porphyrins, particularly those with carboxyl groups, exhibit enhanced stability and binding efficiency, leading to higher hydrogen production rates [78]. Additionally, high-intensity light sources (e.g., 300 W and 450 W Xenon lamps) significantly increase hydrogen yields, emphasizing the importance of photon energy in photocatalysis [77–79,85,86]. Table 2 presents studies on porphyrin-doped TiO_2 photocatalysts for H_2 production.

Table 2. Summary of porphyrin-Doped TiO_2 photocatalysts and their hydrogen production efficiencies.

Entry	Photocatalyst	Porphyrin	Light Source	H_2 Production Efficiency	Reference
1	Pt/TiO_2-Pt-Tc3CP	Pt-Tc3CP	40 W white LED	707 mmol g^{-1} h^{-1}	[82]
2	Pt-TiO_2 NPs-Zn-TM(pCOOH)P	ZnTM(pCOOH)P	40 W white LED	1959 mmol g^{-1} h^{-1}	[83]
3	PdTCP and PtTCP-TiO_2	PdTCP and PtTCP	visible light irradiation ($\lambda > 420$ nm)	30,200 μmol g^{-1}	[52]
4	Pt/TiO_2-THPP-Pd	THPP-Pd	visible light irradiation ($\lambda > 420$ nm)	2025.4 μmol g^{-1} h^{-1}	[13]
5	TiO_2 MS-Cu-TCPP	TCPP-Cu	300 W Xenon	1.3 mmol g^{-1} h^{-1}	[77]
6	Pt/TiO_2-(LG-5)	Zinc–porphyrin dye (LG-5)	450 W Xenon	4196 mmol g^{-1} h^{-1}	[78]
7	Pt-TiO_2-PCT-LG-23	Zinc-PCT-LG-23	300 W Xenon	9793.5 μmol g^{-1} h^{-1}	[79]
8	TiO_2/ZnTmMpHPP + 0.5%Pt	ZnTmMpHPP	300 W Xenon	326.3 μmol h^{-1}	[85]
9	THPP-TiO_2-H NPs	THPP	UV cutoff filter > 400 nm	4.80 mmol g^{-1}	[87]
10	TiO2/Co/PA	TCPP	300 W xenon lamp	4318.4 μmol g^{-1} h^{-1}	[86]

Firstly, the Pt/TiO_2 with Pt-Tc3CP (entry 1, Table 2) system, using platinum tetraphenylporphyrin (Pt-Tc3CP) as a photosensitizer paired with TiO_2, achieved a hydrogen production rate of 707 mmol g^{-1} h^{-1} under a 40 W white LED. This high yield demonstrates the effectiveness of platinum porphyrins in extending TiO_2's light absorption into the visible spectrum. Platinum cocatalysts stabilize photogenerated electrons, enhancing charge separation and reducing recombination losses. This synergy between porphyrin sensitization and platinum's catalytic activity underpins the system's impressive performance [82]. Zn-TM(pCOOH)P on Pt-TiO_2 (entry 2, Table 2) achieves a hydrogen evolution rate of 1959 mmol g^{-1} h^{-1}, highlighting the effectiveness of zinc porphyrins functionalized with carboxyl groups. These groups enhance binding to TiO_2 surfaces, improving light harvesting and charge transfer. The interaction between zinc porphyrins and TiO_2 minimizes energy loss during electron transfer, optimizing performance [83]. The PdTCP and PtTCP-TiO_2 (entry 3, Table 2) configuration delivers an exceptional 30,200 μmol g^{-1} of hydrogen under visible light ($\lambda > 420$ nm). The combined use of palladium (Pd) and platinum (Pt) tetraphenylporphyrins maximizes photocatalytic activity by leveraging Pd and Pt's unique electron affinities, facilitating rapid charge separation. The tetraphenylporphyrin structure aids light absorption and stabilizes photogenerated carriers, positioning this system as one of the most effective in the table [52]. Palladium-substituted tetraphenylporphyrin (THPP-Pd), when combined with Pt/TiO_2 (entry 4, Table 2), achieves a hydrogen evolution rate of 2025.4 μmol g^{-1} h^{-1} under visible light. This setup highlights the importance of palladium's catalytic properties in reducing overpotentials for hydrogen generation while enabling better charge mobility. The palladium component works synergistically with the porphyrin sensitizer and TiO_2 to deliver reliable photocatalytic performance [13]. Copper–tetraphenylporphyrin (TCPP), integrated into TiO_2 modified structures (MS-TiO_2) (entry 5, Table 2), achieves a modest hydrogen production rate of 1.3 mmol g^{-1} h^{-1} under 300 W xenon light. Copper functions as an efficient redox catalyst, aiding electron mobility and improving water-splitting reactions. However, its lower performance compared to Pt and Pd systems indicates the need for optimization in porphyrin–TiO_2 interactions [77]. The zinc–porphyrin dye (LG-5) system (entry 6, Table 2), which achieves 4196 mmol g^{-1} h^{-1} under a 450 W xenon lamp, demonstrates remarkable photocatalytic activity. The strong absorption of visible light by the zinc dye and efficient energy transfer to TiO_2 contribute to the high hydrogen evolution rate, showcasing the potential of zinc porphyrins in high-intensity light applications [78]. Zn-PCT-LG-23 on TiO_2 (entry 7, Table 2) produced 9793.5 μmol g^{-1} h^{-1}, highlighting the impact of advanced porphyrin structural modifications on hydrogen production. These modifications enhance light absorption and charge transfer efficiency, enabling higher performance under 300 W xenon light. The system's success emphasizes the importance of fine-tuning porphyrin structures for optimized photocatalytic activity [79].The ZnTmMpHPP on TiO_2 with 0.5% Pt (entry 8, Table 2) system achieves a moderate hydrogen production rate of 326.3 μmol h^{-1} under xenon light. Although the zinc porphyrin (ZnTmMpHPP) sensitizer contributes to visible light absorption, the efficiency is limited compared to other systems due to weaker porphyrin–TiO_2 interactions and possible electron–hole recombination. The addition of platinum as a cocatalyst stabilizes photogenerated electrons, but further optimization is needed to enhance hydrogen evolution [85].The incorporation of two novel porphyrin-doped TiO_2 systems, THPP-TiO_2-H NPs (entry 9, Table 2) [87] and TiO_2/Co/PA (entry 10, Table 2) [86], underscores significant advancements in addressing TiO_2's limitations for photocatalytic hydrogen production. The THPP-TiO_2-H NPs system (entry 9, Table 2) leverages a hybrid microstructure combining meso-tetra(4-hydroxyphenyl) porphyrin (THPP) with anatase-phase TiO_2, achieving a hydrogen evolution rate of 4.80 mmol g^{-1} h^{-1} under visible light ($\lambda > 400$ nm). This performance arises from the intimate interfacial contact between THPP and TiO_2, which enhances charge transfer efficiency, while the hydrothermal treatment ensures high crystallinity of TiO_2, reducing electron–hole recombination. The system's scalability (up to 500 mL synthesis) further highlights its practical potential. However, its moderate efficiency compared to noble metal-coordinated systems like Zn-TM(pCOOH)P/TiO_2 (entry 2, Table 2: 1959 mmol g^{-1} h^{-1}) or PdTCP/TiO_2 (entry 3, Table 2: 30,200 μmol g^{-1}) suggests limitations in the absence of Pt/Pd co-catalysts, which are critical for stabilizing photogenerated electrons and optimizing light utilization [87]. In contrast, the TiO_2/Co/PA nano–micro hybrid PA (entry 10, Table 2) introduces cobalt ions as interfacial anchors to immobilize J-type porphyrin assembly (PA) derived from CuTCPP, achieving 4318.4 μmol g^{-1} h^{-1} under a 300 W xenon lamp. The cobalt-mediated anchoring minimizes the H-aggregation of PA, ensuring uniform distribution and enhanced electron transfer, while the hybrid's dual functionality—efficient methylene blue degradation alongside H_2 production—demonstrates its versatility. Notably, the system's stability in mixed sacrificial agents (methanol/TEOA) addresses a common challenge in porphyrin-based photocatalysts. Despite these strengths, its performance remains lower than Pt- or Pd-doped systems (e.g., entry 3, Table 2), likely due to the reliance on cobalt rather than noble metals and the use of a xenon lamp rather than optimized visible light sources [86].

Both systems exemplify the critical role of structural engineering (hybrid/nano–micro architectures) and co-catalyst integration (Co^{2+}) in improving charge separation and light absorption. However, they also highlight the persistent efficiency gap between noble metal-free systems and their noble metal-doped counterparts. These findings reinforce the need for future research to optimize porphyrin–metal interactions, explore low-cost co-catalysts (e.g., Ni, Cu), and scale visible light-driven designs to achieve competitive efficiencies while maintaining sustainability. Collectively, these entries advance the field by demonstrating how material innovation and strategic co-catalyst use can balance performance, stability, and practicality in photocatalytic hydrogen production. The versatility of porphyrins in adapting to diverse architectures—from hybrid nanocomposites to hierarchical nano–micro structures—is a key strength highlighted in these systems. For instance, the THPP-TiO_2-H NPs exemplify how porphyrins can be tailored for self-assembly into TiO_2 matrices [87], while the TiO_2/Co/PA showcases their integration into complex heterostructures via metal coordination [86]. This adaptability extends beyond TiO_2-based systems; porphyrins are widely incorporated into metal–organic frameworks (MOFs) [88], nanoparticles (NPs) [89], and core–shell designs, where their tunable electronic properties and strong visible light absorption enhance photocatalytic performance. In MOFs, porphyrins act as light-harvesting nodes [90], while in NPs [91], their conjugation with semiconductors improves charge separation. The reviewed systems align with this broader trend, demonstrating how strategic porphyrin functionalization (e.g., carboxyl or hydroxyl groups) enables robust anchoring and interfacial electron transfer across varied platforms. Reported hydrogen production rates vary widely, from 326.3 μmol h^{-1} to 30,200 μmol g^{-1}, depending on the porphyrin structure, light source, and cocatalysts [52,85]. The differences in photocatalytic performance highlight the need for optimized porphyrin design and integration with TiO_2 to maximize light absorption and minimize recombination losses. High-performing systems, such as PdTCP (entry 3, Table 2) and Zn-TM (pCOOH)P (entry 2, Table 2), are excellent examples due to optimized structural features, effective cocatalysts, and strong porphyrin–TiO_2 interactions [52,83]. In contrast, lower-performing systems suffer from weaker [52,85] binding, suboptimal light absorption, and less efficient cocatalysts. Several key factors influence the efficiency of porphyrin-doped TiO_2 photocatalysts. The molecular structure of the porphyrin (Figure 5) and the presence of functional groups, such as carboxyl, determine binding strength with TiO_2. Cocatalysts, particularly noble metals like Pt and Pd, enhance catalytic activity by stabilizing photogenerated electrons and lowering the energy required for hydrogen evolution. Less effective cocatalysts, such as copper, generally result in reduced performance [77]. Additionally, porphyrins optimized for broad-spectrum light absorption—especially under high-intensity light—demonstrate superior photocatalytic activity. The strength of porphyrin–TiO_2 interactions, often influenced by functionalization, plays a crucial role in ensuring efficient electron transfer and reducing recombination losses. Systems with weak porphyrin–TiO_2 interactions tend to exhibit lower hydrogen production rates. Ultimately, achieving high photocatalytic efficiency requires a balance between effective charge carrier utilization and minimized recombination. Systems that integrate strong porphyrin sensitization, efficient cocatalysts, and structural modifications demonstrate superior performance, underscoring the importance of optimizing both structural (Figure 5) and operational parameters [79,83]. These findings underscore the importance of structural and operational factors in determining the performance of porphyrin-doped TiO_2 photocatalysts.

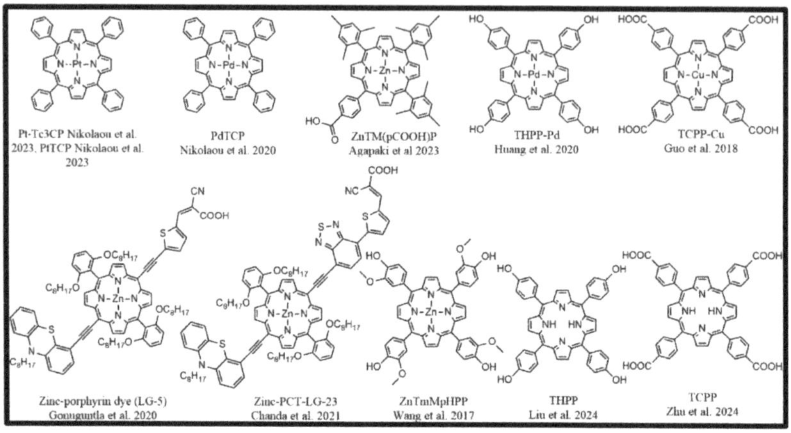

Figure 5. Molecular structure of the porphyrins [13,52,77–79,83,85–87].

4.4. Drawbacks and Challenges of Porphyrin-Doped TiO_2 Systems for Hydrogen Production

Despite the promising benefits, several challenges hinder the maximization of efficiency and scalability in porphyrin-doped TiO_2 systems for hydrogen production. A primary concern is the photostability of porphyrins, which are susceptible to photodegradation (or photobleaching) under extended light exposure. This degradation leads to a decrease in photocatalytic activity over time, limiting the feasibility of these systems for continuous hydrogen production. Strategies for stabilization, such as using metal-centered porphyrins or functionalized porphyrin derivatives, are being explored to improve durability in photocatalytic applications. Another key issue is electron–hole recombination. While porphyrins enhance TiO_2's light absorption, recombination of photogenerated charge carriers remains a challenge. To address this, sacrificial agents or additional cocatalysts are often required, adding complexity and cost to the system. Furthermore, porphyrin molecules tend to aggregate on the TiO_2 surface, potentially reducing light absorption and charge transfer efficiency. Managing porphyrin loading and dispersion on TiO_2 is critical to maintaining optimal performance. Finally, the cost of metal porphyrins, particularly those containing platinum or other noble metals, poses an economic challenge for large-scale applications. Alternatives such as copper or zinc porphyrins are being investigated to reduce costs, although their performance and stability still require further evaluation [77–79,83,85].

5. Conclusions

In this review, we have highlighted recent advances in the use of metal-doped and porphyrin-doped TiO_2 for photocatalytic hydrogen production. Our focus has been on the impact of both monometallic dopants (e.g., Pt, Au, Cu, Ni, Co, Ru, Fe, and Ag) and bimetallic dopants (e.g., Ni-Fe, Ni-C, Nb-Ta, and Ni-Pt). Metal doping has emerged as a promising strategy for enhancing charge carrier separation on the TiO_2 surface and shifting light absorption from the UV to the visible spectrum, thus improving overall photocatalytic efficiency under sunlight. However, a significant challenge remains in the development of low-cost, readily available dopants. While metals like Pt and Au are among the most effective, their high-cost limits scalability. The integration of porphyrins as sensitizers or dopants onto TiO_2 has also shown considerable potential for visible light-driven hydrogen production. Porphyrins' strong visible light absorption and electron transfer capabilities enable TiO_2 to efficiently harvest solar energy and produce hydrogen. Despite this potential, challenges such as porphyrin stability under extended illumination and the need to control molecular aggregation remain. Continued research on functionalized porphyrins and optimized anchoring methods could further enhance the effectiveness and stability of these systems.

6. Future Perspectives and Environmental Considerations

Looking ahead, challenges for metal-doped TiO_2 include economic viability and environmental sustainability. While noble metals like Pt and Au provide superior performance, their high-cost limits scalability. Future research should focus on cost-effective, earth-abundant metals (e.g., Fe, Ni, Cu) and safer, non-toxic alternatives to prevent environmental contamination. Additionally, attention should be given to reducing the environmental impact of metal leaching during photocatalysis, ensuring that these materials are both efficient and sustainable. Developing stable photocatalysts that balance efficiency, cost, and environmental safety will be key for advancing sustainable hydrogen production. To fully realize the potential of TiO_2-based photocatalysts, future studies should focus on creating stable materials with improved surface areas and reduced charge recombination sites. Moreover, optimizing metal and porphyrin deposition methods will be critical for enhancing hydrogen production efficiency and selectivity. Future studies should also prioritize sustainable synthesis methods for porphyrins, ensuring that the materials used are not only effective but also eco-friendly. Expanding the use of bio-inspired or biodegradable porphyrins may provide new, cost-effective solutions compatible with scalable hydrogen production. Finally, alternative sacrificial agents that are both effective and less toxic should be explored to further enhance sustainability in photocatalytic hydrogen production.

Author Contributions: Writing—original draft preparation: D.R.B.; original draft preparation, writing: A.S.; original draft preparation, writing: E.O.; writing—review and editing: A.G.C.; writing—review and editing: R.I.K.; writing—review and editing: A.C.M.; visualization, supervision, original draft preparation, writing: K.L. All authors have read and agreed to the published version of the manuscript.

Funding: This research received no external funding.

Institutional Review Board Statement: Not applicable.

Informed Consent Statement: Not applicable.

Data Availability Statement: No new data were created or analyzed in this study.

Conflicts of Interest: The authors declare no conflicts of interest.

References

1. Lewis, N.S.; Nocera, D.G. Powering the planet: Chemical challenges in solar energy utilization. *Proc. Natl. Acad. Sci. USA* **2006**, *103*, 15729–15735. [CrossRef] [PubMed]
2. Global Energy Crisis. Available online: https://www.iea.org/topics/global-energy-crisis (accessed on 14 January 2025).
3. Zayer Kabeh, K.; Teimouri, A.; Changizian, S.; Ahmadi, P. Techno-economic assessment of small-scale gas to liquid technology to reduce waste flare gas in a refinery plant. *Sustain. Energy Technol. Assess.* **2023**, *55*, 102955. [CrossRef]
4. Yue, M.; Lambert, H.; Pahon, E.; Roche, R.; Jemei, S.; Hissel, D. Hydrogen energy systems: A critical review of technologies, applications, trends and challenges. *Renew. Sustain. Energy Rev.* **2021**, *146*, 111180. [CrossRef]
5. Teets, T.S.; Nocera, D.G. Photocatalytic hydrogen production. *Chem. Commun.* **2011**, *47*, 9268–9274. [CrossRef]
6. Van Hoecke, L.; Laffineur, L.; Campe, R.; Perreault, P.; Verbruggen, S.W.; Lenaerts, S. Challenges in the use of hydrogen for maritime applications. *Energy Environ. Sci.* **2021**, *14*, 815–843. [CrossRef]
7. Wei, D.; Shi, X.; Qu, R.; Junge, K.; Junge, H.; Beller, M. Toward a Hydrogen Economy: Development of Heterogeneous Catalysts for Chemical Hydrogen Storage and Release Reactions. *ACS Energy Lett.* **2022**, *7*, 3734–3752. [CrossRef]
8. Love, J.G.; O'Mullane, A.P.; Boulaire, F.A.; Mackinnon, I.D.R. Impact of fuel cells on hydrogen energy pathways in a sustainable energy economy. *Sustain. Energy Fuels* **2022**, *6*, 4008–4023. [CrossRef]

9. van der Spek, M.; Banet, C.; Bauer, C.; Gabrielli, P.; Goldthorpe, W.; Mazzotti, M.; Munkejord, S.T.; Røkke, N.A.; Shah, N.; Sunny, N.; et al. Perspective on the hydrogen economy as a pathway to reach net-zero CO_2 emissions in Europe. *Energy Environ. Sci.* **2022**, *15*, 1034–1077. [CrossRef]
10. Park, H.; Park, Y.; Kim, W.; Choi, W. Surface modification of TiO_2 photocatalyst for environmental applications. *J. Photochem. Photobiol. C Photochem. Rev.* **2013**, *15*, 1–20. [CrossRef]
11. Orfanos, E.; Ladomenou, K.; Angaridis, P.A.; Papadopoulos, T.; Charalambidis, G.; Vasilopoulou, M.; Coutsolelos, A.G. A stable platinum porphyrin based photocatalyst for hydrogen production under visible light in water. *Sustain. Energy Fuels* **2022**, *6*, 5072–5076. [CrossRef]
12. You, B.; Sun, Y. Innovative Strategies for Electrocatalytic Water Splitting. *Acc. Chem. Res.* **2018**, *51*, 1571–1580. [CrossRef] [PubMed]
13. Huang, L.-Y.; Huang, J.-F.; Lei, Y.; Qin, S.; Liu, J.-M. Porous Hybrid Materials Based on Mesotetrakis(Hydroxyphenyl) Porphyrins and TiO_2 for Efficient Visible-Light-Driven Hydrogen Production. *Catalysts* **2020**, *10*, 656. [CrossRef]
14. Ladomenou, K.; Natali, M.; Iengo, E.; Charalampidis, G.; Scandola, F.; Coutsolelos, A.G. Photochemical hydrogen generation with porphyrin-based systems. *Coord. Chem. Rev.* **2015**, *304–305*, 38–54. [CrossRef]
15. Hausmann, J.N.; Schlögl, R.; Menezes, P.W.; Driess, M. Is direct seawater splitting economically meaningful? *Energy Environ. Sci.* **2021**, *14*, 3679–3685. [CrossRef]
16. Yang, X.; Wang, D. Photocatalysis: From Fundamental Principles to Materials and Applications. *ACS Appl. Energy Mater.* **2018**, *1*, 6657–6693. [CrossRef]
17. Zhu, S.; Wang, D. Photocatalysis: Basic Principles, Diverse Forms of Implementations and Emerging Scientific Opportunities. *Adv. Energy Mater.* **2017**, *7*, 1700841. [CrossRef]
18. Checchetti, A.; Fantini, A. Experimental Determination of Planck's constant using Light Emitting Diodes (LEDs) and Photoelectric Effect. *World J. Chem. Educ.* **2015**, *3*, 87–92. [CrossRef]
19. Jaffe, H.H.; Miller, A.L. The fates of electronic excitation energy. *J. Chem. Educ.* **1966**, *43*, 469. [CrossRef]
20. Belver, C.; Bedia, J.; Gómez-Avilés, A.; Peñas-Garzón, M.; Rodriguez, J.J. Chapter 22—Semiconductor Photocatalysis for Water Purification. In *Nanoscale Materials in Water Purification*; Thomas, S., Pasquini, D., Leu, S.-Y., Gopakumar, D.A., Eds.; Elsevier: Amsterdam, The Netherlands, 2019; pp. 581–651. [CrossRef]
21. Li, S.; Xu, W.; Meng, L.; Tian, W.; Li, L. Recent Progress on Semiconductor Heterojunction-Based Photoanodes for Photoelectrochemical Water Splitting. *Small Sci.* **2022**, *2*, 2100112. [CrossRef]
22. Bahnemann, D.W.; Hilgendorff, M.; Memming, R. Charge Carrier Dynamics at TiO_2 Particles: Reactivity of Free and Trapped Holes. *J. Phys. Chem. B* **1997**, *101*, 4265–4275. [CrossRef]
23. Zhao, D.; Zhuang, Z.; Cao, X.; Zhang, C.; Peng, Q.; Chen, C.; Li, Y. Atomic site electrocatalysts for water splitting, oxygen reduction and selective oxidation. *Chem. Soc. Rev.* **2020**, *49*, 2215–2264. [CrossRef]
24. Goddard, W.A. Electrocatalytic Water Splitting ($H_2O \rightarrow H_2 + ½ O_2$). In *Computational Materials, Chemistry, and Biochemistry: From Bold Initiatives to the Last Mile: In Honor of William A. Goddard's Contributions to Science and Engineering*; Shankar, S., Muller, R., Dunning, T., Chen, G.H., Eds.; Springer International Publishing: Cham, Switzerland, 2021; pp. 1257–1264. [CrossRef]
25. Zhang, W.L.; Wei, W.; Liu, W.; Guan, T.; Tian, Y.; Zeng, H. Engineering the morphology of TiO_2/carbon hybrids via oxidized Ti_3C_2Tx MXene and associated electrorheological activities. *Chem. Eng. J.* **2019**, *378*, 122170. [CrossRef]
26. Wang, H.-X.; Li, X.-X.; Tang, L. Effects of surfactants on the morphology and properties of TiO_2. *Appl. Phys. A* **2020**, *126*, 448. [CrossRef]
27. Najafi, M.; Kermanpur, A.; Rahimipour, M.R.; Najafizadeh, A. Effect of TiO_2 morphology on structure of TiO_2-graphene oxide nanocomposite synthesized via a one-step hydrothermal method. *J. Alloys Compd.* **2017**, *722*, 272–277. [CrossRef]
28. Nasir, J.A.; Rehman, Z.u.; Shah, S.N.A.; Khan, A.; Butler, I.S.; Catlow, C.R.A. Recent developments and perspectives in CdS-based photocatalysts for water splitting. *J. Mater. Chem. A* **2020**, *8*, 20752–20780. [CrossRef]
29. Zaleska, A. Doped-TiO_2: A Review. *Recent Pat. Eng.* **2008**, *2*, 157–164. [CrossRef]
30. Yang, D.-X.; Qu, D.; Miao, X.; Jiang, W.-S.; An, L.; Wen, Y.-J.; Wu, D.-D.; Sun, Z.-C. TiO_2 sensitized by red-, green-, blue-emissive carbon dots for enhanced H_2 production. *Rare Met.* **2019**, *38*, 404–412. [CrossRef]
31. Rajaambal, S.; Sivaranjani, K.; Gopinath, C.S. Recent developments in solar H_2 generation from water splitting. *J. Chem. Sci.* **2015**, *127*, 33–47. [CrossRef]
32. Ye, Z.; Kong, L.; Chen, F.; Chen, Z.; Lin, Y.; Liu, C. A comparative study of photocatalytic activity of ZnS photocatalyst for degradation of various dyes. *Optik* **2018**, *164*, 345–354. [CrossRef]
33. Ajibade, P.A.; Mbuyazi, T.B.; Oluwalana, A.E. Lead Sulphide Nanoparticles as Photocatalyst for the Degradation of Methylene Blue: Effects of pH, Time, Adsorption Kinetics and Recyclability Studies. *J. Inorg. Organomet. Polym. Mater.* **2021**, *31*, 2197–2208. [CrossRef]
34. Luttrell, T.; Halpegamage, S.; Tao, J.; Kramer, A.; Sutter, E.; Batzill, M. Why is anatase a better photocatalyst than rutile?—Model studies on epitaxial TiO_2 films. *Sci. Rep.* **2014**, *4*, 4043. [CrossRef] [PubMed]

35. Bourikas, K.; Kordulis, C.; Lycourghiotis, A. Titanium Dioxide (Anatase and Rutile): Surface Chemistry, Liquid–Solid Interface Chemistry, and Scientific Synthesis of Supported Catalysts. *Chem. Rev.* **2014**, *114*, 9754–9823. [CrossRef]
36. Khlyustova, A.; Sirotkin, N.; Kusova, T.; Kraev, A.; Titov, V.; Agafonov, A. Doped TiO_2: The effect of doping elements on photocatalytic activity. *Mater. Adv.* **2020**, *1*, 1193–1201. [CrossRef]
37. Sultana, M.; Mondal, A.; Islam, S.; Khatun, M.A.; Rahaman, M.H.; Chakraborty, A.K.; Rahman, M.S.; Rahman, M.M.; Nur, A.S.M. Strategic development of metal doped TiO_2 photocatalysts for enhanced dye degradation activity under UV–Vis irradiation: A review. *Curr. Res. Green Sustain. Chem.* **2023**, *7*, 100383. [CrossRef]
38. Cai, Y.; Feng, Y.P. Review on charge transfer and chemical activity of TiO_2: Mechanism and applications. *Prog. Surf. Sci.* **2016**, *91*, 183–202. [CrossRef]
39. Fujishima, A.; Honda, K. Electrochemical Photolysis of Water at a Semiconductor Electrode. *Nature* **1972**, *238*, 37–38. [CrossRef] [PubMed]
40. Bahnemann, D.; Henglein, A.; Lilie, J.; Spanhel, L. Flash photolysis observation of the absorption spectra of trapped positive holes and electrons in colloidal titanium dioxide. *J. Phys. Chem.* **1984**, *88*, 709–711. [CrossRef]
41. Xu, Y.; Qiu, F.; Xu, S.-Q.; Zhu, S.; Zhang, Y.; Liu, H.; Zong, P.; Ma, L.; Hong, K.; Wang, Y. Boosting Photocatalytic Hydrogen Evolution by a Light Coupling and Charge Carrier Confinement Strategy. *Adv. Funct. Mater.* **2025**, *35*, 2417553. [CrossRef]
42. Gao, W.; Lu, J.; Zhang, S.; Zhang, X.; Wang, Z.; Qin, W.; Wang, J.; Zhou, W.; Liu, H.; Sang, Y. Suppressing Photoinduced Charge Recombination via the Lorentz Force in a Photocatalytic System. *Adv. Sci.* **2019**, *6*, 1901244. [CrossRef]
43. Guo, Q.; Zhou, C.; Ma, Z.; Yang, X. Fundamentals of TiO_2 Photocatalysis: Concepts, Mechanisms, and Challenges. *Adv. Mater.* **2019**, *31*, 1901997. [CrossRef]
44. Rahimi, N.; Pax, R.A.; Gray, E.M. Review of functional titanium oxides. I: TiO_2 and its modifications. *Prog. Solid State Chem.* **2016**, *44*, 86–105. [CrossRef]
45. Humayun, M.; Raziq, F.; Khan, A.; Luo, W. Modification strategies of TiO_2 for potential applications in photocatalysis: A critical review. *Green Chem. Lett. Rev.* **2018**, *11*, 86–102. [CrossRef]
46. Augustin, A.; Ganguly, P.; Shenoy, S.; Chuaicham, C.; Pillai, S.C.; Sasaki, K.; Lee, A.F.; Sekar, K. Impact of Hole Scavengers on Efficient Photocatalytic Hydrogen Production. *Adv. Sustain. Syst.* **2024**, *8*, 2400321. [CrossRef]
47. Rozman, N.; Nadrah, P.; Cornut, R.; Jousselme, B.; Bele, M.; Dražić, G.; Gaberšček, M.; Kunej, Š.; Škapin, A.S. TiO_2 photocatalyst with single and dual noble metal co-catalysts for efficient water splitting and organic compound removal. *Int. J. Hydrogen Energy* **2021**, *46*, 32871–32881. [CrossRef]
48. Ding, D.; Liu, K.; He, S.; Gao, C.; Yin, Y. Ligand-Exchange Assisted Formation of Au/TiO_2 Schottky Contact for Visible-Light Photocatalysis. *Nano Lett.* **2014**, *14*, 6731–6736. [CrossRef]
49. Yu, S.; Han, B.; Lou, Y.; Liu, Z.; Qian, G.; Wang, Z. Rational design and fabrication of TiO_2 nano heterostructure with multi-junctions for efficient photocatalysis. *Int. J. Hydrogen Energy* **2020**, *45*, 28640–28650. [CrossRef]
50. Brady, M.D.; Sampaio, R.N.; Wang, D.; Meyer, T.J.; Meyer, G.J. Dye-Sensitized Hydrobromic Acid Splitting for Hydrogen Solar Fuel Production. *J. Am. Chem. Soc.* **2017**, *139*, 15612–15615. [CrossRef]
51. Jaafar, S.N.H.; Minggu, L.J.; Arifin, K.; Kassim, M.B.; Wan, W.R.D. Natural dyes as TIO_2 sensitizers with membranes for photoelectrochemical water splitting: An overview. *Renew. Sustain. Energy Rev.* **2017**, *78*, 698–709. [CrossRef]
52. Nikolaou, V.; Charalambidis, G.; Ladomenou, K.; Nikoloudakis, E.; Drivas, C.; Vamvasakis, I.; Panagiotakis, S.; Landrou, G.; Agapaki, E.; Stangel, C.; et al. Controlling Solar Hydrogen Production by Organizing Porphyrins. *ChemSusChem* **2021**, *14*, 961–970. [CrossRef]
53. Mulewa, W.; Tahir, M.; Amin, N.A.S. MMT-supported Ni/TiO_2 nanocomposite for low temperature ethanol steam reforming toward hydrogen production. *Chem. Eng. J.* **2017**, *326*, 956–969. [CrossRef]
54. Xing, X.; Zhang, M.; Hou, L.; Xiao, L.; Li, Q.; Yang, J. Z-scheme BCN-TiO_2 nanocomposites with oxygen vacancy for high efficiency visible light driven hydrogen production. *Int. J. Hydrogen Energy* **2017**, *42*, 28434–28444. [CrossRef]
55. Muñoz-Batista, M.J.; Rodríguez-Padrón, D.; Puente-Santiago, A.R.; Kubacka, A.; Luque, R.; Fernández-García, M. Sunlight-Driven Hydrogen Production Using an Annular Flow Photoreactor and g-C3N4-Based Catalysts. *ChemPhotoChem* **2018**, *2*, 870–877. [CrossRef]
56. Sun, S. N-Doped TiO_2 Nanobelts with Coexposed (001) and (101) Facets and Their Highly Efficient Visible-Light-Driven Photocatalytic Hydrogen Production. *ACS Appl. Mater.* **2016**, *8*, 18126–18131. [CrossRef]
57. Wu, M.-C.; Hiltunen, J.; Sápi, A.; Avila, A.; Larsson, W.; Liao, H.-C.; Huuhtanen, M., Tóth, G.; Shchukarev, A.; Laufer, N.; et al. Nitrogen-Doped Anatase Nanofibers Decorated with Noble Metal Nanoparticles for Photocatalytic Production of Hydrogen. *ACS Nano* **2011**, *5*, 5025–5030. [CrossRef]
58. Kumaravel, V.; Mathew, S.; Bartlett, J.; Pillai, S.C. Photocatalytic hydrogen production using metal doped TiO_2: A review of recent advances. *Appl. Catal. B Environ.* **2019**, *244*, 1021–1064. [CrossRef]
59. Banerjee, B.; Amoli, V.; Maurya, A.; Sinha, A.K.; Bhaumik, A. Green synthesis of Pt-doped TiO_2 nanocrystals with exposed (001) facets and mesoscopic void space for photo-splitting of water under solar irradiation. *Nanoscale* **2015**, *7*, 10504–10512. [CrossRef]

60. Fontelles-Carceller, O.; Muñoz-Batista, M.J.; Rodríguez-Castellón, E.; Conesa, J.C.; Fernández-García, M.; Kubacka, A. Measuring and interpreting quantum efficiency for hydrogen photo-production using Pt-titania catalysts. *J. Catal.* **2017**, *347*, 157–169. [CrossRef]
61. Fang, J.; Cao, S.-W.; Wang, Z.; Shahjamali, M.M.; Loo, S.C.J.; Barber, J.; Xue, C. Mesoporous plasmonic Au–TiO_2 nanocomposites for efficient visible-light-driven photocatalytic water reduction. *Int. J. Hydrogen Energy* **2012**, *37*, 17853–17861. [CrossRef]
62. Gogoi, D.; Namdeo, A.; Golder, A.K.; Peela, N.R. Ag-doped TiO_2 photocatalysts with effective charge transfer for highly efficient hydrogen production through water splitting. *Int. J. Hydrogen Energy* **2020**, *45*, 2729–2744. [CrossRef]
63. Ouyang, W.; Muñoz-Batista, M.J.; Kubacka, A.; Luque, R.; Fernández-García, M. Enhancing photocatalytic performance of TiO_2 in H_2 evolution via Ru co-catalyst deposition. *Appl. Catal. B Environ.* **2018**, *238*, 434–443. [CrossRef]
64. Díaz, L.; Rodríguez, V.D.; González-Rodríguez, M.; Rodríguez-Castellón, E.; Algarra, M.; Núñez, P.; Moretti, E. M/TiO_2 (M = Fe, Co, Ni, Cu, Zn) catalysts for photocatalytic hydrogen production under UV and visible light irradiation. *Inorg. Chem. Front.* **2021**, *8*, 3491–3500. [CrossRef]
65. David, T.M.; Dev, P.R.; Wilson, P.; Sagayaraj, P.; Mathews, T. A critical review on the variations in anodization parameters toward microstructural formation of TiO_2 nanotubes. *Electrochem. Sci. Adv.* **2022**, *2*, e202100083. [CrossRef]
66. Sadanandam, G.; Lalitha, K.; Kumari, V.D.; Shankar, M.V.; Subrahmanyam, M. Cobalt doped TiO_2: A stable and efficient photocatalyst for continuous hydrogen production from glycerol: Water mixtures under solar light irradiation. *Int. J. Hydrogen Energy* **2013**, *38*, 9655–9664. [CrossRef]
67. Dholam, R.; Patel, N.; Adami, M.; Miotello, A. Hydrogen production by photocatalytic water-splitting using Cr- or Fe-doped TiO_2 composite thin films photocatalyst. *Int. J. Hydrogen Energy* **2009**, *34*, 5337–5346. [CrossRef]
68. Tudu, B.; Nalajala, N.; Prabhakar Reddy, K.; Saikia, P.; Gopinath, C.S. Rationally Designed, Efficient, and Earth-Abundant Ni–Fe Cocatalysts for Solar Hydrogen Generation. *ACS Sustain. Chem. Eng.* **2021**, *9*, 13915–13925. [CrossRef]
69. Pinna, M.; Wei, A.W.W.; Spanu, D.; Will, J.; Yokosawa, T.; Spiecker, E.; Recchia, S.; Schmuki, P.; Altomare, M. Amorphous NiCu Thin Films Sputtered on TiO_2 Nanotube Arrays: A Noble-Metal Free Photocatalyst for Hydrogen Evolution. *ChemCatChem* **2022**, *14*, e202201052. [CrossRef]
70. Gao, D.; Liu, W.; Xu, Y.; Wang, P.; Fan, J.; Yu, H. Core-shell Ag@Ni cocatalyst on the TiO_2 photocatalyst: One-step photoinduced deposition and its improved H_2-evolution activity. *Appl. Catal. B Environ.* **2020**, *260*, 118190. [CrossRef]
71. Ngo, N.M.; Nguyen, M.D.; Tran, H.-V.; Medhi, R.; Lee, J.M.; Lee, T.R. Photocatalytic Hydrogen Generation by Monodisperse TiO_2 Nanoparticles Singly and Dually Doped with Niobium and Tantalum. *ACS Appl. Nano Mater.* **2025**, *8*, 4841–4851. [CrossRef]
72. Afrin, M.F.; Furukawa, M.; Tateishi, I.; Katsumata, H.; Uzzaman, M.; Kaneco, S. Enhanced Photocatalytic Hydrogen Generation from Methanol Solutions via In Situ Ni/Pt Co-Deposition on TiO_2. *J. Compos. Sci.* **2025**, *9*, 68. [CrossRef]
73. Ibrahim, N.S.; Leaw, W.L.; Mohamad, D.; Alias, S.H.; Nur, H. A critical review of metal-doped TiO_2 and its structure–physical properties–photocatalytic activity relationship in hydrogen production. *Int. J. Hydrogen Energy* **2020**, *45*, 28553–28565. [CrossRef]
74. Charisiadis, A.; Nikolaou, V.; Nikoloudakis, E.; Ladomenou, K.; Charalambidis, G.; Coutsolelos, A.G. Metalloporphyrins in bio-inspired photocatalytic conversions. *Chem. Commun.* **2025**, *61*, 4630–4646. [CrossRef]
75. Landrou, G.; Panagiotopoulos, A.A.; Ladomenou, K.; Coutsolelos, A.G. Photochemical hydrogen evolution using Sn-porphyrin as photosensitizer and a series of Cobaloximes as catalysts. *J. Porphyr. Phthalocyanines* **2016**, *20*, 534–541. [CrossRef]
76. Xing, X.; Zhu, H.; Zhang, M.; Hou, L.; Li, Q.; Yang, J. Interfacial oxygen vacancy layer of a Z-scheme BCN–TiO_2 heterostructure accelerating charge carrier transfer for visible light photocatalytic H_2 evolution. *Catal. Sci. Technol.* **2018**, *8*, 3629–3637. [CrossRef]
77. Guo, X.; Li, X.; Qin, L.; Kang, S.-Z.; Li, G. A highly active nano-micro hybrid derived from Cu-bridged TiO_2/porphyrin for enhanced photocatalytic hydrogen production. *Appl. Catal. B Environ.* **2019**, *243*, 1–9. [CrossRef]
78. Gonuguntla, S.; Tiwari, A.; Madanaboina, S.; Lingamallu, G.; Pal, U. Revealing high hydrogen evolution activity in zinc porphyrin sensitized hierarchical porous TiO_2 photocatalysts. *Int. J. Hydrogen Energy* **2020**, *45*, 7508–7516. [CrossRef]
79. Chanda, N.; Koteshwar, D.; Gonuguntla, S.; Bojja, S.; Pal, U.; Giribabu, L. Efficient visible-light-driven hydrogen production by Zn–porphyrin based photocatalyst with engineered active donor–acceptor sites. *Mater. Adv.* **2021**, *2*, 4762–4771. [CrossRef]
80. Gkika, D.A.; Ladomenou, K.; Bououdina, M.; Mitropoulos, A.C.; Kyzas, G.Z. Adsorption and photocatalytic applications of porphyrin-based materials for environmental separation processes: A review. *Sci. Total Environ.* **2024**, *908*, 168293. [CrossRef]
81. Nikoloudakis, E.; López-Duarte, I.; Charalambidis, G.; Ladomenou, K.; Ince, M.; Coutsolelos, A.G. Porphyrins and phthalocyanines as biomimetic tools for photocatalytic H_2 production and CO_2 reduction. *Chem. Soc. Rev.* **2022**, *51*, 6965–7045. [CrossRef]
82. Nikolaou, V.; Agapaki, E.; Nikoloudakis, E.; Achilleos, K.; Ladomenou, K.; Charalambidis, G.; Triantafyllou, E.; Coutsolelos, A.G. Highly efficient light-driven hydrogen evolution utilizing porphyrin-based nanoparticles. *Chem. Commun.* **2023**, *59*, 11256–11259. [CrossRef]
83. Agapaki, E.; Ladomenou, K.; Nikolaou, V.; Coutsolelos, A.G. Efficient solar hydrogen production of zinc trimesityl porphyrin-based photocatalysts. *J. Porphyr. Phthalocyanines* **2023**, *27*, 479–489. [CrossRef]

84. Wang, D.; Niu, L.; Qiao, Z.-Y.; Cheng, D.-B.; Wang, J.; Zhong, Y.; Bai, F.; Wang, H.; Fan, H. Synthesis of Self-Assembled Porphyrin Nanoparticle Photosensitizers. *ACS Nano* **2018**, *12*, 3796–3803. [CrossRef] [PubMed]
85. Wang, Q.; Shi, Y.; Ma, Q.; Gao, D.; Zhong, J.; Li, J.; Wang, F.; He, Y.; Wang, R. A flower-like TiO_2 with photocatalytic hydrogen evolution activity modified by Zn(II) porphyrin photocatalysts. *J. Mater. Sci. Mater. Electron.* **2017**, *28*, 2123–2127. [CrossRef]
86. Zhu, H.; Zhu, W.; Kang, S.-Z.; Qin, L.; Zhang, T.; Li, X. Facile interfacial modification in a novel TiO_2/porphyrin nano-micro hybrid for dual functions of boosting photocatalytic H_2 production and methylene blue degradation. *Appl. Surf. Sci.* **2024**, *644*, 158783. [CrossRef]
87. Liu, Y.; Lv, X.; Zhong, Y.; Wang, G.; Liu, S.; Chen, S.; Qi, C.; He, M.; Shangguan, P.; Luo, Z.; et al. Self-Assembly Regulated Photocatalysis of Porphyrin-TiO_2 Nanocomposites. *Molecules* **2024**, *29*, 3872. [CrossRef]
88. Wang, S.; Hu, G.; Dou, Y.; Li, S.; Li, M.; Feng, H.; Feng, Y.-S. Z-scheme promoted interfacial charge transfer on Cu/In-porphyrin MOFs/CdIn2S4 heterostructure for efficient photocatalytic H_2 evolution. *Sep. Purif. Technol.* **2025**, *354*, 129220. [CrossRef]
89. Myltykbayeva, Z.; López Nieto, J.M.; Moreno-Torralbo, B.M.; Concepción, P.; Abylaikhan, A.; Koca, A. Optimizing the electrocatalytic hydrogen production with vanadyl porphyrin impregnated on mesoporous SiO_2 nanoparticles. *Electrochim. Acta* **2025**, *514*, 145686. [CrossRef]
90. Xiao, L.; Zhang, Q.; Wang, X.; Li, B.; Li, X.; Yang, X.; Li, W.; Zhang, H. Integration of platinum nanoparticles and Pd-porphyrin photosensitiser into a metal–organic framework for effective photocatalytic hydrogen evolution. *J. Colloid Interface Sci.* **2025**, *685*, 165–172. [CrossRef]
91. Van Tran, C.; Dang, T.-D.; Lai, V.D.; Thi Nguyen, G.; Thi Hoa, B.; Nguyen, H.T.; Thi, H.P.N.; Nguyen, T.T.; La, D.D. Self-assembly of porphyrin nanofiber on the surface $CuFe_2O_4$ nanoparticles: A novel photoanode for enhanced photo-electrochemical water splitting. *Fuel* **2025**, *380*, 133196. [CrossRef]

inorganics

MDPI

Article

Solvent-Driven Structural Modulation of Co-Ni_3S_2 and Impact on Electrochemical Water Splitting

Sagar M. Mane [1,†], Komal S. Wagh [1,†], Sangoh Lee [2], Aviraj M. Teli [3], Ganesh T. Chavan [4], Jae Cheol Shin [3,*] and Jaewoong Lee [1,*]

[1] Department of Fiber System Engineering, Yeungnam University, 280 Dehak-Ro, Gyeongsan 38541, Republic of Korea; manesagar99@gmail.com (S.M.M.)
[2] Department of Clothing and Fashion, Yeungnam University, 280 Dehak-Ro, Gyeongsan 38541, Republic of Korea
[3] Division of Electronics and Electrical Engineering, Dongguk University, Seoul Campus, Seoul 04620, Republic of Korea
[4] Department of Civil and Environmental Engineering, Hanyang University ERICA, Ansan 15588, Republic of Korea
* Correspondence: jcshin@dgu.ac.kr (J.C.S.); jaewlee@yu.ac.kr (J.L.)
† These authors contributed equally to this work.

Abstract: Understanding the role of synthesis parameters in tailoring catalyst morphology is crucial for enhancing performance in electrochemical water splitting. This research systematically explores how different solvent environments affect the structural evolution and morphology of cobalt-doped nickel sulfide (Co-Ni_3S_2) nanomaterials. By systematically modifying the solvent environment using ethylene glycol and glycerol, distinct morphologies of Co-Ni_3S_2 were obtained, leading to variations in their electrocatalytic water-splitting performance. The fabricated compounds were thoroughly tested for their catalytic performance in facilitating hydrogen and oxygen evolution processes. Notably, the use of ethylene glycol as a synthesis medium led to the formation of a unique interconnected petal-like structure, significantly improving electrocatalytic activity, as evidenced by low overpotentials of 190.7 mV for HER at 10 mA cm^{-2} and 414 mV for OER at 30 mA cm^{-2}. In contrast, when glycerol was employed as the solvent, the resulting Co-Ni_3S_2 material displayed overpotentials of 223.8 mV and 535 mV for HER and OER, respectively. Eventually, Co-doping was found to enhance the electrocatalytic performance, as pure Ni_3S_2 synthesized under the same solvent conditions exhibited higher overpotentials for both HER and OER. These findings underscore the crucial role of solvent selection in tailoring the structural and functional properties of materials for high-performance electrochemical applications.

Keywords: Co-Ni_3S_2; solvent variation; structural modulation; electrocatalytic water splitting

1. Introduction

The worldwide transition to environmentally friendly and renewable energy solutions has intensified interest in electrochemical water splitting as a viable method for generating hydrogen fuel. A key focus within this approach is the creation of high-performance electrocatalysts that can efficiently facilitate both the hydrogen evolution reaction (HER) and the oxygen evolution reaction (OER), while requiring the least possible energy input [1–4]. In general, the hydrogen evolution reaction (HER) typically occurs at the cathode while the oxygen evolution reaction takes place at the anode, and its overall process in alkaline electrolyte can be expressed by the following equations [5];

$$2H_2O + 2e^- \rightarrow H_2 + 2OH^- E^\circ \quad (1)$$

$$2OH^- \rightarrow \frac{1}{2}O_2 + 2H_2O + 2e^- E^\circ \quad (2)$$

Traditionally, noble metals such as platinum (Pt) for hydrogen evolution reactions and ruthenium dioxide (RuO_2) or iridium dioxide (IrO_2) for oxygen evolution reactions have been regarded as benchmark electrocatalysts due to their exceptional activity and stability. However, their widespread application is hindered by several drawbacks, primarily the limited availability and prohibitive cost of these precious elements. This has spurred intensive research into the development of earth-abundant, cost-effective alternatives that can deliver comparable performance in water-splitting processes [6–8]. Recent advancements have led to comprehensive research into a wide array of materials, including transition metal-based oxides, hydroxides, sulfides, phosphides, nitrides, and carbides, which have emerged as efficient bifunctional electrocatalysts capable of driving both hydrogen evolution and oxygen evolution reactions in integrated water-splitting systems [3,9,10]. These materials have garnered considerable attention due to their intrinsic catalytic activity, structural versatility, and favorable electronic configurations, which collectively contribute to their efficiency in driving the dual electrochemical processes required for sustainable hydrogen production. Among the transition metal compound family, transition metal sulfides (TMS) are considered promising alternatives to precious metal catalysts due to their superior electrical conductivity and catalytic activity [11,12].

In transition metal sulfides, nickel-based sulfides (Ni_xS_y), such as Ni_3S_2, NiS, and NiS_2, feature an extended Ni-Ni bonding network within their crystal structure, which imparts distinct metallic characteristics and enhances electrical conductivity [13,14]. Nonetheless, the electrocatalytic efficiency of nickel sulfides is still hindered by their sparse active site distribution and suboptimal electrical conductivity. Addressing these challenges necessitates continued refinement and engineering of this non-precious metal material to achieve improved catalytic performance. The catalytic efficiency of any material is fundamentally governed by several interrelated parameters: rapid electron transport ensures swift charge transfer during reactions; low energy barriers facilitate favorable reaction kinetics, and a high density of active sites enhances overall reactivity by offering more locations for catalytic events to occur [15]. These intrinsic properties can be systematically optimized through surface and interface engineering techniques. For instance, defect engineering, such as the introduction of vacancies, dislocations, and grain boundaries via plasma treatment, can disrupt lattice symmetry and create localized electronic states that serve as new catalytic centers [16,17]. Compositional doping modifies the electronic structure and conductivity of the catalyst, often introducing new energy levels or shifting the Fermi level to boost charge mobility and adsorption properties [18,19]. Additionally, the construction of heterogeneous structures enables the formation of high-energy interfaces that promote synergistic interactions between different phases, facilitating improved adsorption, electron transfer, and intermediate stabilization [20]. Together, these strategies offer a multi-faceted approach to enhancing catalytic performance at the atomic and nanoscale levels.

In this context, Cui et al. [21] synthesized a bifunctional electrocatalyst based on Mo-doped Ni_3S_2 by adjusting the reaction temperature during the hydrothermal process to obtain different structural morphologies. Among the three resulting structures, the porous nanorod configuration exhibited the most efficient performance for overall water splitting. When integrated with nickel foam as a substrate, the Mo-Ni_3S_2 electrode exhibited low overpotentials, specifically 180 mV for oxygen evolution and 278 mV for hydrogen evolution when tested at a current density of 100 mA cm^{-2}. Kong et al. [22] used interface engineering to develop an N-doped Ni_3S_2/CoS_2 heterostructure catalyst that was directly fabricated on nickel foam through electrochemical deposition and low-temperature vulcanization, forming a highly conductive and interface-rich material. This catalyst demonstrated outstanding bifunctional performance for water splitting, achieving 10 mA cm^{-2} at low overpotentials for both HER (100 mV) and OER (245 mV), and requiring only 1.56 V for overall water electrolysis in alkaline media. MoS_2 was employed as a surface modifier to enhance the water-splitting capabilities of Ni_3S_2 nanosheets deposited on nickel foam, as demonstrated by Song et al. [2]. The disordered amorphous MoS_2 (a-MoS_2) interacted synergistically with the conductive Ni_3S_2 substrate, leading to outstanding bifunctional electrocatalytic activity. Under alkaline conditions, this setup demonstrated highly efficient overall water splitting, operating at a cell voltage of just 1.54 V. It exhibited minimal overpotentials, only 81 mV for the hydrogen evolution reaction (HER) at 10 mA cm^{-2} and 150 mV for the oxygen evolution reaction (OER) at 100 mA cm^{-2}. Yu et al. [6] introduced a method for synthesizing Co and Ce co-doped Ni_3S_2 nanosheets directly on nickel foam (CoCe-Ni_3S_2/NF), aiming to develop a high-performance catalyst for overall water splitting. The dual doping of cobalt and cerium enhances electronic interactions within the material, increasing active site availability and promoting efficient electron transport. The catalyst's ultrathin 2D nanosheet architecture (~10 nm thick) improves electrolyte penetration and facilitates effective gas evolution. The catalyst exhibits remarkable electrocatalytic performance, achieving overpotentials as low as 285 mV for the oxygen evolution reaction (OER) and 193 mV for the hydrogen evolution reaction (HER) at a current density of 50 mA cm^{-2} in 1.0 M KOH. It also maintains stable performance over 100 h and reaches 1.68 V at 20 mA cm^{-2} during overall water splitting. In addition to these studies, numerous efforts have focused on enhancing the bifunctional electrocatalytic activity of Ni_3S_2 by employing strategies such as structural modification, interface engineering, cation doping, and tuning of synthesis conditions. However, the influence of solvent variation as a route for structural modification remains largely unexplored, particularly in the context of bifunctional water-splitting applications of doped and undoped Ni_3S_2. The choice of solvent plays an active role in influencing nucleation dynamics, controlling growth kinetics, and directing phase formation, thereby significantly affecting the physicochemical properties of the synthesized materials. In particular, solvents such as ethylene glycol and glycerol offer complementary physicochemical properties that are highly relevant for nanostructure engineering. Ethylene glycol, with its moderate polarity and reducing ability, promotes uniform nucleation and assists in tuning the electronic structure of the active phase. In contrast, glycerol, owing to its higher viscosity and stronger intermolecular interactions, slows down ion diffusion and crystal growth, thereby enabling better control over particle size, morphology, and surface orientation. Moreover, the abundant hydroxyl groups in both solvents interact with precursor ions and crystal facets, influencing surface energies and stabilizing anisotropic growth. The combined effects of polarity, viscosity, and coordinating ability directly dictate the structural, morphological, and compositional characteristics of the resulting materials, which in turn impact their electrochemical activity and stability in water-splitting applications [23,24].

Building upon the above discussion, this study explores the solvent-driven structural modulation of Co-Ni_3S_2 synthesized under two distinct solvent environments and examines how these structural variations influence electrochemical performance in water-splitting reactions. A two-step solvothermal–hydrothermal synthesis strategy was employed, wherein ethylene glycol and glycerol served as solvents in the first step to facilitate the formation of the Co-Ni hydroxide framework. This intermediate was subsequently subjected to a sulfurization process in the second step to yield the final Co-Ni_3S_2 structure. By elucidating the role of different solvents in tailoring the structural and physicochemical properties of Co-doped Ni-sulfide, this work aims to provide a rational basis for optimizing their catalytic activity in water-splitting applications.

2. Results and Discussion

To investigate the crystallinity of the samples, X-ray diffraction (XRD) analysis was conducted on Co-Ni_3S_2 structures grown on Ni foam. The corresponding diffraction patterns are shown in Figure 1 for Co-doped Ni_3S_2 and Figure S1 (Supplementary Information) for pure Ni_3S_2. The synthesis involved constructing Co–Ni hydroxide frameworks within two separate solvent media: ethylene glycol and glycerol. Following the sulfurization step, the products were categorized as CNS-E.G. and CNS-Glycerol for the Co-doped Ni_3S_2 catalysts, while the corresponding undoped Ni_3S_2 samples were labeled NS-E.G. and NS-Glycerol, respectively. While the diffraction signals from the Ni foam substrate are predominant, distinct and relatively intense peaks attributed to Ni_3S_2 can still be identified, indicating the presence and crystalline nature of the active material. The sharp and prominent diffraction peaks of Ni-foam are observed at 44.9°, 52.3°, and 76.7°, corresponding to the crystallographic planes indexed according to JCPDS card No. 04-0850 [3,7]. The XRD peaks of the material deposited on Ni-foam, identified as Ni_3S_2, were observed at 2θ values of 22.3°, 31.65°, 38.4°, 38.7°, 44.7°, 50.2°, 50.4°, 55.4°, 55.5°, 55.8°, 78.02°, 78.20°, and 78.55°. These diffraction peaks correspond to the (101), (110), (021), (003), (202), (211), (113), (300), (122), (104), (401), (223), and (205) crystal planes of Ni_3S_2, respectively, and are consistent with the standard reference pattern JCPDS card No. 01-085-0775 or 00-044-1418 [9,10,15]. This confirms the formation of Ni_3S_2 with a rhombohedral crystal structure. Successful doping of Co ions was confirmed in CNS-E.G. and CNS-Glycerol samples, as no cobalt sulfide phases were detected in the XRD analysis. However, a closer examination of the diffraction patterns reveals a shift in the Co-Ni_3S_2 reflection, synthesized using ethylene glycol, towards a higher 2θ angle, which was also noted for the pure Ni_3S_2 sample (Figure S1). This shift indicates lattice expansion, typically associated with the solvent variation and incorporation of dopant ions with larger ionic radii, suggesting a higher degree of Co incorporation in the CNS-E.G. sample [7,25]. The increased Co incorporation in the ethylene glycol-synthesized sample, indicated by lattice expansion, is expected to enhance water splitting performance by improving charge transfer and increasing active sites.

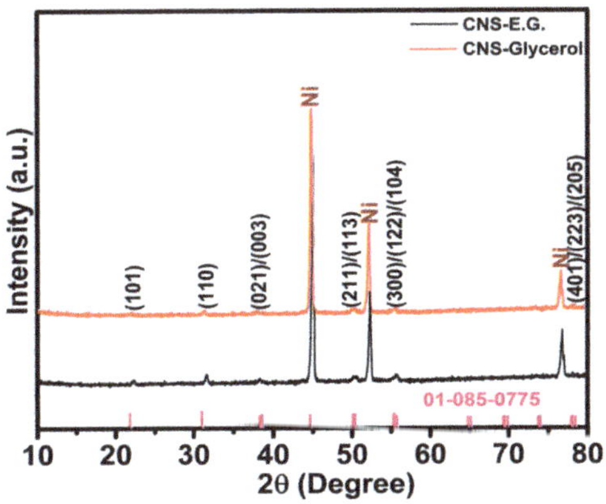

Figure 1. Diffraction patterns of Co-Ni_3S_2 on Ni-foam synthesized using two different solvents (CNS-E.G. and CNS-Glycerol).

Understanding the surface properties of catalysts is vital for improving electrochemical water splitting, since both the hydrogen evolution reaction (HER) and oxygen evolution reaction (OER) are surface-dependent phenomena. To explore these surface-related features, X-ray photoelectron spectroscopy (XPS) was employed to analyze the chemical states and elemental distributions present on the surfaces of the samples. Survey XPS spectra of the CNS-E.G. and CNS-Glycerol samples are illustrated in Figure 2a, indicating the presence of Ni, Co, and S elements in the samples. Figure 2b,c present the deconvoluted Ni 2p spectra for the CNS-E.G. and CNS-glycerol samples, respectively. Although the overall spectral profiles remain similar between the two, noticeable distinctions are evident in the binding energy positions and elusive variations within the peak structures. Both spectra prominently feature the Ni $2p_{3/2}$ and Ni $2p_{1/2}$ spin–orbit components, each accompanied by distinct satellite peaks, signifying the presence of nickel species. The spectral fitting indicates contributions from two oxidation states of nickel: Ni^{2+} and Ni^{3+} [7,14]. Detailed analysis reveals that each spin–orbit component ($2p_{3/2}$ and $2p_{1/2}$) contains two major contributions with the presence of the Ni-metallic peak (Ni^0) shown in the orange color in Figure 2b,c, respectively. The peaks rendered in cyan and olive correspond to Ni^{2+} species, while those in dark yellow and violet are attributed to Ni^{3+}, confirming the coexistence of multiple valence states of nickel in both samples. Additionally, satellite features were identified: the satellite peak associated with the $2p_{3/2}$ level is highlighted in magenta, while the satellite peak related to the $2p_{1/2}$ level appears in navy blue. The binding energies corresponding to the nickel oxidation states were analyzed for both CNS-E.G. and CNS-Glycerol samples. Ni^0 peak was noted at 853 eV for CNS-E.G., and it was at 852.6 eV for the CNS-Glycerol sample. For the Ni^{2+} oxidation state, the CNS-E.G. sample exhibited characteristic peaks at 854.7 eV and 872.8 eV, which are attributed to the Ni $2p_{3/2}$ and Ni $2p_{1/2}$ core levels, respectively. In the case of the CNS-Glycerol sample, these Ni^{2+} peaks were observed at slightly lower energies of 854.5 eV and 872.5 eV for the same core orbitals. Regarding the Ni^{3+} state, the CNS-E.G. sample showed peaks at 855.8 eV ($2p_{3/2}$) and 874.9 eV ($2p_{1/2}$), whereas the CNS-Glycerol counterpart presented these peaks at 855.8 eV and 874.7 eV, indicating a minor shift. Furthermore, satellite peaks corresponding to the Ni $2p_{3/2}$ and $2p_{1/2}$ levels appeared at 860.7 eV and 878.9 eV for the CNS-E.G. sample. In comparison, the CNS-Glycerol sample's satellites were found at slightly lower energies of 860.4 eV and 878.7 eV, respectively [26,27].

As shown in Figure 2d,e, the Co 2p spectra for the CNS-E.G. and CNS-Glycerol samples display two distinct core-level signals, corresponding to $2p_{3/2}$ and $2p_{1/2}$, respectively. Upon deconvolution, the CNS-E.G. spectrum reveals spin–orbit doublets in both core levels (Figure 2d), signifying the presence of both Co^{3+} and Co^{2+} oxidation states along with their associated satellite features. In contrast, the $2p_{3/2}$ core level of the CNS-Glycerol sample exhibits a spin–orbit doublet, as shown in Figure 2e, whereas no such feature is observed in the $2p_{1/2}$ level. This suggests that the integration of cobalt is incomplete when glycerol is used as the solvent, a conclusion that is consistent with observations from the XRD analysis. In the Co 2p XPS analysis, the Co^{3+} species is characterized by distinct peaks appearing in the $2p_{3/2}$ and $2p_{1/2}$ regions, highlighted, respectively, by cyan and gold colors. In contrast, the Co^{2+} state is marked by peaks shown in magenta for the $2p_{3/2}$ level and navy blue for the $2p_{1/2}$ level, which are uniquely observed in the CNS-E.G. sample spectrum. Satellite features related to these core levels are also present, with the $2p_{3/2}$ satellite illustrated in pink and the $2p_{1/2}$ satellite in green. For the CNS-E.G. specimen, the Co^{3+} peaks were found at binding energies of 780.4 eV ($2p_{3/2}$) and 795.5 eV ($2p_{1/2}$), while Co^{2+} peaks appeared at 782.6 eV and 796.3 eV for the same respective levels. The corresponding satellite peaks were located at 785.7 eV ($2p_{3/2}$) and 802.7 eV ($2p_{1/2}$). Meanwhile, in the CNS-glycerol sample, Co^{3+} was detected with peaks at 780.8 eV for $2p_{3/2}$ and 796.7 eV for $2p_{1/2}$. The Co^{2+} peak for the $2p_{3/2}$ level was found at 784.8 eV, coinciding with a satellite peak centered at the same position. An additional satellite peak corresponding to the $2p_{1/2}$ level was observed at 800.1 eV [3,17,19,22,28].

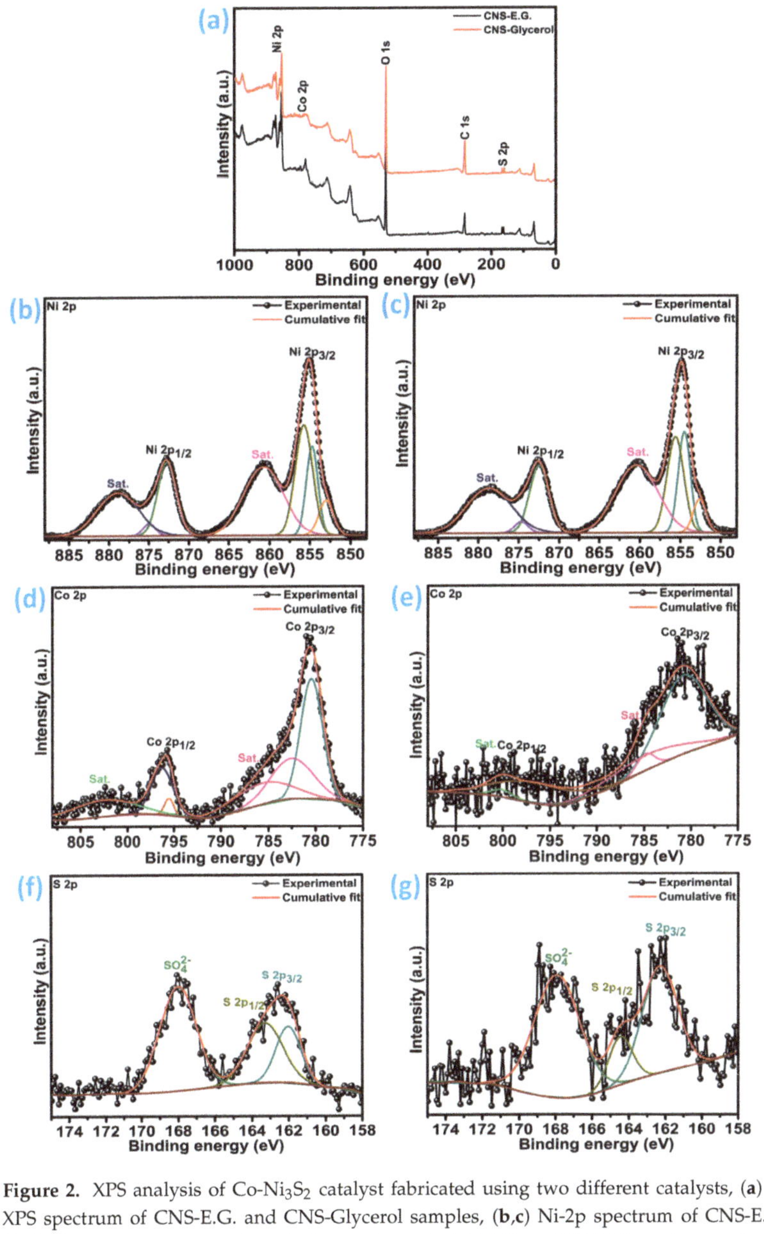

Figure 2. XPS analysis of Co-Ni_3S_2 catalyst fabricated using two different catalysts, (**a**) survey XPS spectrum of CNS-E.G. and CNS-Glycerol samples, (**b**,**c**) Ni-2p spectrum of CNS-E.G. and CNS-Glycerol, respectively, (**d**,**e**) Co-2p spectrum of CNS-E.G. and CNS-Glycerol, respectively, and (**f**,**g**) S-2p spectrum of CNS-E.G. and CNS-Glycerol, respectively.

The sulfur (S 2p) core-level spectra for CNS-E.G. (Figure 2f) and CNS-Glycerol (Figure 2g) display three prominent binding energy peaks, each corresponding to distinct sulfur chemical states. In the CNS-E.G. spectrum, peaks located at 162.0 eV and 163.3 eV, and in the CNS-Glycerol spectrum at 162.3 eV and 164.4 eV, represent the spin–orbit split doublets S $2p_{3/2}$ and S $2p_{1/2}$, respectively. These features are typically linked to sulfur in reduced forms, such as metal sulfides or elemental sulfur [1,2,7]. Additionally, both samples exhibit a peak near 168.0 eV, characteristic of oxidized sulfur species, particularly sulfate ions (SO_4^{2-}) [7,9]. The occurrence of this higher energy peak implies possible surface oxidation or residual contamination from sulfur-containing precursors. Overall, this spectral interpretation sheds light on the sulfur speciation and surface chemical characteristics of the materials under investigation.

Figure 3a–f presents FE-SEM micrographs of Co-Ni_3S_2 structures synthesized on Ni-foam substrates under two different solvent conditions, examined at varying magnifications. In the case of the sample synthesized using ethylene glycol (CNS-E.G.), shown in Figure 3a–c, the surface morphology reveals the formation of ultrathin, curled nanosheets, or "nanopetals". These nanostructures exhibit a high degree of interconnectivity, extending uniformly across the entire Ni-foam surface. The resultant architecture resembles a compact, honeycomb-like network, which is advantageous for maximizing surface area and promoting efficient electron and ion transport pathways. In contrast, the sample prepared using glycerol as the solvent (CNS-Glycerol), depicted in Figure 3d–f, displays a distinctly different morphological profile. Here, the Ni-foam substrate is covered with a densely packed layer of irregularly shaped particles. These particles lack a defined geometry and vary in size, yet they are evenly distributed across the substrate. Over time, their aggregation leads to the development of a slab-like or sheet-like morphology, indicating a different nucleation and growth mechanism compared to the CNS-E.G. sample. These contrasting surface features highlight the significant influence of solvent type on the microstructural evolution of Co-Ni_3S_2 during synthesis, potentially affecting the material's electrochemical properties and overall performance in device applications. This structural variation is also noted for the pure Ni_3S_2 samples fabricated with ethylene glycol and glycerol, as noted in Figure S2a–f, respectively. Comparing the morphology of pure Ni_3S_2 and Co-doped Ni_3S_2, it is evident that Co-doping induces significant structural transformations. In ethylene glycol, pure Ni_3S_2 shows only the initiation of nanosheet growth, whereas Co-doped Ni_3S_2 develops into highly interconnected and uniformly distributed nanosheet networks. In glycerol, pure Ni_3S_2 primarily forms agglomerated nanoparticles, while Co-doping drives the evolution of more organized sheet-like assemblies. The EDS spectra of CNS-E.G. and CNS-Glycerol are presented in Figure 3g,h, respectively, while the corresponding spectra for NS-E.G. and NS-Glycerol are shown in Figure S2g,h. Subsequently, elemental mapping was performed for the CNS-E.G. and CNS-Glycerol samples, as shown in Figure 4a–d and Figure 5a–d, respectively. EDS spectra (Figure S2g,h) of Ni_3S_2 confirm the constitution of Ni and S elements. Furthermore, EDS with elemental mapping of Co-doped Ni_3S_2 samples confirms the presence of Ni, Co, and S elements. This indicates the successful formation of the target materials Ni_3S_2 and Co-doped Ni_3S_2.

Transmission electron microscopy (TEM) was carried out to gain deeper insight into the nanoscale structure of the CNS-E.G. sample, complementing the surface-level observations obtained from FE-SEM. As presented in Figure 6a–d, the TEM micrographs confirm the formation of extremely thin structures that adopt a curled or petal-like morphology. These petals extend laterally over several hundred nanometers, but due to their ultrathin nature, they appear highly transparent to the electron beam. The curled and wrinkled edges, clearly visible in Figure 6a,b, demonstrate the intrinsic flexibility of the nanopetals, a feature that is advantageous for creating a large electrochemically accessible surface area. Such wrinkling not only prevents dense stacking but also exposes a greater number of edge sites, which can act as highly active regions for ion adsorption and charge transfer. Closer examination at higher magnifications (Figure 6c,d) reveals the presence of nanoscale pores distributed across the surface of the petals. The coexistence of ultrathin dimensions with internal porosity provides a hierarchical architecture: the thin sheets shorten electron and ion transport distances, while the pores facilitate electrolyte penetration and enhance accessibility of the internal active regions. This unique combination of curling, porosity, and ultrathin sheet morphology indicates a highly open and accessible network that is expected to significantly improve electrochemical kinetics.

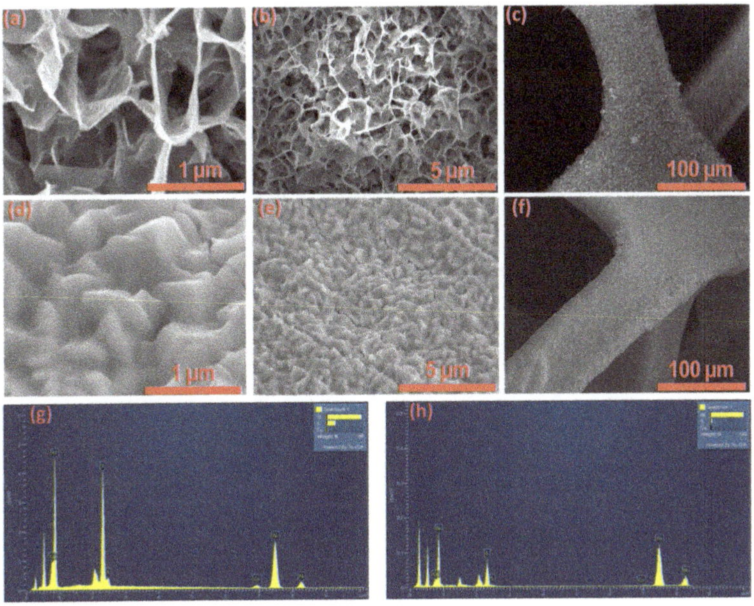

Figure 3. FE-SEM and EDS analysis of Co-Ni_3S_2 catalyst fabricated using two different catalysts, (**a–c**) FE-SEM images of CNS-E.G. at different magnifications, (**d–f**) FE-SEM images of CNS-Glycerol at different magnifications, (**g**,**h**) EDS spectrum of CNS-E.G. and CNS-Glycerol, respectively.

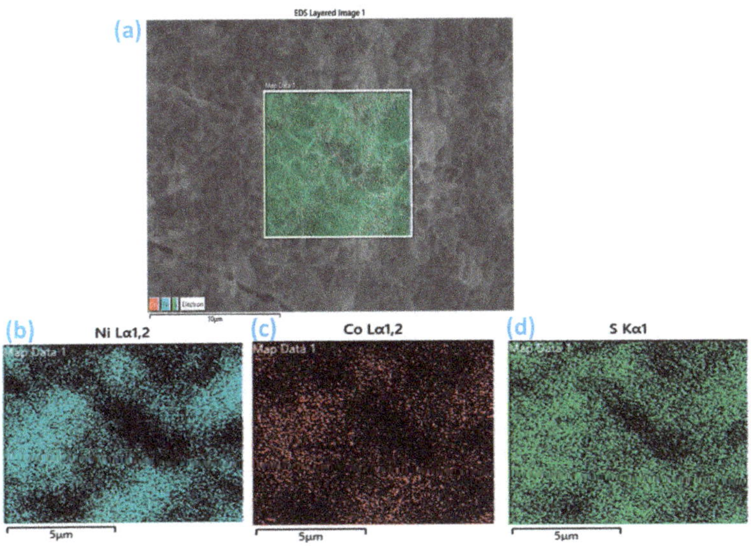

Figure 4. Elemental mapping of CNS-E.G. catalyst, (**a**) overall mapping, (**b**) Ni-element, (**c**) Co-element, and (**d**) S-element.

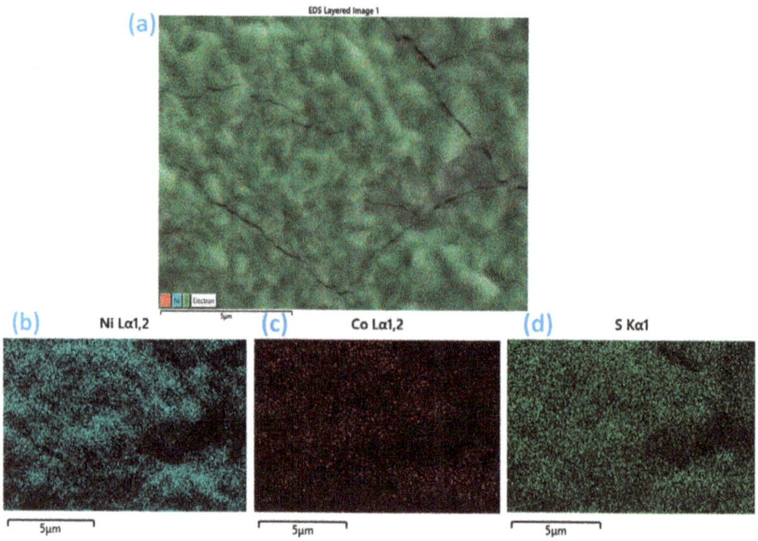

Figure 5. Elemental mapping of CNS-Glycerol catalyst, (**a**) overall mapping, (**b**) Ni-element, (**c**) Co-element, and (**d**) S-element.

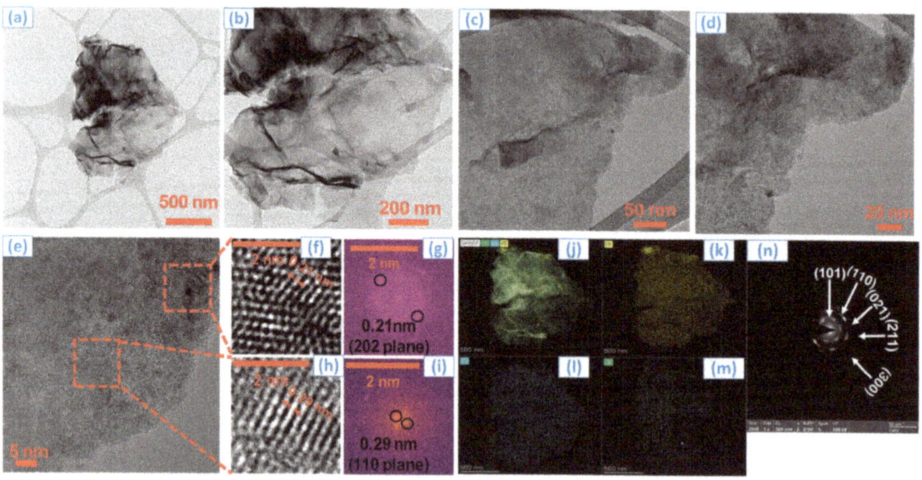

Figure 6. TEM analysis of CNS-E.G. sample, (**a**–**d**) TEM images of CNS-E.G. at different magnifications, (**e**) HRTEM image.

High-resolution TEM (HRTEM) image, as illustrated in Figure 6e, provides more definitive insights into the crystallinity of the nanosheets. Well-resolved lattice fringes are visible, with an interplanar spacing consistent with the (202) and (110) planes of the Co-doped Ni_3S_2 phase, which are represented in Figure 6f,h, respectively, whereas the FFT patterns of both planes were illustrated in Figure 6g,i. This confirms the successful incorporation of Co into the Ni_3S_2 lattice. The presence of coherent lattice fringes throughout large regions suggests high crystallinity, which is favorable for efficient charge transport. The elemental mapping results obtained from TEM, shown in Figure 6j–m, clearly verify that the nanopetals structures are uniformly composed of nickel (Ni), cobalt (Co), and sulfur (S). This elemental distribution is consistent with the findings from other physicochemical characterization techniques, further confirming the successful formation of the intended

Co-doped Ni_3S_2 composition. Selected-area electron diffraction (SAED) pattern represented in Figure 6n collected from the nanosheets displays a set of well-defined diffraction rings/spots corresponding to the characteristic planes of Co-Ni_3S_2, in good agreement with XRD analysis. The polycrystalline nature, as indicated by the diffraction rings, further supports the nanosheet growth mechanism, in which interconnected crystalline domains are assembled into extended networks.

The hydrogen evolution reaction (HER) activities of the Ni_3S_2 and Co-doped Ni_3S_2 catalysts, NS-E.G., NS-Glycerol, CNS-E.G., and CNS-Glycerol, were evaluated through analysis of their polarization curves. To confirm the involvement of the Ni-foam current collector in the overall process, the polarization curve was obtained using the same experimental procedure and compared accordingly. Electrochemical measurements were performed using a conventional three-electrode configuration immersed in a 1.0 M KOH solution. From the linear sweep voltammetry (LSV) results depicted in Figure 7a, the CNS-E.G. catalyst demonstrated a notably lower onset potential at 149.5 mV, whereas the CNS-Glycerol counterpart showed a higher onset potential of 182.1 mV. For the undoped Ni_3S_2 samples, the onset potential values are comparatively higher than those observed for the Co-incorporated counterparts. Specifically, the sample synthesized in ethylene glycol (NS-E.G.) exhibits an onset potential of 243.7 mV, while the sample prepared in glycerol (NS-Glycerol) shows an onset potential of 269.2 mV. This decrease in onset potential for CNS-E.G. is attributed to the ethylene glycol solvent, which enhances cobalt incorporation and fosters the development of a porous honeycomb-like architecture, thereby boosting catalytic efficiency for HER. The overpotentials corresponding to current densities of 10, 30, and 50 mA cm^{-2} were also determined to compare the catalysts' performance under different operating conditions, as presented in Figure 7b. At these current densities, CNS-E.G. required overpotentials of 190.7, 245.6, and 276.8 mV, respectively. In contrast, CNS-Glycerol exhibited higher overpotentials of 223.8, 307.8, and 348.9 mV at the same current levels. For the pristine NS-E.G. sample, the recorded overpotentials are 267.7, 341.7, and 373.4 mV at the corresponding current densities. In contrast, the NS-Glycerol variant exhibits the largest overpotentials among all tested samples, reaching 291.9, 373.1, and 401.8 mV under the same conditions. For the pure Ni-foam current collector, the measured overpotentials are 329.2 mV at a current density of 10 mA cm^{-2} and 525.7 mV at 30 mA cm^{-2}. However, when the current density is increased to 50 mA cm^{-2}, the overpotential exceeds the measurable range, indicating significant polarization and reduced electrochemical performance. These variations in overpotential with increasing current density clearly illustrate the inherent limitations of bare Ni-foam in sustaining high-rate electrochemical reactions, as depicted in Figure 7a. These findings highlight that increasing the current density during HER leads to a pronounced increase in overpotential for all catalysts, with CNS-E.G. consistently outperforming CNS-Glycerol and two pure samples, NS-E.G. and NS-Glycerol. Specifically, the CNS-Glycerol sample shows an increase in overpotential by approximately 55.9%, whereas the CNS-Ethylene Glycol (CNS-E.G.) sample displays a comparatively lower increase of about 45.2% with an increase in current density from 10 to 50 mA cm^{-2}. NS-E.G. has the lowest change among all samples, with 39.5% and NS-glycerol exhibits a change of 40.5%. This suggests a stronger dependence of the CNS-Glycerol sample on the applied current density, potentially reflecting differences in intrinsic conductivity, active surface area, or catalytic efficiency between the two samples under high reaction rates.

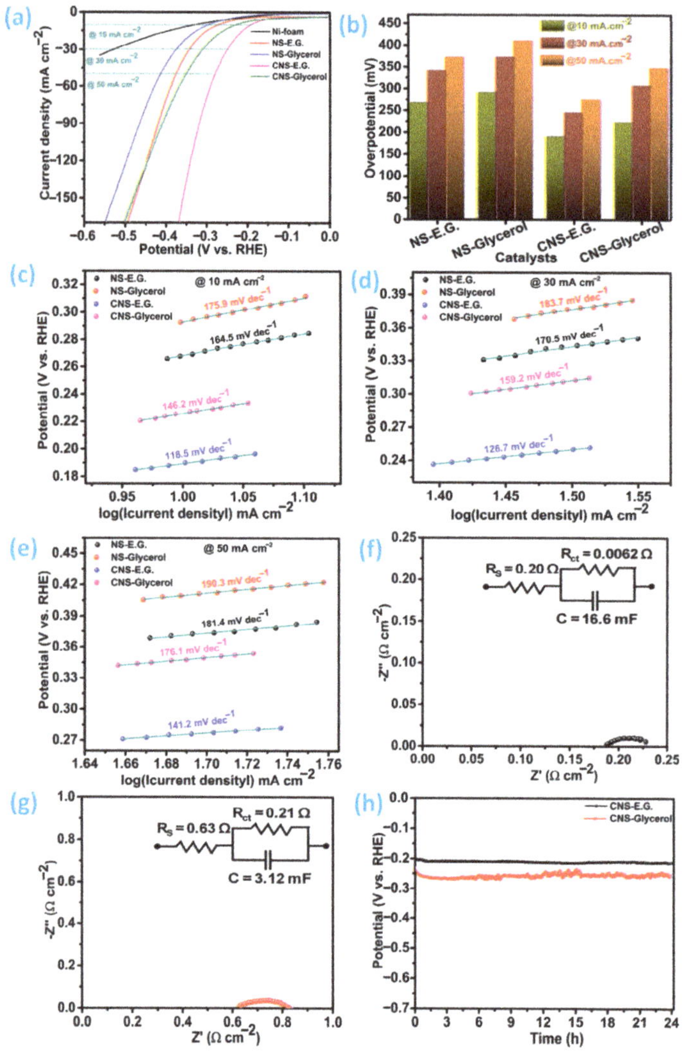

Figure 7. HER activity performance of the Co-Ni_3S_2 catalyst fabricated using two different solvents, (**a**) LSV curve measured at 5 mV s^{-1}, (**b**) estimated overpotential at different current densities, (**c–e**) Tafel slopes at 10, 30, and 50 mA cm^{-2}, respectively, (**f**,**g**) Nyquist plots of CNS-E.G. and CNS-Glycerol, respectively, with equivalent circuit inset, (**h**) HER stability at 10 mA cm^{-2} of CNS-E.G. and CNS-Glycerol.

To thoroughly understand the catalytic reaction kinetics, Tafel slope measurements were conducted at three separate current densities, 10, 30, and 50 mA cm^{-2}, as depicted in Figure 7c–e. The Tafel slope serves as a vital indicator of the electrochemical reaction rate at the electrode surface, where lower values correspond to enhanced reaction efficiency and quicker electron transfer. In this investigation, the CNS-E.G. catalyst consistently exhibited reduced Tafel slope values compared to the CNS-Glycerol, NS-E.G., and NS-Glycerol catalysts across all tested current densities, highlighting its improved kinetic behavior. The CNS-E.G. catalyst exhibited Tafel slopes of 118.5, 126.7, and 141.2 mV dec^{-1} at current densities of 10, 30, and 50 mA cm^{-2}, respectively. In contrast, the CNS-Glycerol variant demonstrated comparatively higher values of 146.2, 159.2, and 176.1 mV dec^{-1} under the same testing conditions. Under identical operating parameters, the NS-E.G. sample recorded slopes of 164.5, 170.5, and 181.4 mV dec^{-1}, whereas the NS-Glycerol catalyst yielded the steepest slopes overall, with values of 175.9, 183.7, and 190.3 mV dec^{-1}. Lower Tafel slopes indicate faster reaction kinetics and more efficient charge-transfer at the electrode. The measured slopes follow the order (best → worst): CNS-E.G. (118.5–141.2 mV dec^{-1}) < CNS-Glycerol (146.2–176.1 mV dec^{-1}) < NS-E.G. (164.5–181.4 mV dec^{-1}) < NS-Glycerol (175.9–190.3 mV dec^{-1}). These findings suggest that the hydrogen evolution reaction (HER) on CNS-E.G. proceeds more efficiently via the Volmer–Heyrovsky pathway in alkaline medium [2,3,18]. The noticeable difference in Tafel slopes between the two catalysts highlights the enhanced charge transfer efficiency and lower energy barrier for the electrochemical reaction in the CNS-E.G. system [3]. This suggests that the structural and compositional features imparted by the ethylene glycol treatment contribute positively to its electrocatalytic performance, particularly under increasing current densities.

Electrochemical impedance spectroscopy (EIS) measurements were performed over a frequency spectrum ranging between 100 Hz and 100 kHz. The excitation voltage applied matched the overpotential recorded at a current density of 10 mA cm^{-2}. The resulting Nyquist plots for the CNS-E.G. and CNS-Glycerol catalysts are presented in Figure 7f,g and Figure S3a,b for NS-E.G. and NS-Glycerol, respectively, all catalysts exhibiting characteristic semicircular profiles. For quantitative analysis, the impedance spectra were modeled using an equivalent circuit, which is displayed as an inset within each Nyquist plot. In such representations, the diameter of the semicircle is directly correlated with the charge transfer resistance (Rct), a critical descriptor of how readily electrons are transported across the catalyst/electrolyte interface. A reduction in Rct corresponds to an accelerated interfacial electron exchange, thereby indicating superior catalytic kinetics. Since Rct is a rate-determining factor for electrochemical reactions, a lower value is indicative of faster charge transfer dynamics and more efficient catalytic activity. Among the catalysts investigated, CNS-E.G. exhibits the lowest Rct of 0.006 $\Omega \cdot cm^2$, highlighting its superior interfacial conductivity and catalytic efficiency. In contrast, CNS-Glycerol presents a substantially larger Rct of 0.21 $\Omega \cdot cm^2$, while the non-doped variants NS-E.G. and NS-Glycerol display intermediate values of 0.13 and 0.15 $\Omega \cdot cm^2$, respectively. This marked reduction in Rct for CNS-E.G. highlights the beneficial role of cobalt incorporation and the ethylene glycol–assisted synthesis route in facilitating charge transfer during the hydrogen evolution reaction (HER). In addition to Rct, the series resistance (Rs) was extracted from the high-frequency intercept of the Nyquist plots. This parameter reflects the combined ohmic contributions from the electrolyte, electrode material, and contact resistance. The Rs values determined were 0.68 $\Omega \cdot cm^2$ for NS-E.G., 0.92 $\Omega \cdot cm^2$ for NS-Glycerol, 0.20 $\Omega \cdot cm^2$ for CNS-E.G., and 0.63 $\Omega \cdot cm^2$ for CNS-Glycerol. The markedly reduced Rs of CNS-E.G. further supports the superior ionic and electronic transport pathways within this catalyst system. Taken together, the concurrent minimization of both R_{ct} and R_s in CNS-E.G., coupled with its reduced Tafel slope, provides compelling evidence for its enhanced catalytic efficiency in the hydrogen evolution reaction.

Furthermore, the long-term durability of the catalysts was a key parameter in evaluating their overall catalytic efficiency. To examine this, both catalysts were subjected to a continuous stability test over 24 h under a fixed current density of 10 mA cm^{-2}, as depicted in Figure 7h. Despite showing comparable trends in stability, the CNS-Glycerol catalyst experienced a more pronounced decline, retaining only 90.7% of its initial HER activity. In contrast, the CNS-E.G. catalyst demonstrated better durability, maintaining approximately 95.58% of its original performance over the same period. The structural and morphological features of the CNS-E.G. and CNS-Glycerol catalysts were examined following their assessment for hydrogen evolution reaction (HER) performance. X-ray diffraction (XRD) analysis and field-emission scanning electron microscopy (FE-SEM) were employed for this purpose. As depicted in Figure 8a, the XRD profiles reveal the diffraction characteristics of both catalyst types. Corresponding surface morphologies captured through FE-SEM are shown in Figure 8b,c for CNS-E.G. and CNS-Glycerol, respectively. While the fundamental crystalline architecture remained stable, a minor decline in peak intensities was noted. This attenuation is likely due to enhanced exposure of the underlying nickel foam, possibly caused by ongoing electrochemical cycling.

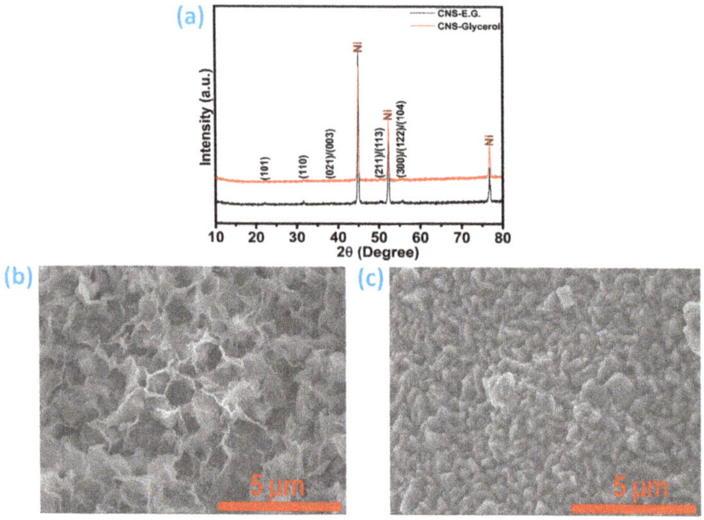

Figure 8. Phase and structure changes after HER stability, (**a**) diffraction patterns after HER activity, (**b**,**c**) FE-SEM images of CNS-E.G. and CNS-Glycerol after HER stability.

Among the reactions that constitute water splitting, the oxygen evolution reaction (OER) is notably the most demanding. This complexity arises from a multi-step mechanism that includes four coupled proton-electron transfers, alongside the slow kinetics associated with forming the O-O bond. Enhancing the catalytic activity for OER is crucial to reducing the overpotential, thereby boosting the overall efficiency of electrochemical water-splitting devices [9]. To evaluate the catalysts' effectiveness in driving OER, polarization measurements were performed using a three-electrode setup at a controlled scan rate of 5 mV per second. Figure 9a presents the comparative results for the NS-E.G., NS-Glycerol, CNS-E.G., and CNS-Glycerol samples. Consistent with the observations from the hydrogen evolution reaction (HER), CNS-E.G. exhibits superior OER activity, displaying a lower overpotential than the CNS-Glycerol and other pure catalysts, NS-E.G., and NS-glycerol. Figure 9b summarizes the overpotential values measured at current densities of 30 and 50 mA·cm^{-2}. The Co-doped Ni_3S_2 synthesized using ethylene glycol as a solvent (CNS-E.G.) demonstrates relatively low overpotentials of 414 mV and 466 mV at 30 and 50 mA·cm^{-2}, respectively. In comparison, the catalyst prepared from glycerol under similar conditions (CNS-Glycerol) exhibits significantly larger overpotentials of 535 mV and 643 mV at the same current densities. In the absence of Co incorporation, the ethylene glycol-derived Ni_3S_2 (NS-E.G.) records overpotentials of 468 mV and 547 mV at 30 and 50 mA·cm^{-2}. These values are slightly higher than those obtained for CNS-E.G., confirming the beneficial role of Co-doping, yet remain lower than the glycerol-derived samples. For the oxygen evolution reaction (OER) performance, only the Ni-foam substrate exhibits measurable catalytic activity. At a current density of 30 mA cm^{-2}, the system demonstrates an overpotential of approximately 810 mV. When the applied current density is increased to 50 mA cm^{-2}, the required overpotential correspondingly rises to around 890 mV. This progressive increase indicates the intrinsic limitations of the Ni-foam surface in facilitating efficient oxygen evolution at higher reaction rates. The elevated overpotential at larger current densities suggests sluggish charge-transfer kinetics and a higher energy barrier for OER, emphasizing the need for surface modification or catalyst incorporation to enhance the electrochemical activity and lower the energy losses associated with the reaction. Among all catalysts studied, the non-Co-doped glycerol-derived Ni_3S_2 (NS-Glycerol) displays the poorest activity, with overpotentials reaching 580 mV and 683 mV at 30 and 50 mA cm^{-2}, respectively.

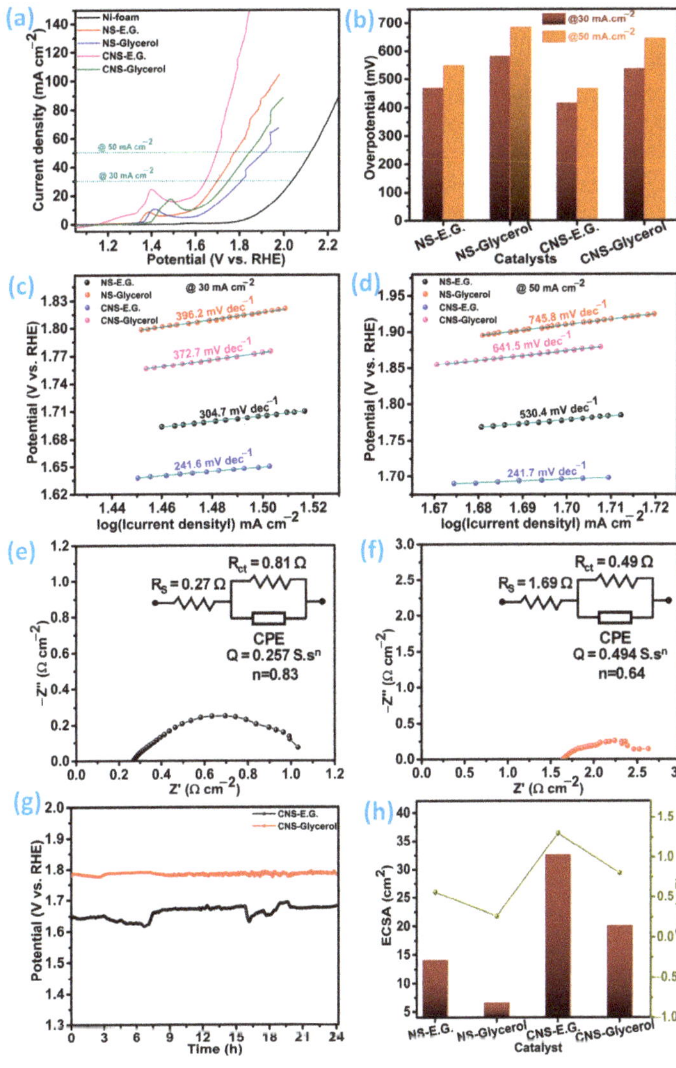

Figure 9. OER activity performance of the Ni_3S_2 and Co-Ni_3S_2 catalyst fabricated using two different solvents, (**a**) LSV curve measured at 5 mV s^{-1}, (**b**) estimated overpotential at different current densities, (**c**,**d**) Tafel slopes at 30 and 50 mA cm^{-2}, respectively, (**e**,**f**) Nyquist plots of CNS-E.G. and CNS-Glycerol, respectively, with equivalent circuit inset, (**g**) OER stability at 30 mA cm^{-2} of CNS-E.G. and CNS-Glycerol, and (**h**) electrochemical surface area and double layer capacitance of each electrocatalyst.

These results clearly indicate that (i) Co incorporation significantly lowers the overpotential, enhancing the catalytic activity, and (ii) the ethylene glycol–derived materials outperform the glycerol-derived ones, regardless of Co-doping, suggesting a favorable structural or electronic contribution from the ethylene glycol–based synthesis route. The enhanced OER activity of the CNS-E.G. catalysts is attributed to the effective incorporation of Co atoms and the strong bonding between these metal cations and the electronegative S atoms, which promotes improved electron delocalization. This modification altered the binding characteristics with the OH^* intermediate, leading to enhanced oxygen evolution reaction (OER) activity [2]. Moreover, in the case of the CNS-E.G. catalyst, there is no significant increase in overpotential when the current density is raised from 30 mA cm^{-2} to 50 mA cm^{-2}. This behavior suggests that, with appropriate structural and compositional modifications, such catalysts can be optimized for commercial electrocatalytic applications, where minimal overpotential at higher current densities is a key requirement [6].

The reaction kinetics of the oxygen evolution reaction (OER) can be systematically evaluated through Tafel slope analysis across varying current densities, as depicted in Figure 9c,d. A comparative evaluation of the catalysts demonstrates that CNS-E.G. consistently achieves smaller Tafel slopes than CNS-Glycerol, NS-E.G., and NS-Glycerol across all investigated current densities, signifying more favorable reaction kinetics. At 30 and 50 mA cm^{-2}, the Tafel slopes for CNS-E.G. were measured at 241.6 and 241.7 mV dec^{-1}, respectively. In contrast, CNS-Glycerol showed a pronounced rise in slope values at the same current densities, reaching 372.7 and 641.5 mV dec^{-1}. The NS-E.G. catalyst also performed better than CNS-Glycerol, recording 304.7 and 530.4 mV dec^{-1} under 30 and 50 mA cm^{-2}, whereas NS-Glycerol exhibited the poorest performance, with the highest slopes of 396.2 and 745.8 mV dec^{-1}. This significant disparity underscores the enhanced oxygen evolution reaction (OER) kinetics provided by CNS-E.G. Notably, the Tafel slope for CNS-E.G. remains nearly constant between the tested current densities, unlike CNS-Glycerol, NS-E.G., and NS-Glycerol catalysts, which display a sharp increase. The stability of the slope in CNS-E.G. indicates that its electrochemical reaction pathway is relatively unaffected by variations in current density. This implies a stable rate-determining step and unaltered surface reaction conditions, which are critical for robust and efficient catalyst performance under varying operational conditions [17]. Moreover, the dramatic increase in Tafel slope for CNS-Glycerol, NS-E.G., and NS-Glycerol at higher current densities may indicate a shift in the rate-determining step or degradation of active surface sites, both of which could compromise catalytic efficiency. In contrast, the electrochemical stability of CNS-E.G. underscores its potential as a more reliable and durable catalyst for practical OER applications.

Electrochemical Impedance Spectroscopy (EIS) was employed to analyze the charge transfer resistance (R_{ct}) associated with the catalysts. The electrochemical impedance spectroscopy (EIS) profiles for the cobalt-doped Ni_3S_2 (CNS) and pristine Ni_3S_2 (NS) electrodes synthesized with ethylene glycol (E.G.) and glycerol as precursors are presented in Figure 9e,f for CNS, and in Figure S3c,d for NS. These Nyquist plots provide a direct comparison of their interfacial charge-transport properties. For the CNS-E.G. electrode, the fitted equivalent circuit analysis revealed a charge-transfer resistance (R_{ct}) of 0.81 $\Omega \cdot cm^{-2}$ along with a series resistance (R_s) of 0.27 $\Omega \cdot cm^{-2}$. The relatively small semicircle diameter and low R_s value reflect highly favorable charge-transfer dynamics at the electrode-electrolyte boundary, suggesting efficient electronic conduction and ion diffusion. On the other hand, the CNS-Glycerol electrode displayed a lower R_{ct} of 0.49 $\Omega \cdot cm^{-2}$ but a substantially larger R_s of 1.69 $\Omega \cdot cm^{-2}$. Although the decreased R_{ct} indicates improved electron-transfer kinetics relative to CNS-E.G., the higher ohmic resistance points to inferior intrinsic conductivity or weaker electrolyte penetration, which counteracts the kinetic benefit. For comparison, the undoped Ni_3S_2 electrodes show poorer transport characteristics overall. The NS-E.G. electrode exhibited R_{ct} and R_s values of 0.56 and 1.40 Ω cm^{-2}, respectively, while the NS-Glycerol electrode yielded even higher resistances of 2.20 and 2.10 Ω cm^{-2}. These results confirm that cobalt incorporation markedly enhances both interfacial kinetics and bulk conductivity compared with the pristine sulfide. The superior performance of CNS-E.G. can be rationalized by considering its physicochemical attributes. The ethylene glycol–assisted synthesis may promote the formation of a more conductive surface network, optimized crystallinity, or favorable electronic states, which collectively reduce interfacial resistance. In contrast, the glycerol-based route seems to introduce structural or morphological features that increase series resistance despite enabling faster charge transfer.

The durability of the Co-incorporated catalysts under oxygen evolution reaction (OER) operation was systematically examined by conducting a continuous electrolysis test for 24 h at a constant current density of 30 mA cm^{-2}, as presented in Figure 9g. Both catalytic systems retained their activity with minimal degradation, indicating strong electrochemical stability during prolonged OER operation. Quantitatively, the CNS-E.G. catalyst displayed a slight reduction of 2.31% in its initial catalytic performance after the 24 h test, corresponding to a retained activity of 97.69%. In contrast, the CNS-Glycerol catalyst showed an even lower performance decay of only 0.24%, maintaining 99.76% of its original activity. To further investigate structural resilience after long-term operation, post-stability characterization was performed using X-ray diffraction (XRD) and field-emission scanning electron microscopy (FE-SEM). The diffraction profiles (Figure 10a) confirmed that the crystalline framework of both catalysts remained largely preserved, with only negligible variations in peak intensity or position, suggesting limited structural reorganization during operation. Complementary FE-SEM micrographs of CNS-E.G. (Figure 10b) and CNS-Glycerol (Figure 10c) corroborated this observation, showing that the surface morphology and nanoscale architecture were largely retained, with only minor textural modifications.

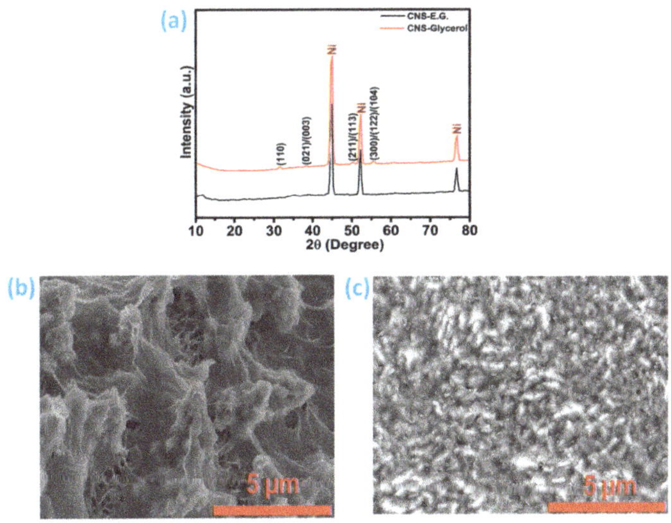

Figure 10. Phase and structure changes after OER stability, (**a**) diffraction patterns after OER activity, (**b**,**c**) FE-SEM images of CNS-E.G. and CNS-Glycerol after OER stability.

The catalytic efficiency of an electrocatalyst is predominantly dictated by the fraction of its surface that actively engages in electrochemical transformations. Importantly, this is not equivalent to the total geometric or physical surface area of the material. Instead, the critical parameter is the electrochemically accessible and catalytically active portion of the surface, which is generally expressed as the Electrochemical Surface Area (ECSA) [7,11]. This parameter provides a direct measure of the number of surface-active sites available for charge transfer and reaction kinetics during operation. To evaluate the ECSA of the prepared catalysts, NS-E.G., NS-Glycerol, CNS-E.G., and CNS-Glycerol cyclic voltammetry (CV) experiments were conducted, with the corresponding profiles presented in Figure S4a–d. From these CV curves, the double-layer capacitance (C_{dl}) values were determined, as depicted in Figure S5a–d. Since C_{dl} arises from the non-faradaic charging of the electrode-electrolyte interface, it scales proportionally with the accessible surface area. Therefore, C_{dl} values are widely used as a proxy for quantifying the ECSA of nanostructured catalysts. The extracted C_{dl} and ECSA values are summarized in Figure 9h. Among all tested samples, the CNS-E.G. catalyst demonstrated the highest double-layer capacitance (1.3 mF) and the largest electrochemical surface area (32.5 cm^2), indicating a superior density of electrochemically active sites. The performance sequence, based on both Cdl and ECSA, follows the order: CNS-E.G. (1.3 mF, 32.5 cm^2) > CNS-Glycerol (0.8 mF, 20.01 cm^2) > NS-E.G. (0.56 mF, 13.96 cm^2) > NS-Glycerol (0.26 mF, 6.54 cm^2). This trend highlights the significant role of Co-doping and the solvent environment (ethylene glycol vs. glycerol) in enhancing the effective surface exposure of active sites, thereby improving the catalytic behavior.

Following individual evaluations of the hydrogen evolution reaction (HER) and oxygen evolution reaction (OER) activities, a dual-electrode configuration was constructed to examine the overall water-splitting capabilities of each catalyst. Linear sweep voltammetry (LSV) was conducted at a scan rate of 5 mV s^{-1} for both the CNS-E.G.//CNS-E.G. and CNS-Glycerol//CNS-Glycerol cell systems, with the corresponding performance curves depicted in Figure 11a. These LSV profiles were utilized to determine the bifunctional catalytic efficiency, primarily through analysis of overpotentials. The electrolyzer incorporating CNS-E.G. electrodes displayed superior performance by requiring a lower voltage to sustain a current density of 10 mA cm^{-2}, reaching this benchmark at 1.537 V. Conversely, the CNS-Glycerol-based system demanded a higher potential of 1.674 V to achieve the same current density, indicating an approximate 8% increase. As shown in Figure 11b, both systems exhibit a trend of increasing overpotential with rising current density. For instance, the CNS-E.G.-configured electrolyzer reached 30 mA cm^{-2} at 1.717 V and 50 mA cm^{-2} at 1.866 V. Under identical conditions, the CNS-Glycerol system required 1.837 V and 1.994 V to reach 30 and 50 mA cm^{-2}, respectively, underscoring its relatively higher energy demand. Collectively, the Co-doped Ni_3S_2 catalyst demonstrates excellent bifunctional electrocatalytic performance, often surpassing or matching the effectiveness of other Ni_3S_2-based catalysts previously documented. A detailed comparison of performance metrics is summarized in Table 1. From the comparative evaluation, it becomes evident that the Co-doped Ni_3S_2 (CNS-E.G.) catalyst synthesized using ethylene glycol as solvent in this study demonstrates relatively modest activity toward the oxygen evolution reaction (OER) when benchmarked against pristine Ni_3S_2-based systems and state-of-the-art noble metal catalysts such as Pt/C (20 wt.% %) [29] and RuO_2 [30]. Despite this limitation on the anodic side, the catalyst exhibits highly competitive hydrogen evolution reaction (HER) performance. In fact, the HER activity of the CNS material is on par with, or in several cases superior to, previously reported electrocatalysts such as Ni_3S_2-$CoMoS_x$/NF [9], Co-Ni_3S_2/NF [11], Mo-Ni_3S_2 [13], and Ni_3S_2/NF [15]. Moreover, when implemented in a two-electrode configuration for overall water splitting, the CNS-E.G. electrolyzer surpasses the efficiency of a number of reported Ni_3S_2-based electrolytic systems. Specifically, its performance exceeds that of N-Ni_3S_2/VS_2//N-Ni_3S_2/VS_2 [1], Ni_3S_2/NF//Ni_3S_2/NF [8], Cd-Ni_3S_2/NF//Cd-Ni_3S_2/NF [12], Mo-Ni_3S_2//Mo-Ni_3S_2 [13],

Fe-Mo-Ni_3S_2//Fe-Mo-Ni_3S_2 [18], and N-Ni_3S_2/CoS_2/NF//N-Ni_3S_2/CoS_2/NF [22]. This outcome underscores the favorable synergistic effects introduced by cobalt incorporation and the use of ethylene glycol solvent, which promotes enhanced HER activity and improved bifunctionality, ultimately translating into superior water-splitting efficiency compared to many analogous Ni_3S_2-based architectures. Chronopotentiometric tests of the CNS-E.G.//CNS-E.G. and CNS-Glycerol//CNS-Glycerol electrolyzers were done as illustrated in Figure 11c. In both electrolyzers, the curve initially reaches the overpotential and then continues to increase steadily over time. The CNS-E.G. electrocatalyzer exhibited a 2.5% performance loss, indicating 97.5% stability. In comparison, the CNS-Glycerol electrocatalyzer showed a higher performance loss of 10.3%, corresponding to 89.7% stability relative to its original performance after 24 h. The catalyst prepared in an ethylene glycol medium (designated as CNS-E.G.) exhibits significantly enhanced electrocatalytic efficiency in various applications, namely, the hydrogen evolution reaction (HER), oxygen evolution reaction (OER), and comprehensive bifunctional catalysis when evaluated against the version synthesized using glycerol (CNS-Glycerol). This enhanced activity is likely attributable to differences in physicochemical properties as noted through various characterizations.

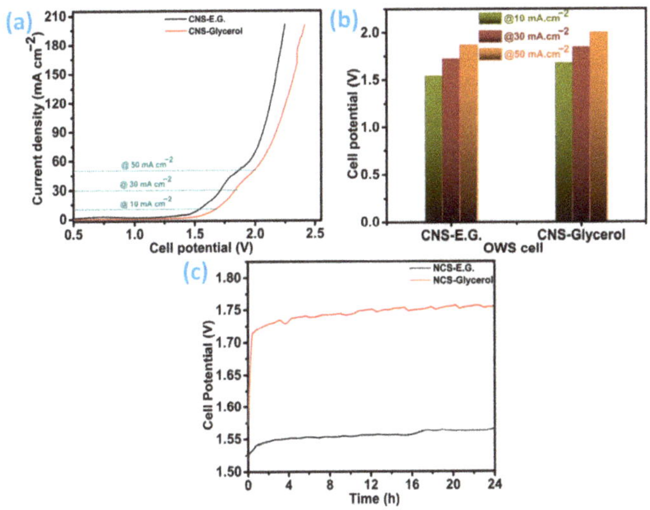

Figure 11. Overall water splitting performance of the Co-Ni_3S_2 catalyst fabricated using two different solvents, (**a**) LSV curve of CNS-E.G.//CNS-E.G. and CNS-Glycerol//CNS-Glycerol electrolyzer at 5 mV s^{-1}, (**b**) cell potential of electrolyzer at different current densities, (**c**) stability curves of CNS-E.G.//CNS-E.G. and CNS-Glycerol//CNS-Glycerol electrolyzer at 10 mA cm^{-2}.

Table 1. Comparative analysis of electrocatalytic performance of Ni_3S_2-based catalysts.

Catalyst	HER Overpotential (mV)	OER Overpotential (mV)	Cell Voltage (V)	Reference
N-Ni_3S_2/VS_2	151@10 mA	227@10 mA	1.648@10 mA	[1]
MoS_2-Ni_3S_2/NF	81@10 mA	150@100 mA	1.54@100 mA	[2]
3D Ni_3S_2/Cu–NiCo LDH	156@10 mA and 304@100 mA	119@10 mA and 218@100 mA	1.75@100 mA	[3]
CoCe-Ni_3S_2/NF	193@50 mA	285@50 mA	1.68@20 mA	[7]
Ni_3S_2/NF	89@10 mA	242@10 mA	1.577@10 mA	[8]
Ni_3S_2-$CoMoS_x$/NF	234@10 mA	90@10 mA	1.52@10 mA	[9]

Table 1. *Cont.*

Catalyst	HER Overpotential (mV)	OER Overpotential (mV)	Cell Voltage (V)	Reference
$Co\text{-}Ni_3S_2/NF$	240@10 mA	120@10 mA	1.69@ 50 mA	[11]
$Cd\text{-}Ni_3S_2/NF$	140@10 mA	197@10 mA	1.54@10 mA	[12]
$Mo\text{-}Ni_3S_2$	212@10 mA	260@10 mA	1.67@10 mA	[13]
$Co\text{-}N\text{-}Ni_3S_2/NF$	215@50 mA	285@50 mA	1.5@10 mA	[14]
Ni_3S_2/NF	223@10 mA	260@10 mA	--	[15]
Co-doped Ni_3S_2	62@10 mA	297@20 mA	1.52@10 mA	[17]
$Fe\text{-}Mo\text{-}Ni_3S_2$	74@10 mA	80@10 mA	1.60@10 mA	[18]
Mo-doped Ni_3S_2	278@100 mA	180@100 mA	1.53@10 mA	[21]
$N\text{-}Ni_3S_2/CoS_2/NF$	100@10 mA	245@10 mA	1.56@10 mA	[22]
Mo-S-Ni	78@10 mA	260@10 mA	1.53@10 mA	[27]
Pt/C (20 wt.%)	43@10 mA	--	--	[29]
RuO_2	--	199@10 mA	--	[30]
$Co\text{-}Ni_3S_2$	190.7@10 mA	414@ 30 mA	1.537@10 mA	This work

3. Experimental Section

3.1. Materials

All chemical reagents used in this study including nickel nitrate hexahydrate ($Ni(NO_3)_2{\cdot}6H_2O$), cobalt nitrate hexahydrate ($Co(NO_3)_2{\cdot}6H_2O$), sodium sulfide nonahydrate ($Na_2S{\cdot}9H_2O$), ethylene glycol ($C_2H_6O_2$), glycerol ($C_3H_8O_3$), urea (CH_4N_2O), and ammonium fluoride (NH_4F) were obtained from Sigma-Aldrich, based in Seoul, South Korea. All chemicals had a purity greater than 99% and were used without further purification.

3.2. Fabrication of $Co\text{-}Ni_3S_2$ Electrocatalyst

A two-stage synthetic route combining solvothermal and hydrothermal techniques was employed to fabricate $Co\text{-}Ni_3S_2$ films directly on nickel foam substrates. Prior to the deposition process, the Ni foam was thoroughly cleaned by sequential ultrasonication in hydrochloric acid, ethanol, and deionized (DI) water for 10 min each to eliminate surface contaminants. In the initial solvothermal phase, a precursor solution was prepared using 60 mL of either ethylene glycol or glycerol as the solvent. Initially, 2 mmol of nickel nitrate hexahydrate ($Ni(NO_3)_2{\cdot}6H_2O$) and 1 mmol of cobalt nitrate hexahydrate ($Co(NO_3)_2{\cdot}6H_2O$) were introduced into a reaction vessel under constant agitation. Following this, 4 mmol of urea (CH_4N_2O) along with 6 mmol of ammonium fluoride (NH_4F) were added to the solution, which was further stirred for 15 min to ensure homogeneity. This precursor mixture was then sealed in a stainless-steel autoclave lined with Teflon and subjected to hydrothermal conditions at 100 °C for 12 h to facilitate the growth of a cobalt-nickel hydroxide framework. Upon completion of the reaction and natural cooling to room temperature, the solid material was isolated, thoroughly washed with deionized water and ethanol, and subsequently dried. For the sulfidation step, a separate solution was prepared by dissolving 8 mmol of sodium sulfide nonahydrate ($Na_2S{\cdot}9H_2O$) in 60 mL of deionized water with 15 min of stirring. The dried Co-Ni hydroxide material from the previous step was immersed in this sulfide-containing solution. The mixture was then transferred into a clean Teflon-lined autoclave, sealed, and placed in an oven at 160 °C for 6 h to complete the sulfidation process. Following the reaction, the resultant $Co\text{-}Ni_3S_2$ films were extracted, thoroughly rinsed with ethanol and DI water to eliminate residual impurities, and dried for further characterization. For the fabrication of pure Ni_3S_2 on Ni-foam, similar steps were utilized without the addition of Co.

3.3. Physicochemical Characterization

The electrocatalysts prepared using various solvent systems were thoroughly examined to assess their physicochemical characteristics through multiple analytical techniques. Crystalline phase composition was determined via X-ray diffraction (XRD) utilizing an X'Pert Pro instrument (Malvern Panalytical, Worcestershire, UK). To explore the elemental oxidation states and surface composition, X-ray photoelectron spectroscopy (XPS) was conducted using a K-Alpha analyzer (Thermo Fisher Scientific, Seoul, South Korea). Morphological and microstructural features were characterized using field-emission scanning electron microscopy (FE-SEM, model S-4800; Hitachi, Tokyo, Japan), which also facilitated energy-dispersive X-ray spectroscopy (EDS) and spatial distribution mapping of elements.

3.4. Electrocatalytic Measurements

The electrocatalytic characteristics of both catalyst samples were thoroughly analyzed using a VersaSTAT3 electrochemical workstation. Their activity in facilitating the oxygen evolution reaction (OER) and hydrogen evolution reaction (HER) was studied using a standard three-electrode configuration. In this setup, a nickel foam electrode with an exposed surface area of 1 cm^2, coated with the catalyst material, was used as the working electrode. A graphite sheet acted as the counter electrode, and an Ag/AgCl electrode served as the reference. To evaluate the bifunctional electrocatalytic capability, symmetric full-cell assemblies were constructed using two identical catalyst-coated electrodes (e.g., CNS-E.G. paired with CNS-E.G.). All electrochemical assessments were conducted in a 1 mol L^{-1} potassium hydroxide (KOH) alkaline solution. Linear sweep voltammetry (LSV) was employed to probe catalytic activity, with measurements taken at a scan rate of 5 mV s^{-1}. Electrochemical impedance spectroscopy (EIS) was performed at the overpotential required to sustain a current density of 10 mA cm^{-2}. To examine durability and long-term functionality, chronopotentiometric measurements were carried out under constant current operation at 30 mA cm^{-2} for OER, 10 mA cm^{-2} for HER, and complete water splitting.

4. Conclusions

This study highlights the pivotal influence of solvent environment on the morphological engineering and electrochemical performance of nickel sulfide (Ni_3S_2) and cobalt-doped nickel sulfide (Co-Ni_3S_2) nanostructures for water splitting applications. By employing ethylene glycol and glycerol as solvent systems, we successfully tailored the nanostructure of Ni_3S_2 and Co-Ni_3S_2 to yield morphologies with distinct catalytic behaviors. A distinct petal-shaped network was synthesized using ethylene glycol as the medium, leading to enhanced electrocatalytic efficiency. This structure exhibited notably low overpotentials, measuring 190.7 mV for the hydrogen evolution reaction (HER) and 1.537 V for overall water splitting. In contrast, the glycerol-derived material, while still active, showed comparatively higher overpotentials. These findings establish a clear link between solvent-driven morphology modulation and catalytic efficiency, emphasizing the strategic importance of solvent selection in catalyst design. This study enhances insight into how structural features influence the properties of transition metal sulfides and provides a strategic approach for designing efficient electrocatalysts aimed at sustainable hydrogen generation.

Supplementary Materials: The following supporting information can be downloaded at: https://www.mdpi.com/article/10.3390/inorganics13110359/s1.

Author Contributions: Conceptualization, S.M.M.; Methodology, S.M.M. and K.S.W.; Software, S.M.M.; Validation, K.S.W., G.T.C. and J.C.S.; Formal analysis, K.S.W., S.L., A.M.T. and G.T.C.; Investigation, S.M.M., K.S.W., A.M.T., G.T.C. and J.L.; Resources, S.L. and J.C.S.; Data curation, S.M.M., K.S.W., A.M.T. and G.T.C.; Writing—original draft, S.M.M.; Writing—review & editing, J.L.; Visualization, J.C.S. and J.L.; Supervision, J.C.S. and J.L. All authors have read and agreed to the published version of the manuscript.

Funding: This work was supported by the Industrial Technology Innovation Program (RS-2024-00433858, Development of natural fiber-reinforced composite materials and components with a porosity of 2% or less for large cargo ship rotor sails) funded By the Ministry of Trade, Industry & Energy (MOTIE), Korea.

Institutional Review Board Statement: Not applicable.

Informed Consent Statement: Not applicable.

Data Availability Statement: The original contributions presented in this study are included in the article. Further inquiries can be directed to the corresponding authors.

Conflicts of Interest: The authors declare no conflict of interest.

References

1. Zhong, X.; Tang, J.; Wang, J.; Shao, M.; Chai, J.; Wang, S.; Yang, M.; Yang, Y.; Wang, N.; Wang, S.; et al. 3D Heterostructured Pure and N-Doped Ni_3S_2/VS_2 Nanosheets for High Efficient Overall Water Splitting. *Electrochim. Acta* **2018**, *269*, 55–61. [CrossRef]
2. Song, S.; Wang, Y.; Li, W.; Tian, P.; Zhou, S.; Gao, H.; Tian, X.; Zang, J. Amorphous MoS_2 Coated Ni_3S_2 Nanosheets as Bifunctional Electrocatalysts for High-Efficiency Overall Water Splitting. *Electrochim. Acta* **2020**, *332*, 135454. [CrossRef]
3. Jia, L.; Du, G.; Han, D.; Hao, Y.; Zhao, W.; Fan, Y.; Su, Q.; Ding, S.; Xu, B. Ni_3S_2/Cu-NiCo LDH Heterostructure Nanosheet Arrays on Ni Foam for Electrocatalytic Overall Water Splitting. *J. Mater. Chem. A Mater.* **2021**, *9*, 27639–27650. [CrossRef]
4. Jiang, J.; Su, H.; Song, S.; Liu, W.; Li, N.; Gao, Y.; Ge, L. W-Doped $FeNi_2S_4/Ni_3S_2$/NF with Interfacial Effect as Efficient Bifunctional Electrocatalyst for Overall Water Splitting. *Nano Res.* **2023**, *16*, 12116–12125. [CrossRef]
5. Mosallaei, H.; Hadadzadeh, H.; Ensafi, A.A.; Mousaabadi, K.Z.; Weil, M.; Foelske, A.; Sauer, M. Evaluation of HER and OER electrocatalytic activity over RuO_2-Fe_2O_3 nanocomposite deposited on HrGO nanosheets. *Int. J. Hydrogen Energy* **2023**, *48*, 1813–1830. [CrossRef]
6. Zhai, P.; Zhang, Y.; Wu, Y.; Gao, J.; Zhang, B.; Cao, S.; Zhang, Y.; Li, Z.; Sun, L.; Hou, J. Engineering Active Sites on Hierarchical Transition Bimetal Oxides/Sulfides Heterostructure Array Enabling Robust Overall Water Splitting. *Nat. Commun.* **2020**, *11*, 5462. [CrossRef]
7. Yu, Z.; Song, S.N.; Dong, Y.W.; Gao, X.T.; Jiang, Q.F.; Zhao, Z. Cobalt and Cerium Dual-Doped Nickel Sulfide Nanostructures as a Bifunctional Catalyst for Overall Water Splitting. *ACS Appl. Nano Mater.* **2025**, *8*, 894–903. [CrossRef]
8. Zhang, J.; Li, Y.; Zhu, T.; Wang, Y.; Cui, J.; Wu, J.; Xu, H.; Shu, X.; Qin, Y.; Zheng, H.; et al. 3D Coral-Like Ni_3S_2 on Ni Foam as a Bifunctional Electrocatalyst for Overall Water Splitting. *ACS Appl. Mater. Interfaces* **2018**, *10*, 31330–31339. [CrossRef]
9. Zhao, L.; Ge, H.; Zhang, G.; Wang, F.; Li, G. Hierarchical Ni_3S_2-$CoMoS_x$ on the Nickel Foam as an Advanced Electrocatalyst for Overall Water Splitting. *Electrochim. Acta* **2021**, *387*, 138538. [CrossRef]
10. Li, X.; Shang, X.; Rao, Y.; Dong, B.; Han, G.Q.; Hu, W.H.; Liu, Y.R.; Yan, K.L.; Chi, J.Q.; Chai, Y.M.; et al. Tuning Crystal Phase of NiS_x through Electro-Oxidized Nickel Foam: A Novel Route for Preparing Efficient Electrocatalysts for Oxygen Evolution Reaction. *Appl. Surf. Sci.* **2017**, *396*, 1034–1043. [CrossRef]
11. Song, S.; Wang, Y.; Li, W.; Tian, P.; Zhou, S.; Gao, H.; Tian, X.; Zang, J. Co-Doped Ni_3S_2 Hierarchical Nanoarrays Derived from Zeolitic Imidazolate Frameworks as Bifunctional Electrocatalysts for Highly Enhanced Overall-Water-Splitting Activity. *J. Alloys Compd.* **2020**, *827*, 154299. [CrossRef]
12. Yan, H.; Deng, R.; Zhang, S.; Yao, H.; Duan, J.; Bai, H.; Li, Y.; Liu, R.; Shi, K.; Ma, S. Cd Doped Ni_3S_2 Nanosheet Arrays Grown on Nickel Foam as Highly Efficient and Robust Bifunctional Electrocatalysts for Alkaline Overall Water Splitting. *J. Alloys Compd.* **2023**, *954*, 170072. [CrossRef]
13. Wu, C.; Liu, B.; Wang, J.; Su, Y.; Yan, H.; Ng, C.; Li, C.; Wei, J. 3D Structured Mo-Doped Ni_3S_2 Nanosheets as Efficient Dual-Electrocatalyst for Overall Water Splitting. *Appl. Surf. Sci.* **2018**, *441*, 1024–1033. [CrossRef]
14. Du, X.; Ma, G.; Zhang, X. Cobalt and Nitrogen Co-Doped Ni_3S_2 nanoflowers on Nickel Foam as High-Efficiency Electrocatalysts for Overall Water Splitting in Alkaline Media. *Dalton Trans.* **2021**, *50*, 8955–8962. [CrossRef]
15. Huang, C.; Miao, X.; Pi, C.; Gao, B.; Zhang, X.; Qin, P.; Huo, K.; Peng, X.; Chu, P.K. Mo $_2$C/VC Heterojunction Embedded in Graphitic Carbon Network: An Advanced Electrocatalyst for Hydrogen Evolution. *Nano Energy* **2019**, *60*, 520–526. [CrossRef]

16. Feng, L.L.; Yu, G.; Wu, Y.; Li, G.D.; Li, H.; Sun, Y.; Asefa, T.; Chen, W.; Zou, X. High-Index Faceted Ni_3S_2 Nanosheet Arrays as Highly Active and Ultrastable Electrocatalysts for Water Splitting. *J. Am. Chem. Soc.* **2015**, *137*, 14023–14026. [CrossRef]
17. Tong, X.; Li, Y.; Pang, N.; Qu, Y.; Yan, C.; Xiong, D.; Xu, S.; Wang, L.; Chu, P.K. Co-Doped Ni_3S_2 Porous Nanocones as High-Performance Bifunctional Electrocatalysts in Water Splitting. *Chem. Eng. J.* **2021**, *425*, 130455. [CrossRef]
18. Wang, T.; Li, B.; Wang, P.; Xu, M.; Wang, D.; Wang, Y.; Zhang, W.; Qu, C.; Feng, M. Modulation of Electronic Structure of Ni_3S_2 via Fe and Mo Co-Doping to Enhance the Bifunctional Electrocatalytic Activities for HER and OER. *J. Colloid. Interface Sci.* **2024**, *672*, 715–723. [CrossRef] [PubMed]
19. Sun, A.; Wei, C.; Guo, J.; Zhang, X. Co-Doped Ni_3S_2 Ultrathin Nanosheets for Efficient Oxygen Evolution Catalysis. *Mater. Lett.* **2021**, *299*, 130069. [CrossRef]
20. Ma, X.; Yang, J.; Xu, X.; Yang, H.; Peng, C. NiSe/Ni_3Se_2 on Nickel Foam as an Ultra-High-Rate HER Electrocatalyst: Common Anion Heterostructure with Built-in Electric Field and Efficient Interfacial Charge Transfer. *RSC Adv.* **2021**, *11*, 34432–34439. [CrossRef]
21. Cui, Z.; Ge, Y.; Chu, H.; Baines, R.; Dong, P.; Tang, J.; Yang, Y.; Ajayan, P.M.; Ye, M.; Shen, J. Controlled Synthesis of Mo-Doped Ni_3S_2 Nano-Rods: An Efficient and Stable Electro-Catalyst for Water Splitting. *J. Mater. Chem. A Mater.* **2017**, *5*, 1595–1602. [CrossRef]
22. Kong, Q.; Fan, N.; Chen, S.; Wu, X.; Liu, L.; Lang, R.; Gao, Z.; Guan, H.; Dong, C.; Chen, G. Interface Engineering of N-Doped Ni_3S_2/CoS_2 Heterostructures as Efficient Bifunctional Catalysts for Overall Water Splitting. *J. Electroanal. Chem.* **2021**, *895*, 115516. [CrossRef]
23. Schabikowski, M.; Stróż, A.; Kruk, A. Beyond Oxidation: Engineering Functional Anodised Metal Matrices Through Molecular and Surface Modifications. *Int. J. Mol. Sci.* **2025**, *26*, 7809. [CrossRef]
24. Cai, J.; Zhu, H. Surface-engineered nanostructured high-entropy alloys for advanced electrocatalysis. *Commun. Mater.* **2025**, *6*, 118. [CrossRef]
25. Ai, T.; Wang, H.; Bao, W.; Feng, L.; Zou, X.; Wei, X.; Ding, L.; Deng, Z.; Rao, B. Fe-V Synergistic Doping Effect of Hierarchical Ni_3S_2 Oblate-Nanorod Arrays for Efficient Electrocatalytic Oxygen Evolution Reaction. *Chem. Eng. J.* **2022**, *450*, 138358. [CrossRef]
26. Ha, Q.N.; Susanto Gultom, N.; Yeh, C.H.; Kuo, D.H. One-Pot Synthesized Li, V Co-Doped Ni_3S_2 Nanorod Arrays as a Bifunctional Electrocatalyst for Industrialization-Facile Hydrogen Production via Alkaline Exchange Membrane Water Electrolysis. *Chem. Eng. J.* **2023**, *472*, 144931. [CrossRef]
27. Wang, C.; Shao, X.; Pan, J.; Hu, J.; Xu, X. Redox Bifunctional Activities with Optical Gain of Ni_3S_2 Nanosheets Edged with MoS_2 for Overall Water Splitting. *Appl. Catal. B* **2020**, *268*, 118435. [CrossRef]
28. Roy, A.; Kang, K.M.; Nah, Y.C.; La, M.; Choi, D.; Wang, S. Improved electrocatalytic water oxidation with cobalt hydroxide nano-flakes supported on copper-modified nickel foam. *Electrochim. Acta* **2021**, *383*, 138368. [CrossRef]
29. Jiang, B.; Liu, S.; Cheng, L.; Zhou, L.; Cui, H.; Liu, M.; Wen, M.; Wang, C.; Wang, W.; Li, S.; et al. Mass synthesis of Pt/C catalysts with high Pt loading for low-overpotential hydrogen evolution. *Int. J. Hydrogen Energy* **2024**, *58*, 268–278. [CrossRef]
30. Zhao, Z.L.; Wang, Q.; Huang, X.; Feng, Q.; Gu, S.; Zhang, Z.; Xu, H.; Zeng, L.; Gu, M.; Li, H. Boosting the oxygen evolution reaction using defect-rich ultra-thin ruthenium oxide nanosheets in acidic media. *Energy Environ. Sci.* **2020**, *13*, 5143–5151. [CrossRef]

inorganics

MDPI

Article

Electrochemical and Gas-Solid Hydrogen Storage Properties of a Multi-Metal Magnesium-Based Alloy Obtained by Ball Milling

Eli Grigorova [1,*], Gülhan Çakmak [2], Hakan Yüce [2] and Pavel Markov [1]

[1] Institute of General and Inorganic Chemistry, Bulgarian Academy of Sciences, Acad. G. Bonchev Str, bl.11, 1113 Sofia, Bulgaria; pvlmarkov@svr.igic.bas.bg

[2] Department of Metallurgical and Materials Engineering, Muğla Sıtkı Koçman University, Kötekli, 48000 Muğla, Turkey; gulhancakmak@mu.edu.tr (G.Ç.); hakanyuce2@posta.mu.edu.tr (H.Y.)

* Correspondence: egeorg@svr.igic.bas.bg

Abstract: The hydrogen sorption and electrochemical properties of the alloy $Mg_{50}Ni_{12.5}Al_{12.5}V_{12.5}Fe_{12.5}$ synthesized by ball milling under the protected atmosphere of argon for 50 h in a planetary ball mill are investigated. The significantly fast rate of absorption reaction is observed along with the hydrogen absorption capacity of 2.04 wt.% H_2 at temperatures 200 and 300 °C and at a pressure of 1 MPa. Even at room temperature, the absorption capacity is relatively high, and it is about 1.6 wt.% H_2. The alloy ball milled for 50 h and the alloy after cycling and hydrogenation were characterized by X-ray diffraction analyses, SEM, and TEM. The prepared alloy was tested as an anode in a Ni/MH battery in a 6 M KOH electrolyte. Galvanostatic and potentiostatic discharge modes were employed, revealing activation after the third cycle and giving a discharge capacity of 257 mAh/g.

Keywords: hydrogen storage; metal hydrides; Mg-based alloys; Ni-MH batteries

1. Introduction

Hydrogen is an important and efficient future energy carrier, but its application remains challenging due to the safety issues related to its storage. Metal hydrides and especially Mg-based hydrides are good alternatives for solid-state hydrogen storage, because they provide reversibility, increased safety, and high volumetric densities and have an affordable price [1–5]. They can be used in various stationary and mobile applications, such as tanks for solid hydrogen storage, heat storage systems, fuel cells, and batteries. One advancing and powerful way of optimizing hydrogen storage alloys is the innovative method of machine learning [6]. However, as the authors in this study admitted, more efforts are necessary to achieve better predictive accuracy of the models. Preparing Mg-based alloys with other metals that are capable of storing hydrogen and also have high catalytic activity or better anti-corrosion resistance is an approach to improve their performance as hydrogen storage media and as anodes in Ni-MH batteries [2–8]. Among the many hydrogen storage alloys that can be used as anodes in Ni-MH batteries, MgNi amorphous alloys prepared by ball milling are very promising because of their low cost and high hydrogen storage capacity. The main drawback of these alloys is the degradation rate of the electrochemical capacity, which inhibits their application. In some published results, it is shown that the most effective method to improve the cycling life and the capacity retention is partial substitution with various metals such as Al, Ni, V, Y, Ti, Cr, Zr, B, Co, Cr, Ta, Pd, and Ce [9–22]. J. Huang et al. have proved that the fast degradation of electrochemical capacity of MgNi amorphous alloys is mainly attributed to hydrogen

https://doi.org/10.3390/inorganics13090299

induced crystallization [16]. Adding other alloying metals to MgNi led to less hydrogen-induced crystallization and an improved capacity retention rate. The substitution of Mg by Ti seems to have a better effect on corrosion. It was claimed that the formation of titanium oxide can hinder further oxidation of Mg, which led to an improvement in electrochemical characteristics [17]. A comparable effect has the Al substitution for Mg in $Mg_{1-x}Al_xNi$ alloys [19]. By these substitutions, the MgNi alloys have shown improvements in their electrochemical properties, particularly in their discharge capacity and cycling stability. Another approach to improving the electrochemical properties of the MgNi alloy is by adding some compounds such as FeB [23] or CoB [24]. The effect of Gd in the Mg-Ni alloys on the microstructural and electrochemical properties was investigated by Chen W. et al. [25]. The reported maximum discharge capacities of $Mg_{71}Ni_{29}$ and $Mg_{72.6}Ni_{25.3}Gd_{1.1}$ are 134.3 mAh/g and 91.9 mAh/g, respectively. The substitution of Ni with Gd in the Mg-Ni alloy positively impacts cycling stability and corrosion resistance but lowers the electrochemical capacity and weakens characteristic peaks of Mg_2Ni and strengthens those of $MgNi_2$.

More detailed investigations of hydrogen storage properties of nanostructured multicomponent metal Mg-based alloys regarding higher storage capacity, fast absorption/desorption kinetics at lower pressure, and better electrochemical properties as an anode in NiMH batteries are needed. Taking into consideration the published results discussed above, we aimed at obtaining a MgNi alloy with a BCC structure where nickel is partially substituted by vanadium, iron, and aluminum. The absence of rare-earth metals reduces cost and environmental concerns, while the use of abundant Mg increases the material's practical relevance for large-scale applications. These results distinguish our work from prior studies on Mg–Ni substitutions, as the performance features observed here are directly related to the multicomponent design. The chosen method of synthesis is by grinding under argon in a planetary mill. This is not only because of the positive effects of ball milling but also because of the low evaporation temperature of magnesium, which makes it difficult to use melting. Therefore, the corresponding alloy with composition by the formula $Mg_{50}Ni_{12.5}Al_{12.5}V_{12.5}Fe_{12.5}$ was synthesized by ball milling in a planetary mill for 50 h under argon, and its hydrogen storage properties were studied at different temperatures. The phase composition of the alloy was monitored by X- ray diffraction, SEM, and TEM analyses. The electrochemical discharge capacity and capacity retention rate in 6 M KOH electrolyte of this alloy were also tested.

2. Results and Discussion

2.1. X-Ray Diffraction Analyses

In Figure 1, X-ray diffraction patterns of the alloy are presented after 0, 5, 10, 20, 30, and 50 h of ball milling. Gradual broadening of the diffraction peaks combined with a decrease in the peaks' intensity and also the disappearance of some of them, especially those of Mg, as well as those related to Al, V, and Fe metals, as a function of grinding time is observed. The ball-milled mixture after 50 h contains a Mg, Ni-type BCC structure and a V BCC-type structure and along with some traces of Mg_3AlNi_2, which is comparable to the background and starts to appear after 30 h of milling. The Mg_2Ni phase is not observed even after 50 h of ball milling treatment.

To determine the crystallite sizes of ball-milled powders, Scherrer's formula is used. The crystallite sizes in nm after different times of ball milling are presented in Table 1. A significant decrease in the crystallite size is observed with the increase of ball milling time for all metals.

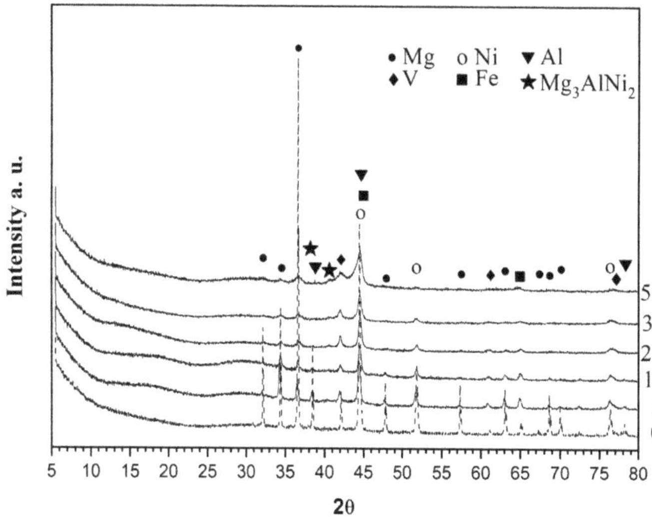

Figure 1. XRD diffraction patterns of the $Mg_{50}Ni_{12.5}Al_{12.5}V_{12.5}Fe_{12.5}$ ball milled from bottom up 0, 5, 10, 20, 30, and 50 h under argon.

Table 1. Crystallite size in nm calculated via Sherrer formula after different durations of ball milling.

Ball Milling, h	Mg	Ni	Al	V
10	38.4	32.6	42.2	30.0
30	16.2	17.3	19.8	17.5
50	10.8	12.1	7.5	9.3

The alloy Mg_3AlNi_2 can be synthesized by melting or ball milling and annealing [26–28] and absorbs hydrogen reversibly and faster and has better electrochemical properties than the Mg_2Ni alloy electrode. A. Parente et al. [26] synthesized nanocrystalline materials Mg_mAlNi_n (m, n $\leq$ 3) by ball milling, made XRD Rietveld refinements in a thermodynamics study with PCT diagrams at different temperatures, and proposed a reaction for the disproportion of the Mg_2AlNi_2 phase with hydrogen. The resulting products are MgH_2, Mg_2NiH_4, and $Mg_{0.5}AlNi_{1.5}$. These authors used a vibratory high energy mill Spex 8000 (SPEX samplePrep, Metuchen, NJ, USA) and the duration of their ball milling experiments was shorter and 30 h in total.

2.2. Gas Hydrogen Storage Properties

The first two cycles of hydrogenation are activation cycles during which slower kinetics and a capacity below 1 wt.% are detected. Usually, the Mg-based materials for hydrogen storage need activation, but the alloy prepared by us is activated relatively fast. The following cycles of hydrogenation/dehydrogenation showed that the $Mg_{50}Ni_{12.5}Al_{12.5}V_{12.5}Fe_{12.5}$ has a very rapid rate of absorption reaction at 200 and 300 °C with the maximum obtained hydrogen absorption capacity value of 2.04 wt.% H_2. The alloy shows similar kinetics of the absorption reaction at both temperatures at the beginning of the process, but the reaction of hydrogen absorption is a bit faster at 200 °C after a very short period of up to 0.5 min. (Figure 2a). After that, with the progress of the absorption process, the rate at 200 °C is slower than at 300 °C. On the other hand, even at room temperature, the absorption capacity is quite high and around 1.6 wt.% H_2. In fact, more than 90% of the maximum absorption capacity is obtained after 2 min of hydrogenation at 200 °C and 1 MPa. The alloy is subjected in total to 20 cycles of hydrogenation/ dehydrogenation at different temperatures.

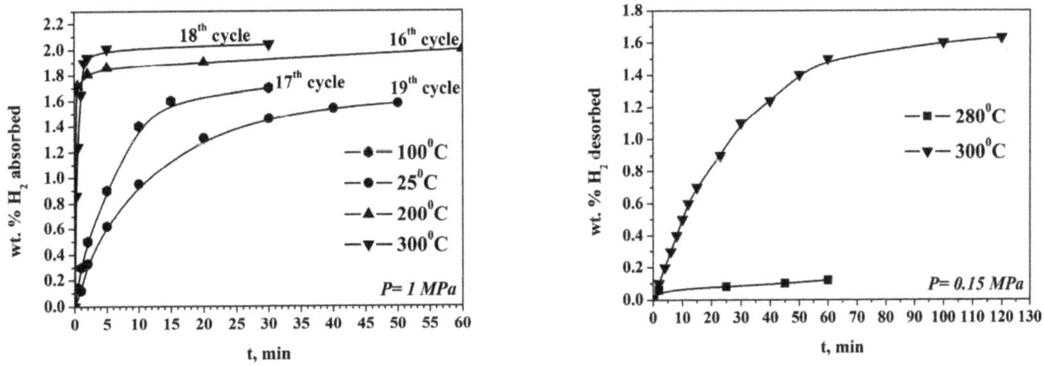

(a) (b)

Figure 2. Hydrogen absorption (**a**) and desorption (**b**) curves of the $Mg_{50}Ni_{12.5}Al_{12.5}V_{12.5}Fe_{12.5}$ ball milled 50 h under argon.

The obtained hydrogen absorption capacity is comparable with some published results of similar materials. The absorption capacity of $Mg_{50}Ni_{50}$ is 1.85 wt.% at 100 °C, and the composite $Mg_{50}Ni_{50}$- 25 wt.% Mg_2Ni absorbs 1.47 wt.% H_2 at 90 °C. The presented hydrogen absorption curves for these alloys showed faster kinetics, but at a hydrogen pressure of 2 MPa [29] and even a similar one at 3.3 MPa [30].

The desorption reaction is faster at 300 °C compared with that at 280 °C, but even after 2 h only 1.6 wt.% H_2 is desorbed. Decreasing the temperature by only 20 degrees Celsius leads to even greater degradation in the desorption kinetics, and a very small amount of hydrogen is desorbed. At 260 °C and 0.15 MPa after 60 min., hydrogen desorption has not been detected (Figure 2b).

The hydrogen absorption and desorption were analyzed using the Johnson–Mehl-Avrami–Kolmogorov equation (JMAK). The constant n called the Avrami exponent reflects the constancy of the nucleation rate, the dimension of nuclei growth, and the rate controlling step. The Avrami exponent can be determined by the slope of the line described by $\ln[-\ln(1 - \alpha)] = \ln t$. The fraction of hydrogenation and dehydrogenation at a certain time α was transformed into normalized values (between 0 to 1). Figure 3a,b show the JMAK calculation results of the alloy's hydrogen absorption and desorption at different temperatures. The R^2 values of the JMAK curve fit for hydrogen absorption range from 0.87 to 0.99 and from 0.97 to 0.99 for hydrogen desorption. A good correlation of the fitting results indicates the values of R^2 0.99 at 100 °C and 25 °C for hydrogen absorption. The lower value of R^2 0.87 for the fits of the two faster absorption processes at higher temperatures at 300 °C and 200 °C has comparatively poor fits. The obtained values for the hydrogen absorption of the Avrami exponent n are 0.5, 0.9, 1.2, and 1 for 200°, 300°, 100°, and 25°, respectively. The fastest hydrogen absorption at the first stage of the reaction is observed at 200 °C, and n at this temperature has the lowest value indicating one dimensional diffusion-controlled growth with a decreasing nucleation rate. For the hydrogen desorption process, the values of the Avrami exponent are increased. The slower hydrogen desorption at 280 °C has a value of n 1.8, and at 300 °C the value is 1.2, indicating nucleation and a two-dimensional growth mechanism for desorption.

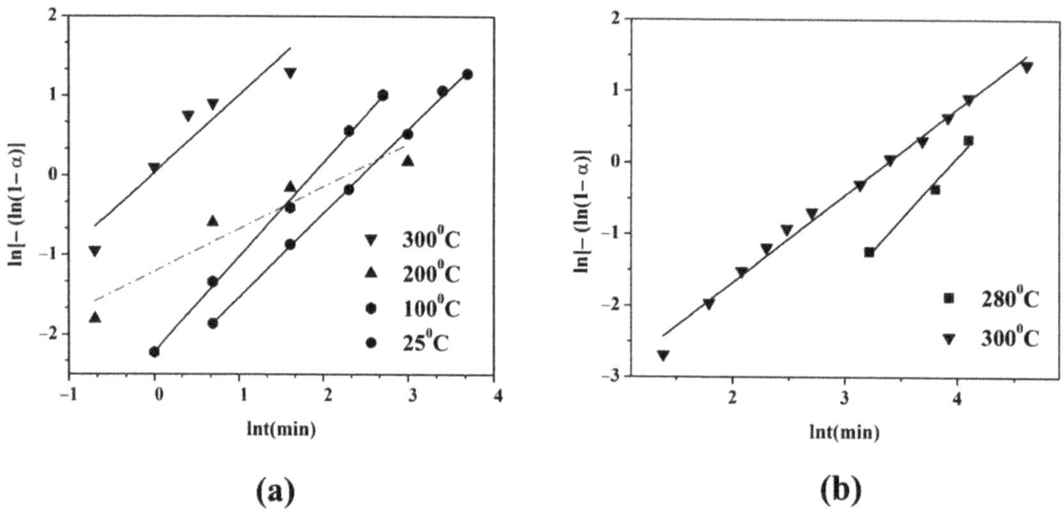

Figure 3. JMFK fit curves of the $Mg_{50}Ni_{12.5}Al_{12.5}V_{12.5}Fe_{12.5}$: (**a**) hydrogen absorption and (**b**) hydrogen desorption.

After hydrogenation at 200 °C and 1MPa for one hour, at the 20th cycle the sample is characterized by XRD and TEM. In Figure 4, the XRD diffraction patterns of the alloy after 50 h of ball milling and after hydrogenation are presented. The phase composition includes MgH_2, VH_x, and $AlFe_{0.23}Ni_{0.77}$ alloy. The ternary hydride Mg_2NiH_4 of the intermetallic Mg_2Ni is not observed.

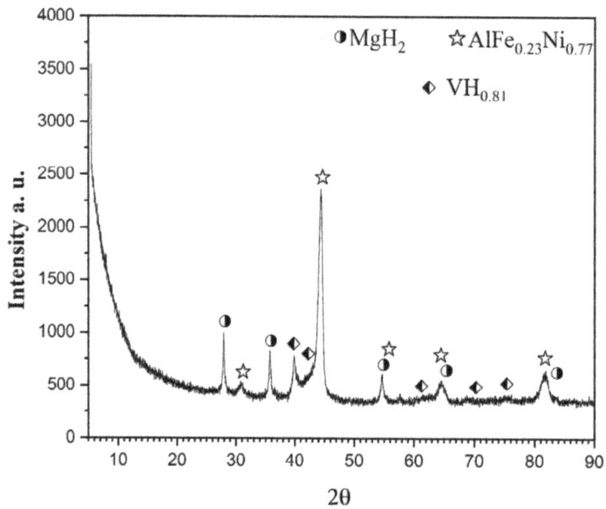

Figure 4. XRD diffraction patterns of $Mg_{50}Ni_{12.5}Al_{12.5}V_{12.5}Fe_{12.5}$ ball milled 50 h under argon and after hydrogenation at 1 MPa and 200 °C.

Despite the good hydrogen absorption rate, the absence of Mg_2Ni after milling and Mg_2NiH_4 after hydrogenation has a negative effect on the desorption rate and temperature of the alloy. The intermetallic Mg_2Ni improves the hydrogen sorption rate and lowers the desorption temperature, so it is of critical importance for the hydrogen storage properties of the Mg-based materials. To improve the hydrogen desorption properties, future experiments will consider adding some catalysts, an excess of magnesium, and changes in the production procedure of the alloy.

2.3. SEM and TEM Characterization of the $Mg_{50}Ni_{12.5}Al_{12.5}V_{12.5}Fe_{12.5}$ After Ball Milling and Hydrogenation

The SEM images are presented in Figure 5a,b. A higher fraction of small particles about a few micrometers in size is observed after hydrogenation. After ball milling for 50 h under argon, there are also some particles with a diameter of about 5 μm or even less than that size, but the majority of them are about 30–40 μm to 150 μm, and some agglomerates are also visible. From Figure 5, it can be concluded that the particles of the alloy ball milled for 50 h have a cracked non-smooth surface and irregular shapes. The process creates fine particles through mechanical fracturing but also causes these smaller particles to cold-weld and agglomerate into larger ones, resulting in a size distribution with both small and large particles. This mechanochemical process also generates strain and microstructural defects on the surface, but it may also promote oxidation and the formation of a magnesium oxide layer. After hydrogenation, the particles have a sponge-like surface and reduced size.

Figure 5. SEM images of the $Mg_{50}Ni_{12.5}Al_{12.5}V_{12.5}Fe_{12.5}$: (**a**) ball milled 50 h and (**b**) after hydrogenation at 200 °C and 1 MPa.

In Figures 6 and 7, brightfield micrographs, SAED, and HRTEM for 50 h ball-milled powder under argon and after hydrogenation at 200 °C and 1 MPa are presented. TEM analyses confirm nanosized microstructures of both samples after milling and hydrogenation as well as the phase composition detected by X-ray diffraction analyses.

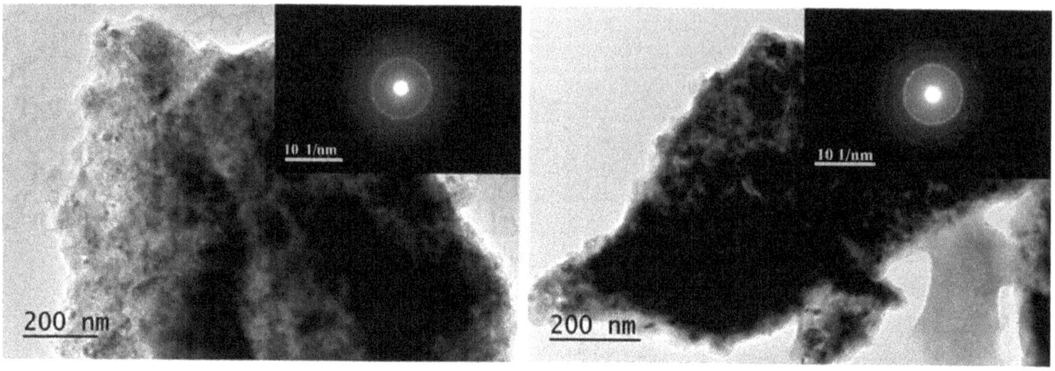

Figure 6. Brightfield micrographs and SAED are presented for $Mg_{50}Ni_{12.5}Al_{12.5}V_{12.5}Fe_{12.5}$ **left**— ball-milled for 50 h and **right**—after hydrogenation at 200 °C and 1 MPa.

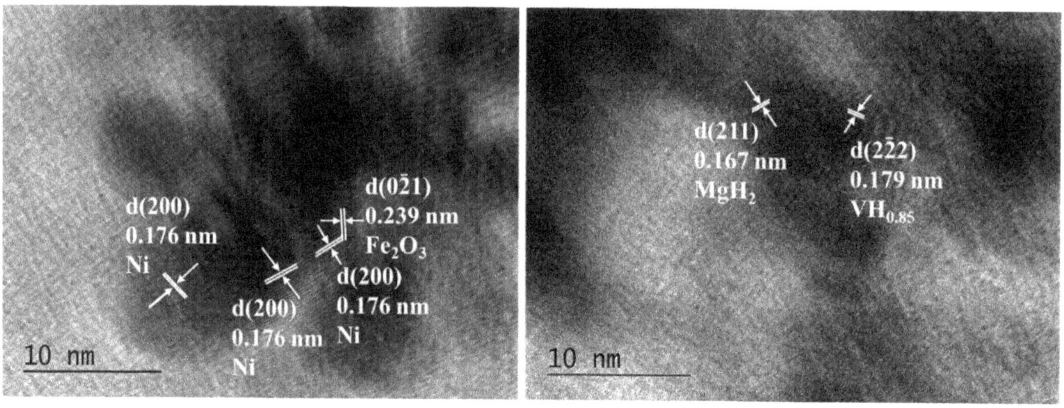

Figure 7. HTREM images of $Mg_{50}Ni_{12.5}Al_{12.5}V_{12.5}Fe_{12.5}$ after 50 h ball milling on the **left** and after hydrogenation at 200 °C and 1 MPa on the **right**.

The particles have an irregular shape and are overlapping. Some nanocrystalline areas are observed (Figure 6). The SAED patterns after 50 h of ball milling in Figure 6 are associated with the metals Mg, Ni, Fe, Al, and V. All patterns observed after hydrogenation by SAED in Figure 6 on the right belong to MgH_2. In the HTREM image for the 50 h ball-milled sample, the interplanar distance of the lattice fringes of 0.176 nm belongs to the (200) planes of Ni. Another lattice fringe is that of Fe_2O_3 $d_{(021)}$ = 0.179 nm (Figure 7 left). The right HTREM image in Figure 7 is for the alloy after hydrogenation, and the lattice fringes of 0.167 nm belong to (211) of MgH_2 and $d_{(222)}$ = 0.179 nm of VH_x. In other obtained SAED results for the alloy after hydrogenation, the patterns of $AlFe_{0.23}Ni_{0.77}$ alloy and MgO are also observed.

Some oxidized areas, e.g., MgO and Fe_2O_3 phases, are detected by TEM in both powders after 50 h of ball milling and hydrogenation, but this could be explained by the copper grid preparation and transfer into the TEM holder when the samples are exposed briefly to air. The phase Mg_3AlNi_2 is not observed by TEM characterization probably because of its low amount and the locality of this method, and after hydrogenation, if the disproportionation reaction occurs as proposed by A. Parente et al. [26], it would be difficult to detect such traces of the phases.

2.4. Electrochemical Properties and Cycle Stability

In this study, to optimize the performance of the anode material that we prepared for Ni-MH batteries, determine a safe charge–discharge range, and ensure its stability during this process, an extensive literature review was conducted initially to understand the electrochemical behavior of similar materials [31–35]. The primary aim of this literature review was to identify the potential ranges based on the data obtained to prevent damage to the alloy electrode during overcharging or discharging. Subsequently, the electrochemical stability limits of our alloy electrode were determined at a cutoff potential of −0.9 V using techniques such as Cyclic Voltammetry (CV) through these comprehensive approaches. These methods enable the identification of potential ranges where stable electrochemical reactions occur without triggering undesirable side reactions or electrode degradation. To evaluate the electrochemical characteristics of the $Mg_{50}Ni_{12.5}Al_{12.5}V_{12.5}Fe_{12.5}$ alloy electrode synthesized through a 50 h ball milling process, galvanostatic and potentiostatic discharge modes were employed. The galvanostatic discharge curve, presented in Figure 8, shows the electrode's performance when charged at a 1C rate and discharged at a 0.2 C rate, resulting in a maximum discharge capacity of 257 mAh/g.

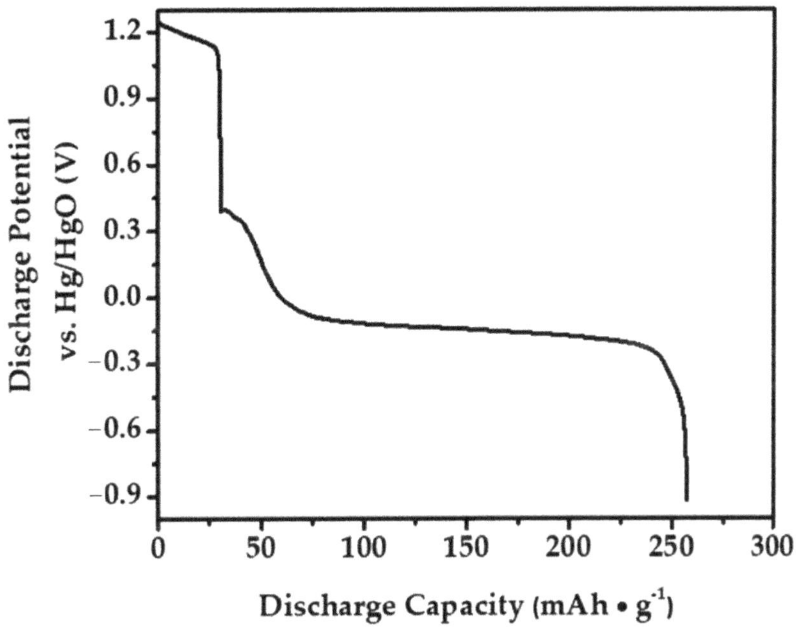

Figure 8. The maximum discharge capacity of the $Mg_{50}Ni_{12.5}Al_{12.5}V_{12.5}Fe_{12.5}$ alloy electrode at the 4th cycle.

Electrochemical cycling tests were performed on three independently prepared electrodes to ensure reproducibility. The maximum discharge capacities obtained for the first cycle were 257.0, 256.5, and 255.5 mAh g^{-1}, respectively. These values are in close agreement, giving an average of 256.3 $\pm$ 0.8 mAh g^{-1} (SD, n = 3) with a maximum deviation of only 1.5 mAh g^{-1}. This narrow variation confirms the high reproducibility of the electrochemical performance. Accordingly, the capacities presented in Figures 9 and 10 represent the mean values of three electrodes, and the corresponding error bars indicate ± standard deviation.

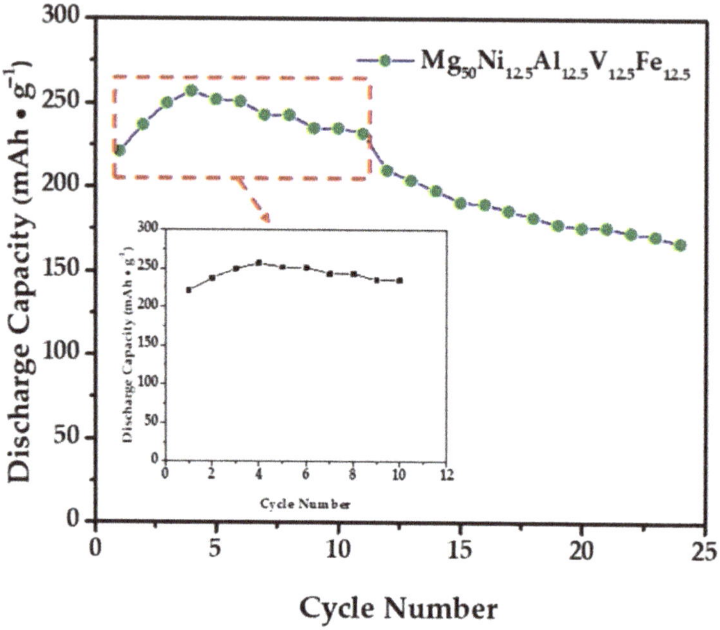

Figure 9. The maximum discharge capacity $Mg_{50}Ni_{12.5}Al_{12.5}V_{12.5}Fe_{12.5}$ alloy electrode at the 4th cycle. The dashed red box marks the activation region (cycles 1–4).

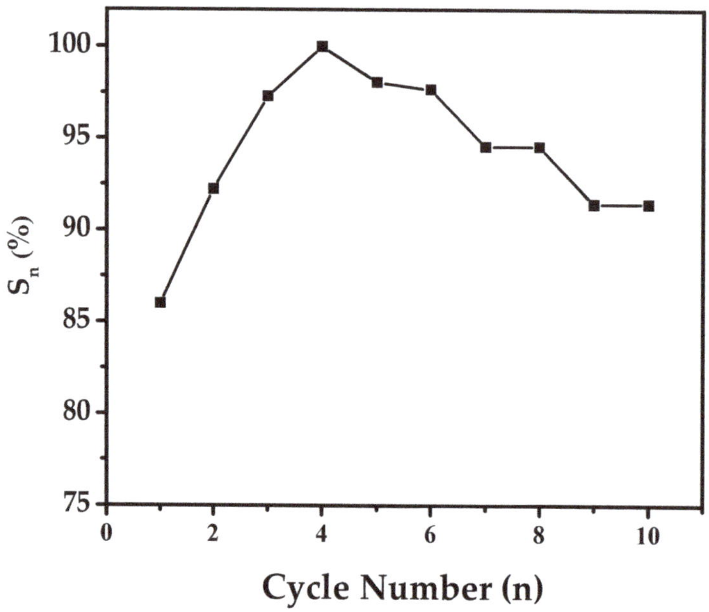

Figure 10. Capacity retention rate (Sn) and cycle number (n) of $Mg_{50}Ni_{12.5}Al_{12.5}V_{12.5}Fe_{12.5}$ after 50 h of ball milling.

The alloy electrode can become fully activated after the 3rd cycle, as shown in Figure 9. The variations in the discharge capacities of $Mg_{50}Ni_{12.5}Al_{12.5}V_{12.5}Fe_{12.5}$ alloys with the cycle number are illustrated in Figure 9, indicating that the alloy can attain its maximum discharge capacities at the fourth cycle. The reason for such easy activation of this multicomponent system can be attributed to the formation of phase boundaries, which are essential diffusion tunnels for hydrogen atoms [36]. Therefore, the easy attainment of the maximum capacity after the third cycle can be attributed to the beneficial effects of these multicomponent and multiphase structures.

The electrochemical test results for the $Mg_{50}Ni_{12.5}Al_{12.5}V_{12.5}Fe_{12.5}$ alloy electrode's 24-cycle performance show that the alloy achieved a maximum discharge capacity of 257 mAh/g at the fourth cycle. As can be seen in Figure 9, at the 10th cycle, the alloy electrode provided a maximum discharge capacity of 235 mAh/g, and at the end of the 24th cycle, the discharge capacity of our alloy electrode is 167 mAh/g. This value corresponds to an approximate capacity retention rate of 65%.

Although the alloy maintains a stable discharge capacity during the first 10 cycles, a pronounced decline occurs thereafter, leading to ~65% retention at cycle 24. This trend is consistent with the behavior reported by Iwakura et al. [37], where post-cycling degradation of Mg–Ni alloys was attributed to oxidation of the Mg-rich surface and structural instability of the hydride phase. Their results (Figure 9) [35] demonstrate that a rapid drop in capacity can occur even in the absence of direct surface analysis such as electrochemical impedance spectroscopy (EIS), underscoring the intrinsic susceptibility of Mg-based alloys to corrosion in an alkaline environment. Therefore, the retention loss observed in our Mg–Ni–Al–V–Fe can be rationalized within this well-established degradation framework.

The cycle stability of the electrode is an important factor in understanding how well the battery holds up after activation. A critical factor here is the capacity retention rate (Sn), calculated as $Sn = Cn/Cmax \times 100\%$. In this formula, Cmax is the highest discharge capacity, and Cn is the capacity at a certain number of charge/discharge cycles.

Figure 10 shows the capacity retention rates (Sn) in percent for the alloy electrode against cycle numbers. The fluctuating curves offer insights into how the discharge capacity undergoes changes due to oxidation during charge/discharge cycles.

Remarkably, the alloy demonstrates a strong cycle stability of 91.43% by the 10th cycle. It is worth highlighting that our observed capacity retention rate stands out compared to findings in the literature. For instance, in the study by Y. Zhang et al. [36], where they explored the electrochemical properties of a distinct alloy electrode specifically, $Mg_{50-x}Ti_xNi_{45}Al_3Co_2$, they noted a capacity retention rate of 73% at the 10th cycle. This observed capacity loss was attributed to the formation of Mg $(OH)_2$ on the surface of the electrode. The capacity fading observed during cycling can be reasonably attributed to pulverization and oxidation-driven surface degradation. Although no direct surface analysis was conducted in this study, the interpretation is consistent with well-documented corrosion mechanisms of Mg-based alloys in an alkaline environment. Similar oxidation-induced particle fragmentation and capacity attenuation have been reported by Jain et al. [3]. Therefore, the degradation pathway proposed here should be regarded as a possible explanation supported by established literature rather than direct experimental verification.

In the literature, the discharge capacity of Mg-Ni-based alloys has been reported over a wide range with a substantial decrease during cycling (Table 2). For example, Huang et al. achieved a discharge capacity of 488.65 mAh/g in the $Mg_{55}Ni_{45}$ alloy, although this value decreased to around 110 mAh/g after cycling [20]. The $La_{1.5}Mg_{0.5}Ni_7$ alloy, produced by mechanical alloying, exhibited a remarkable discharge capacity of 304 mAh/g [38]. In this study, the $Mg_{50}Ni_{12.5}Al_{12.5}V_{12.5}Fe_{12.5}$ demonstrated a discharge capacity of 257 mAh/g, which is competitive within the mid-range when compared to the literature.

Table 2. Comparison between the received electrochemical capacity and performance in this study and some published data of other Mg-based materials.

Composition	Electrochemical Capacity (mAh/g)	Notes/Cycling Performance	References
$Mg_{50}Ni_{12.5}Al_{12.5}V_{12.5}Fe_{12.5}$	257	235 mAh/g, 10 cycles	This work
$Mg_{1-x}Ti_{x+y}Ni_y$ (x = 0.15, 0.2; y = 0, 0.05, 0.1)	542–442	≈350 mAh/g, 20 cycles	[11]
$Mg_{1-x}Al_xNi$ (x = 0.1, 0.2)	404, 350	234 and 231 mAh/g, 20 cycles	[12]
$Mg_{50}Ni_{50}$, $Mg_{45}Pd_5Ni_{50}$	445, 315.8	≈60 and 150 mAh/g, 20 cycles	[13]
$Mg_{55}Ni_{45}$, $Mg_{55}Pd_3Ni_{42}$ and $Mg_{55}Pd_4Ni_{41}$	488.65, 543 and 564.6	≈110 mAh/g, 13 cycles 181 and 287 mAh/g, 50 cycles	[22]
MgNi–FeB 15h MA	317	≈200 mAh/g, 50 cycles	[23]
MgNi-CoB	254.9–440	113.1–174.8 mAh/g, 30 cycles	[24]
$Mg_{71}Ni_{29}$ and $Mg_{72.6}Ni_{25.3}Gd_{1.1}$	134.3 and 91.9	≈30 mAh/g, 10 cycles	[25]
$La_{2-x}Mg_xNi_7$(x = 0.25, 0.5, 0.75)	304–220	225–150 mAh/g, 30 cycles	[38]

3. Materials and Methods

Argon with a purity of 99.999% and hydrogen with a purity of 99.99% were used for these experiments. Mg was purchased from Strem chemicals (Newburyport, MA 01950-4098, USA) with a purity of 99% and of size 50 mesh. Other metals were purchased from Sigma Aldrich (St. Louis, MO, USA) as follows: Ni powder 150 µm with a purity of 99.9%, V powder 325 mesh with a 99.5% purity, Al powder 60 µm with a purity of 99.9%, and Fe powder 60 µm with a purity of 99%. The metal mixture of 3 g with a composition corresponding to the formula $Mg_{50}Ni_{12.5}Al_{12.5}V_{12.5}Fe_{12.5}$ was milled under argon in a planetary mono mill Pulverisette 6 Fritsch (Fritsch, Idar-Oberstein, Germany) with a rotation speed of 300 rpm, stainless steel balls with diameters 10 mm, and a ball to powder weight ratio of 15:1 for a total duration of 50 h. After 5, 10, 20, 30, 40, and 50 h of ball milling, the stainless steel vial with a volume of 80 mL was opened in a glove box under argon and some amount of the powder was taken for X-ray diffraction analyses using a Powder X-ray Diffractometer Bruker D8 Advance with a LynxEye detector and Cu kα radiation (Bruker, Karlsruhe, Germany). After 50 h of ball milling under argon and after hydrogenation at 200 °C and 1 MPa, the samples were characterized by TEM HR STEM JEOL JEM 2100 with a GATAN Orius 832 SC1000 CCD Camera (JEOL, Tokyo, Japan). Scanning electron microscopy (SEM) images were obtained with a JEOL JSM 7100 F microscope(JEOL, Tokyo, Japan).

The hydrogen sorption properties of the alloy were studied at 25, 100, 200, and 300 °C and a pressure of 1 MPa for absorption and 0.15 MPa for desorption and at 300, 280, and 260 °C using a volumetric Sievert's type apparatus.

The electrochemical properties and kinetics characteristics were studied by using a three-electrode system. The working electrode was made by mixing the alloy powder ground under 30–60 microns with high purity nickel powder at a ratio of 3:1 and then cold pressing it into 10 mm diameter pellets under 15 MPa. The electrode pellets then were wrapped with foam nickel. Finally, the electrodes were immersed in a 6 M KOH solution for 1 day to be completely wetted before the electrochemical measurement. The $Ni(OH)_2$ counter electrode and a Hg/HgO reference electrode were used to set up a three-electrode cell in the 6 M KOH solution. The amount of nickel hydroxide was four times the weight of the alloy placed as the working electrode. The discharge capacity of the working electrode was calculated based on the theoretical capacity of $(NiOH)_2$, reported to be 289 mAh/g [39]. The discharge capacity was calculated using the amount of the active material. Electrochemical measurements were carried out on three independently fabricated electrodes under identical conditions. The discharge capacities reported in this work are expressed as mean ± standard deviation (SD). The three-electrode cell was monitored by a Bio-Logic SP-300 galvanostat/potentiostat (Seyssinet-Pariset, France). The system was charged/discharged at a 0.2 C rate in order to fully activate the electrodes. After activation, each electrode was charged at a 1 C rate for 5 h followed by 30 min rest and then discharged at a 0.25 C rate to the cut-off potential of −0.9 V vs. the Hg/HgO reference electrode.

4. Conclusions

The X-ray diffraction patterns of the ball-milled alloy $Mg_{50}Ni_{12.5}Al_{12.5}V_{12.5}Fe_{12.5}$ for 50 h under argon showed the presence of Mg, Al, Ni, Fe, and some minor quantity of the partially disordered CsCl-type cubic structure Mg_3AlNi_2 phase. TEM analyses confirmed the results from X-ray diffraction analyses and revealed nanosized microstructures for the ball-milled alloy and for the hydrogenated one. After hydrogenation of this alloy, MgH_2, VH_x, Fe, Al, and Ni were detected. Mg_2Ni was not formed after ball milling nor was Mg_2NiH_4 formed after hydrogenation. The reversible hydrogen storage capacity of this nanosized material $Mg_{50}Ni_{12.5}Al_{12.5}V_{12.5}Fe_{12.5}$ was due mainly to magnesium. The nonstoichiometric vanadium hydride and other metals along with Mg_3AlNi_2 favor the hydrogen sorption properties. This metallic powder showed very fast absorption kinetics and resistance to oxidation, which is confirmed by the results from the electrochemical testing. It appears that the effects of ball milling such as decreasing particles size and introducing many defects also contributed to these good absorption kinetics properties. The kinetics of desorption was not as fast, and only a small quantity of hydrogen was desorbed at the lower temperature 280 °C. To improve the desorption characteristics of this type of alloy in order for it to be more attractive for hydrogen storage applications, changes in the conditions of synthesis and optimization of the composition are being considered for the future.

The alloy exhibited a maximum electrochemical discharge capacity of 257 mAh/g. Despite not surpassing the elevated values reported in previous studies for similar materials [10–13,21], this capacity is comparable to results from other sources [23–25,38] and showed a better capacity retention rate than the materials published in [9,10,16,20,22] and a comparable retention rate to that in [18]. Notably, the alloy demonstrated stability during electrochemical cycling, making it a promising candidate for Ni-MH battery applications.

Author Contributions: Conceptualization, E.G. and G.Ç.; methodology, E.G. and P.M.; validation, E.G. and G.Ç.; formal analysis, E.G. and G.Ç.; investigation, E.G., G.Ç., H.Y., and P.M.; resources, E.G. and G.Ç.; data curation, E.G.; writing—original draft preparation, E.G.; writing—review and editing, G.Ç., H.Y., and P.M.; visualization, E.G., G.Ç., H.Y., and P.M.; supervision, E.G. and G.Ç.; project administration, E.G. and G.Ç.; funding acquisition, E.G. and G.Ç. All authors have read and agreed to the published version of the manuscript.

Funding: The authors gratefully acknowledge the Bulgarian Academy of Sciences under the project IC-TR/02/2022-2023 and TUBITAK (The Scientific and Technological Research Council of Turkey) project № 121N774 for the support of this work. Some research equipment of Distributed Research Infrastructure INFRAMAT, part of Bulgarian National Roadmap for Research Infrastructures, supported by Bulgarian Ministry of Education and Science under project D01-322/30.11.2023 were used for this investigation.

Institutional Review Board Statement: Not applicable.

Informed Consent Statement: Not applicable.

Data Availability Statement: The original contributions presented in the study are included in the article, and further inquiries can be directed to the corresponding author.

Acknowledgments: The authors are grateful to Loïc Joanny and Francis Gouttefangeas for SEM experiments at CMEBA platform (ScanMAT, UAR 2025 University of Rennes-CNRS).

Conflicts of Interest: The authors declare no conflicts of interest.

References

1. Rampai, M.M.; Mtshali, C.B.; Seroka, N.S.; Khotseng, L. Hydrogen production, storage, and transportation: Recent advances. *RSC Adv.* **2024**, *14*, 6699–6718. [CrossRef]
2. Yartys, V.A.; Lototskyy, M.V.; Akiba, E.; Albert, R.; Antonov, V.E.; Ares, J.R.; Zhu, M. Magnesium based materials for hydrogen-based energy storage: Past, present and future. *Int. J. Hydrogen Energy* **2019**, *44*, 7809–7859. [CrossRef]
3. Jain, I.P.; Lal, C.; Jain, A. Hydrogen storage in Mg: A most promising material. *Int. J. Hydrogen Energy* **2010**, *35*, 5133–5144. [CrossRef]
4. Xu, Y.; Zhou, Y.; Li, Y.; Hao, Y.; Wu, P.; Ding, Z. Magnesium-Based Hydrogen Storage Alloys: Advances, Strategies, and Future Outlook for Clean Energy Applications. *Molecules* **2024**, *29*, 2525. [CrossRef]
5. Ren, L.; Li, Y.; Zhang, N.; Li, Z.; Lin, X.; Zhu, W.; Lu, C.; Ding, W.; Zou, J. Nanostructuring of Mg-based hydrogen storage materials: Recent advances for promoting key applications. *Nano-Micro Lett.* **2023**, *15*, 93. [CrossRef]
6. Çolak, A.B. An innovative study on high entropy energy storage Mg-Y-Ni-cu systems: Machine learning-driven optimization of electrical cycling in Ni-MH battery alloys. *J. Energy Storage* **2025**, *107*, 114958. [CrossRef]
7. Song, J.; Chen, J.; Xiong, X.; Peng, X.; Chen, D.; Pan, F. Research advances of magnesium and magnesium alloys worldwide in 2021. *J. Magnes. Alloys* **2022**, *10*, 863–898. [CrossRef]
8. Shang, Y.; Pistidda, C.; Gizer, G.; Klassen, T.; Dornheim, M. Mg-based materials for hydrogen storage. *J. Magnes. Alloys* **2021**, *9*, 1837–1860. [CrossRef]
9. Ruggeri, S.; Roue, L.; Huot, J.; Schulz, R.; Aymard, L.; Tarascon, J.M. Properties of mechanically alloyed Mg-Ni-Ti ternary hydrogen storage alloys for Ni-MH batteries. *J. Power Sources* **2002**, *112*, 547–556. [CrossRef]
10. Abe, T.; Inoue, S.; Mu, D.; Hatano, Y.; Watanabe, K. Electrochemical studies of the effect of surface modification of amorphous MgNi electrodes by carbon or Ni. *J. Alloys Compd.* **2003**, *349*, 279–283. [CrossRef]
11. Anik, M. Effect of Titanium additive element on the discharging behavior of MgNi Alloy electrode. *Int. J. Hydrogen Energy* **2011**, *36*, 15075–15080. [CrossRef]
12. Anik, M.; Özdemir, G.; Kücükdeveci, N.; Baksan, B. Effect of Al, B, Ti and Zr additive elements on the electrochemical hydrogen storage performance of MgNi alloy. *Int. J. Hydrogen Energy* **2011**, *36*, 1568–1577. [CrossRef]
13. Zhuang, X.; Zhang, Y.; Zhu, Y.; Qu, Y.; Zhan, L.; Wan, N.; Li, L. The effects of Pd and/or Zr additives on the structures and cyclic stabilities of Mg50Ni50-based electrode alloys. *Int. J. Hydrogen Energy* **2015**, *40*, 2768–2774. [CrossRef]
14. Küçküdeveci, N.; Erdoğan, I.A.; Aybar, A.B. Effects of Zr addition on electrochemical characteristics of MgTiMnNi hydrogen storage alloys. *Int. J. Hydrogen Energy* **2023**, *48*, 18753–18760. [CrossRef]
15. Li, P.; Wang, H.; Jiang, W.; Liu, J.; Ouyang, L.; Zhu, M. Promoting the cycling stability of amorphous MgNi-based alloy electrodes by mitigating hydrogen-induced crystallization. *Int. J. Hydrogen Energy* **2021**, *46*, 6701–6708. [CrossRef]
16. Huang, J.; Ouyang, L.; Wang, H.; Liu, J.; Zhu, M.; Fang, F.; Sun, D. Hydrogenation and crystallization of amorphous phase: A new mechanism for the electrochemical capacity and its decay in milled MgNi alloys. *Electrochim. Acta* **2019**, *305*, 145–154. [CrossRef]
17. Han, S.-C.; Lee, P.S.; Lee, J.-Y.; Züttel, A.; Schlapbach, L. Effects of Ti on the cycle life of amorphous MgNi-based alloy prepared by ball milling. *J. Alloys Compd.* **2000**, *306*, 219–226. [CrossRef]
18. Küçükdeveci, N.; Akay Erdoğan, I.; Binal Aybar, A.; Anik, M. Electrochemical hydrogen storage properties of mechanically alloyed Mg0.8Ti0.2-xMnxNi (x = 0, 0.025, 0.05, 0.1) type alloys. *Int. J. Hydrogen Energy* **2022**, *47*, 2511–2519. [CrossRef]
19. Wang, J.F.; Ding, P.D.; Gao, S.; Pan, F.S.; Tang, A. Effect of Al content on hydrogen storage properties of amorphous $Mg_{1-x}Al_xNi$ (x = 0, 0.1, 0.2, 0.3) alloy. *Mater. Sci. Forum* **2009**, *610–613*, 946–950. [CrossRef]
20. Rongeat, C.; Grosjean, M.H.; Ruggeri, S.; Dehmas, M.; Bourlot, S.; Marcotte, S.; Roué, L. Evaluation of different approaches for improving the cycle life of MgNi-based electrodes for Ni-MH batteries. *J. Power Sources* **2006**, *158*, 747–753. [CrossRef]
21. Huang, J.; Wang, H.; Ouyang, L.; Liu, J.; Zhu, M. Reducing the electrochemical capacity decay of milled Mg-Ni alloys: The role of stabilizing amorphous phase by Ti-substitution. *J. Power Sources* **2019**, *438*, 226984. [CrossRef]
22. Huang, J.; Li, J.; Cai, Y.; Zhao, S. Enhancing the cycle stability of milled Mg-Ni alloys: The role of Pd substitution on reversible electrochemical hydrogenation/dehydrogenation reactions. *Electrochem. Commun.* **2025**, *170*, 107860. [CrossRef]
23. Zhao, M.; Feng, Y.; Jiao, L.; Yuan, H.; Zhou, X.; Jang, M.-B. Preparation and electrochemical characteristics of MgNi-FeB alloys. *Int. J. Hydrogen Energy* **2007**, *32*, 3915–3920. [CrossRef]

24. Guo, M.; Wang, H.; Liu, J.; Ouyang, L. Improving hydrogen-induced crystallization and electrochemical hydrogen storage properties of MgNi amorphous alloy with CoB addition. *J. Non-Cryst. Solids* **2022**, *588*, 121646. [CrossRef]
25. Chen, W.; Sha, J.; Li, L.; Bao, J.; Yang, Y.; Qiao, M.; Tian, J.; Zhang, Z.J. The effect of Gd on the microstructure and electrochemical properties of Mg-Ni-based alloys. *Alloys Compd.* **2024**, *981*, 173638. [CrossRef]
26. Parente, A.; Nale, A.; Catti, M.; Kopnin, E.; Caracino, P. Hydrogenation properties of Mg_2AlNi_2 and mechanical alloying in the Mg-Al-Ni system. *J. Alloys Compd.* **2009**, *477*, 420–424. [CrossRef]
27. Yang, X.; Zhu, Y.; Zhang, J.; Zhang, Y.; Liu, Y.; Lin, H.; Wang, T.; Li, L. Effect of partial substitution of Ti for Al on the phase structure and electrochemical hydrogen storage properties of Mg_3AlNi_2 alloy. *J. Alloys Compd.* **2018**, *746*, 421–427. [CrossRef]
28. Sheng- Long, L.; Fu Kai, H.; Wen- Chi, C.; Chih- Kuang, L.; Jing- Chie, L. Influence of Mg_3AlNi_2 content on the cycling stability of Mg_2Ni-Mg_3AlNi_2 hydrogen storage alloy electrodes. *Intermetallics* **2011**, *19*, 1953–1958.
29. Guzmán, D.; Ordoñez, S.; Fernández, J.F.; Sánchez, C.; Serafini, D.; Rojas, P.A.; Aguilar, C.; Tapia, P. Effect of amorphous $Mg_{50}Ni_{50}$ on hydriding and dehydriding behavior of Mg_2Ni alloy. *Mater. Charact.* **2011**, *62*, 442–450. [CrossRef]
30. Shao, H.; Asano, K.; Enoki, H.; Akiba, E. Preparation and hydrogen storage properties of nano-structured Mg-Ni BCC alloys. *J. Alloys Compd.* **2009**, *477*, 301–306. [CrossRef]
31. Zhang, Y.; Chen, L.-X.; Lei, Y.-Q.; Wang, Q.-D. The reduction of cycling capacity degradation of Mg–Ni-based electrode alloys by Fe substitution. *Int. J. Hydrogen Energy* **2002**, *27*, 501–506. [CrossRef]
32. Young, K.; Fetcenko, M.A.; Li, F.; Ouchi, T. Structural, thermodynamic, and electrochemical properties of $Ti_xZr_{1-x}(VNiCrMnCoAl)_2$ C14 Laves phase alloys. *J. Alloys Compd.* **2008**, *464*, 238–247. [CrossRef]
33. Etiemble, A.; Rousselot, S.; Guo, W.; Idrissi, H.; Roué, L. Influence of Pd addition on the electrochemical performance of Mg-Ni-Ti-Al-based metal hydride for Ni-MH batteries. *Int. J. Hydrogen Energy* **2013**, *38*, 7169–7177. [CrossRef]
34. Tan, S.; Shen, Y.; Onur Şahin, E.; Noréus, D.; Öztürk, T. Activation behavior of an AB2 type metal hydride alloy for NiMH batteries. *Int. J. Hydrogen Energy* **2016**, *41*, 9948–9953. [CrossRef]
35. Wang, L.; Young, K.; Meng, T.; Ouchi, T.; Yasuoka, S. Partial substitution of cobalt for nickel in mixed rare earth metal based superlattice hydrogen absorbing alloy–Part 1 structural, hydrogen storage and electrochemical properties. *J. Alloys Compd.* **2016**, *660*, 407–415. [CrossRef]
36. Zhang, Y.; Wei, X.; Gao, J.; Hu, F.; Qi, Y.; Zhao, D. Electrochemical hydrogen storage behaviors of as-milled Mg-Ti-Ni-Co-Al-based alloys applied to Ni-MH battery. *Electrochim. Acta* **2020**, *342*, 136123. [CrossRef]
37. Iwakura, H.; Inoue, H.; Matsuoka, M.; Tamura, S. Electrochemical and structural properties of Mg–Ni hydrogen storage alloys. *J. Less-Common Met.* **1991**, *172–174*, 1051–1060.
38. Werwiński, M.; Szajek, A.; Marczyńska, A.; Smardz, L.; Novak, M.; Jurczyk, M. Effect of substitution La by Mg on electrochemical and electronic properties in $La_{2-x}Mg_xNi_7$ alloys: A combined experimental and ab initio studies. *J. Alloys Compd.* **2018**, *763*, 951–959. [CrossRef]
39. Ash, B.; Nalajala, V.S.; Popuri, A.K.; Subbaiah, T.; Minakshi, M. Perspectives on nickel hydroxide electrodes suitable for rechargeable batteries: Electrolytic vs. chemical synthesis routes. *Nanomaterials* **2020**, *10*, 1878. [CrossRef] [PubMed]

inorganics

MDPI

Article

Structure and Electrochemical Performance of Glasses in the Li_2O-B_2O_3-V_2O_5-MoO_3 System

Margarita Milanova [1,*], Xinhao Yang [2,3], Pamela Vargas [2], Nataly Carolina Rosero-Navarro [2], Ruzha Harizanova [4], Bojidar Jivov [5], Lyubomir Aleksandrov [1], Reni Iordanova [1], Maya Shopska [6] and Savina Koleva [7]

1 Institute of General and Inorganic Chemistry, Bulgarian Academy of Sciences, Acad. Georgi Bonchev Str., Bld. 11, 1113 Sofia, Bulgaria; lubomirivov@gmail.com (L.A.); reni@svr.igic.bas.bg (R.I.)
2 Instituto de Cerámica y Vidrio, Kelsen 5, 28049 Madrid, Spain; xinhao@icv.csic.es (X.Y.); pamela@icv.csic.es (P.V.); rosero@icv.csic.es (N.C.R.-N.)
3 Faculty of Sciences, Universidad de Autónoma de Madrid, 28049 Madrid, Spain
4 Department of Physics, University of Chemical Technology and Metallurgy, 8 Kliment Ohridski Boulevard, 1756 Sofia, Bulgaria; rharizanova@uctm.edu
5 Bulgarian Academy of Sciences, Institute of Metal Science, Equipment and Technologies with Hydro- and Aerodynamics Centre "Acad. A. Balevski", 67 "Shipchenski Prohod" Blvd., 1574 Sofia, Bulgaria; b_jiv@abv.bg
6 Institute of Catalysis, Bulgarian Academy of Sciences, Acad. Georgi Bonchev Str., Bld. 11, 1113 Sofia, Bulgaria; shopska@ic.bas.bg
7 Faculty of Chemistry and Pharmacy, Sofia University "St. Kliment Ohridski", 1 James Bourchier Blvd, 1164 Sofia, Bulgaria; savi.koleva02@gmail.com
* Correspondence: margi@svr.igic.bas.bg

Abstract: Applying the melt quenching method (cooling rate 10^1–10^2 K/s), new multi-component vanadate glasses were synthesized, containing different amounts of MoO_3 at the expense of B_2O_3 with the composition $20Li_2O$:$(30 - x)B_2O_3$:$50V_2O_5$:$xMoO_3$, x = 10, 20 mol%. The obtained samples were characterized by X-ray diffraction, infrared spectroscopy, differential scanning calorimetry and impedance spectroscopy. The density of the glasses was measured by the Archimedes method, on the basis of which the physicochemical parameters molar volume, oxygen molar volume and oxygen packing density were calculated. It was found that the replacement of B_2O_3 with MoO_3 leads to changes in electrical conductivity, which are a consequence of the increase in non-bridging oxygen atoms in the amorphous structure. The electrochemical characterization of the $20Li_2O$:$(30 - x)B_2O_3$:$50V_2O_5$:$20MoO_3$ glass obtained was performed by assembling an all-solid-state cell, employing $20Li_2O$:$(30 - x)B_2O_3$:$50V_2O_5$:$20MoO_3$ glass as a cathode active material. The obtained results show that the studied glass compositions are interesting in view of their potential application as cathode materials in all-solid-state lithium-ion batteries.

Keywords: glasses; structure; density; all-solid-state battery

1. Introduction

Currently, various types of batteries are widely used to power [1,2] consumer electronics, mobile devices, lighting components, medical equipment, etc. Batteries are used [1] in lifting and transport equipment (electric trucks, stackers, etc.), hybrid cars, electric cars, ships, and others. Stationary energy storage systems provide an alternative source of electricity for industrial and domestic users. At the same time, due to the high reliability [1,2] of lithium-ion batteries, their further development and the development of new generation samples are of great interest [3–14]. Many research groups [15–32] have studied

 https://doi.org/10.3390/inorganics13090285

the structure, physicochemical properties and electrochemical behavior of glass materials synthesized in oxide systems with the participation of various glass formers. In the Li_2O-P_2O_5-MoO_3 system, distinct regions (with different ratios and concentrations of oxide components) characterized by dominant ionic or electronic conductivity of the synthesized glasses were established [17,18,21]. The change in DC and AC conductivity [21] was studied by changing the composition of a series of phosphate glasses yLi_2O–$(1 - y)[0.35(MoO_3)_2$–$0.65(P_2O_5)]$. It was proven that modification of glassy materials from the Li_2O-P_2O_5-MoO_3 system [25], by introducing low concentrations of LiCl, causes a change in the conductivity mechanism from mixed (ionic–electronic) to predominantly ionic. In the case of glasses [19] with the composition xLi_2O–$(100 - 2x)V_2O_5$–xP_2O_5 (x = 15, 25, 35, 40 and 45 mol%), a transition from predominantly electronic to ionic conductivity was observed at a concentration of 40 mol% Li_2O. In glassy samples with the composition $50V_2O_5$–$(50 - x)P_2O_5$–xA_2O (A = Li, Na, K), a decrease in DC conductivity was observed with an increase in the content of alkali oxides [24]. The effect of crystallization on the ionic and electronic conductivity of glasses prepared in the Li_2O–V_2O_5–P_2O_5 system was reported in [23]. A significant increase in electronic conductivity has been observed for a crystallized sample having composition $50Li_2O$–$25V_2O_5$–$25P_2O_5$. The nanostructured features, as well as the ionic conductivity of $Li_{1.3}Nb_{0.3}V_{1.7}(PO_4)_3$ glass–ceramic materials, that may find application as cathode materials in battery production, were presented in [4]. At low V_2O_5 concentrations ($0.05 \le x \le 0.5$), it was discovered that the added divanadium pentoxide acts as a modifier that promotes the formation of boron-oxygen tetrahedra for glasses with compositions xV_2O_5–$(1 - x)B_2O_3$. When vanadium oxide content is higher ($x > 0.5$), a glass network composed of vanadium-oxygen tetrahedra is formed. Vanadium oxide acts as a glass former in the binary systems $xMeO$–$(100 - x)V_2O_5$ (where Me = Ba, Mg, and Sr) and creates an amorphous network of vanadium-oxygen polyhedra [20]. Vanadium oxide is mainly a modifier with the formation of VO_4 tetrahedra for $(100 - x)[B_2O_3{\bullet}Li_2O]$–$xV_2O_5$ glasses, at low x values [16]. With increasing content of V_2O_5, its structural role changes to a glass former and the formation of VO_5 trigonal bipyramids occurs [16]. A study on the local structure [28] of glasses from the Li_2O–B_2O_3–V_2O_5 system has been carried out, in which the co-ordination of boron (3 and 4) and vanadium (5 and 6) was identified, and a structural model of the amorphous network was proposed. An increase in the fraction of tetracoordinated boron was observed with increasing vanadium concentration. Glass samples [26] with compositions $0.3Li_2O$–$(0.7 - x)B_2O_3$–xV_2O_5 and x = 0.3, 0.325, 0.35, 0.375, 0.4, 0.425, 0.45 and 0.475 were obtained, where the boron anomaly typical for borate glasses with a high Li_2O content was not detected. The dominant role of electronic conductivity, which increases exponentially (7.4×10^{-5} S cm^{-1} at x = 0.475) with temperature and with the increase of vanadium content, has been confirmed [26]. The structure and electrochemical activity of glasses with compositions $20Li_2O$-$30V_2O_5$-$(50 - x)SiO_2$-xB_2O_3 mol.% (x = 10, 20, 30, 40) were investigated [5]. For the composition of $20Li_2O$-$30V_2O_5$-$40SiO_2$-$10B_2O_3$, the highest content of fraction V^{4+}, increased electronic conductivity, and higher charge–discharge capacity were recorded compared to other samples. The reduction of the particle size by grinding (while preserving the amorphous structure of the material) stabilizes the charge–discharge cycle in the studied operating mode, lowers the impedance [5], facilitates the migration of Li^+ ions and increases the conductivity. The possible states of vanadium from V^{5+} to V^{2+} allow [15] the realization of a complete charge–discharge cycle with adequate capacity under certain structural conditions. In most cases, vanadium ions exist in the form of V^{3+}, V^{4+} and V^{5+} with different coordination environments [30]. In

glassy materials, vanadium ions exhibit mainly +4 and +5 oxidation states. V^{4+} ions with four-coordination form tetrahedral structural units, while V^{5+} ions exist in the amorphous network with coordination 4, 5, and 6, allowing the formation of tetrahedra, bipyramids, and octahedra, respectively. DC conductivity in the interval 318 K–473 K was investigated for series of oxide glasses with compositions $xMoO_3 + (30 - x)(V_2O_5) + 70(B_2O_3)$ [29]. The values were recorded to increase with the concentration of MoO_3 up to x = 0, 15 and then decrease at x > 0.15. With increasing MoO_3 concentration, a non-linear change in density and molar volume was found. Mott's variable-range hopping model has been shown to allow adequate interpretation [29] of DC conductivity data obtained at $T < \theta_D/2$.

Several previous studies have discussed the potential of glassy materials as solid-state battery electrolytes [33] or interfacial layers [34] due to their favorable stability against lithium metal, environmental durability, and moderate ionic conductivity. In particular, B_2O_3-based glasses, such as $Li_2O–V_2O_5–SiO_2–B_2O_3$, have been explored as promising cathode materials [5]. For example, composite electrodes based on $V_2O_5–LiBO_2$ glass and reduced graphene oxide (RGO) have demonstrated a high energy density of approximately 1000 Wh/kg, while maintaining a specific capacity of around 300 mAh/g over the first 100 cycles [3]. Building on this understanding, we designed an all-solid-state composite cathode by mixing the synthesized glass powder with a sulfide-based solid electrolyte and conductive carbon black [35–37]. The sulfide electrolyte, known for its high lithium-ion conductivity, mechanical softness, and excellent processability [38–40], plays a dual role. It minimizes interfacial resistance and mechanical mismatch between the electrode and electrolyte, and it provides efficient ionic pathways throughout the cathode. The composite can be fabricated without high-temperature sintering. This simplifies the manufacturing process and improves interfacial compatibility between the components.

The aim of the present work is to study the synthesis conditions, structure and impedance characteristics of new glassy materials from the Li_2O-B_2O_3-V_2O_5-MoO_3 system. The electrochemical performance of one selected glass composition was also investigated by assembling an all-solid-state cell, using the glass sample obtained as an active material.

2. Results and Discussion

2.1. XRD and DCS Studies

The phase formation of selected compositions from the Li_2O-B_2O_3-V_2O_5-MoO_3 system was studied by X-ray diffraction analysis (XRD). The first batch of compositions with reduced V_2O_5 content yielded opaque and visually non-uniform glasses (Table 1) (Figure 1a). According to the XRD data (Figure 1b), glass-crystalline material exhibiting an amorphous halo and diffraction peaks corresponding to the $Li_{0.66}V_2O_5$ (PDF # 01-074-0054) [41] and to the LiV_3O_8 (PDF # 01-072-1193) [41] was obtained from the ternary composition in the second series containing 50 mol% of V_2O_5 without MoO_3 (Table 2). The addition of 10 and 20 mol% MoO_3 results in the formation of visually uniform X-ray amorphous samples (Figure 1b), while from the quaternary composition possessing higher MoO_3 content of 30 and 40 mol%, glass-crystalline specimens containing $LiVMoO_6$ (PDF # 01-082-8687) [41] as a crystalline phase were quenched (Figure 1c).

Thermal parameters of the obtained two homogeneous glass samples $20Li_2O:50V_2O_5:(30 - x) B_2O_3:xMoO_3$, x = 10, 20 mol% were established by differential scanning calorimetry (DSC). Figure 2 compares the DSC curves of the two homogeneous glass samples x = 10 and x = 20 obtained.

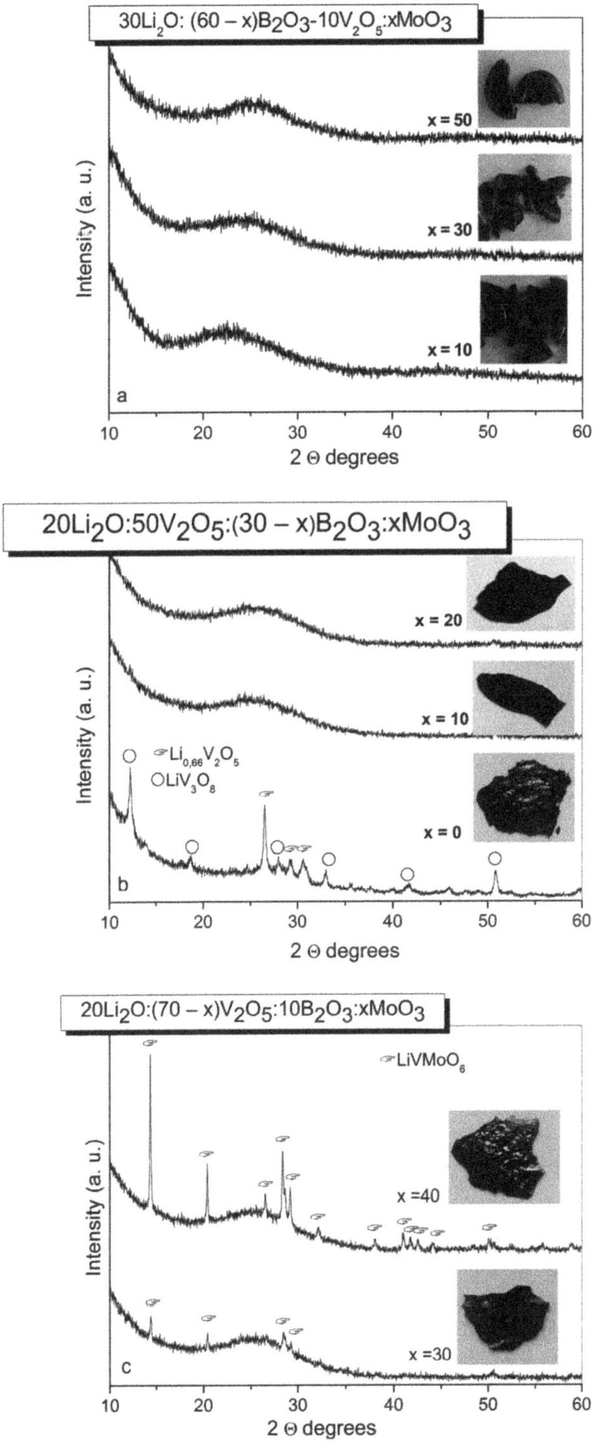

Figure 1. XRD patterns of glass (**a**,**b**) and glass–crystalline materials (**b**,**c**), obtained by quenching of the melts with various compositions from the Li_2O-B_2O_3-V_2O_5-MoO_3 system.

Table 1. Nominal compositions, experimental conditions applied and visual observations of the quenched samples.

Sample	Li_2O (mol%)	B_2O_3 (mol%)	V_2O_5 (mol%)	MoO_3 (mol%)	Melting Temperature (°C)	Colling Rate K/s	Visual Assessment
1.	30	10	10	50	850	$\leq 10^1–10^2$	Hygroscopic/Non homogeneous glass
2.	30	30	10	30	850	$\leq 10^1–10^2$	Hygroscopic/Non homogeneous glass
3.	30	50	10	10	850	$\leq 10^1–10^2$	Hygroscopic/Non homogeneous glass

Table 2. Nominal compositions, experimental conditions applied and visual observations of the quenched samples.

Sample	Li_2O (mol%)	B_2O_3 (mol%)	V_2O_5 (mol%)	MoO_3 (mol%)	Melting Temperature (°C)	Colling Rate K/s	Visual Assessment
1.	20	20	50	-	950	$10^1–10^2$	Glass + crystals
2.	20	20	50	10	950	$10^1–10^2$	Non hygro-scopic/Homogeneous glass
3.	20	10	50	20	950	$10^1–10^2$	Non hygro-scopic/Homogeneous glass
4.	20	10	40	30	950	$10^1–10^2$	Glass + crystals
5.	20	10	30	40	950	$10^1–10^2$	Glass + crystalls

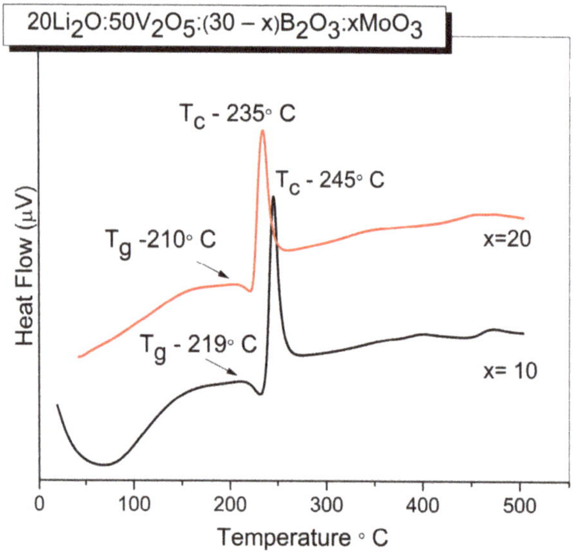

Figure 2. DSC curves of glasses with the compositions $20Li_2O:50V_2O_5:(30 - x) B_2O_3:xMoO_3$, x = 10, 20 mol% obtained by melt quenching method.

The humps correspond to the glass transition temperature T_g, while the exothermic peaks are due to the glass crystallization temperature T_c. Their values and the calculated glass thermal stability $\Delta T = T_c - T_g$ are listed in Table 3. As observed, T_g and T_c slightly decrease with the increasing MoO_3 content. Bearing in mind that the glass transition temperature is very sensitive to any alteration in the coordination number of network-forming atoms (i.e., short-range order) and to the formation of non-bridging oxygens [42,43], we can explain a slight reduction of T_g with the change in the short-range order and with slight depolymerization of the amorphous network (increasing number of non-bridging oxygens) with compositional changes. In addition, the values of T_g were correlated with average single bond enthalpy *EB* of glasses using the following relationship proposed in [14,15]:

$$EB = \frac{20E_{Li\text{-}O} + 50E_{V\text{-}O} + xE_{Mo\text{-}O} + (30 - x)E_{B\text{-}O}}{100} \tag{1}$$

where $E_{Li\text{-}O}$, $E_{V\text{-}O}$, $E_{Mo\text{-}O}$ and $E_{B\text{-}O}$ are the bond dissociation energies for the single bond Li-O, V-O, Mo-O and B-O, respectively, and the values of *EB* calculated are also included in Table 3. The higher *EB* value of the glass x = 10 can be connected with the difference in bond dissociation energies of metal oxide. Since the B-O bond enthalpy is 806 kJ mol^{-1}, exceeding Mo-O bond enthalpy (607 kJ mol^{-1}), the average single bond energy decreases with the addition of MoO_3 at the expense of B_2O_3, because of the formation of weaker Mo-O at the expense of stronger B-O bonds. Both glasses are characterized with a low thermal stability criterion ΔT of 25 °C that remains constant with the composition, indicating low thermal stability and poor glass-forming ability of the compositions [44].

Table 3. Values of glass transition temperature T_g, crystallization temperatures (T_c), thermal stability ΔT and average single-bond enthalpy *EB* of investigated glasses.

Sample ID	T_g/°C	T_c/°C	ΔT/°C	*EB*/KJ mol^{-1}
x = 10	219	245	24	612
x = 20	210	235	25	592

2.2. IR Spectral Analysis

The structure of the two glasses obtained (x = 10 and x = 20) was investigated by IR spectroscopy (Figure 3a). The IR spectra of both glasses are extremely similar, as can be seen in the figure, and exhibit IR activity in three spectral regions: 1600–1150 cm^{-1}, 1150–850 cm^{-1}, and 850–400 cm^{-1}. To obtain more precise structural information, the spectra were deconvoluted and are displayed in Figure 3b,c.

In this way, in the first spectral region between 1600–1150 cm^{-1} several peaks were observed due to the stretching vibration of BO_3 groups in meta-, pyro-, orthoborate structures (bands in the range 1196–1461 cm^{-1}) and in superstructural units (BO_3+BO_4) (band at 1266–1270 cm^{-1}) [45,46]. In the second spectral range 1150–850 cm^{-1}, bands related to the vibrations of BO_4 tetrahedral units (bands at 1090 and at 1072 cm^{-1}) [47], VO_4 (band at 649–651 cm^{-1}), VO_5 (bands at 1000–993 cm^{-1}) and VO_6 (band at 932–916 cm^{-1}), as well as MoO_4 (bands at 910 cm^{-1}) and MoO_6 (bands at 923–916 cm^{-1} and at 980–971 cm^{-1}) structural groups are situated [48–50]. Me-O-Me, Me = V, Mo bonding is also identified, manifested by the presence of the band at 885 cm^{-1} [51]. The third spectral region from 850–400 cm^{-1} contains bands connected with the bending vibrations of B-O-B bonds in meta-, pyro- and superstructural borate units (band at 697 cm^{-1}), V-O-V stretching vibrations (band at 767–764 cm^{-1}), bending, asymmetric and symmetric stretching mode of $V(Mo)_2O_2$ entities present in $V(Mo)_2O_8$ units (bands at 468–472 cm^{-1}, at 573 cm^{-1} and at 728 cm^{-1}, respectively), and also vibrations of VO_4 tetrahedral groups (band at 649–651 cm^{-1}) [45–50]. The detailed assignments of the bands observed in the deconvoluted IR spectra of the present glasses are summarized in Table 4. With increasing MoO_3 content, MoO_4→MoO_6 transformation (disappearance of the band at 910 cm^{-1} and appearance of the band at 448 cm^{-1}), and Me-O-Me, Me = V and Mo bonding (increased intensity of the band at 885 cm^{-1}) take place [51,52]. On the other hand, some depolymerization of the borate oxygen network and formation of orthoborate BO_3^{3-} groups also occurs, evidenced by the appearance of the band at 1196 cm^{-1} and increased intensity of the band at 1253 cm^{-1}, both characteristic for the BO_3^{3-} units [45] in the spectrum of glass x = 20.

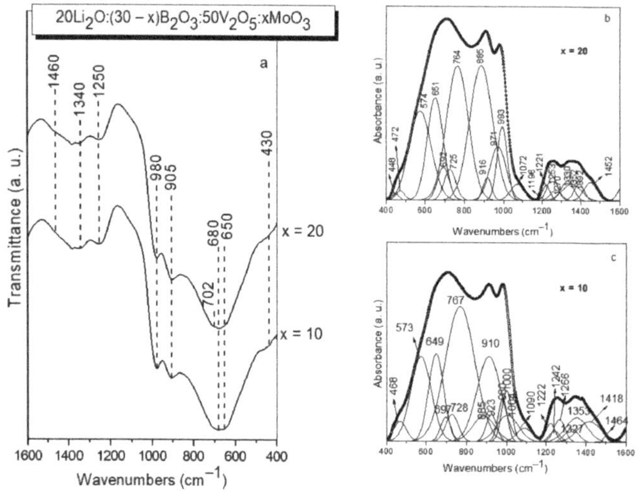

Figure 3. IR spectra (**a**) and deconvoluted IR spectra (**b**,**c**) of glasses with the compositions $20Li_2O{:}50V_2O_5{:}(30 - x)\ B_2O_3{:}xMoO_3$, x = 10, 20 mol% obtained by melt quenching method.

Table 4. Infrared bands (in cm^{-1}) and their assignments for glasses $20Li_2O{:}50V_2O_5{:}(30 - x)\ B_2O_3{:}xMoO_3$, x = 10, 20 mol%.

Infrared Bands Position (cm^{-1})	Assignment	Ref.
468–472	$\delta\ V_2O_2\ (VO_5)$	[49]
573	$\nu_s\ (Me_2O_8)$, Me = V/Mo	[51]
649–651	$\nu_{as}\ (VO_4)$	[48]
697	$\nu_3\ (MoO_4)$	[52]
728	$\nu_s\ (Me_2O_8)$, Me = V/Mo	[51]
767–764	ν (V-O-V)	[49]
885	ν (Me-O-Me), (Mo/VO_6) Me = V/Mo	[51]
910	$\nu_1\ (MoO_4)$	[52]
923–916	ν (Me = O), (Mo/VO_6) Me = V/Mo	[48,52]
980–971	ν (Mo = O), MoO_6	[50]
1000–993	ν (V = O), (VO_5)	[49]
1090–1072	$\nu\ BØ_4^-$	[47]
1196; 1242–1253	$\nu_3\ BO_3^{3-}$	[45]
1221; 1237–1330	B-O-B stretch in pyroborate units, $BØO_2^{2-}$;B-O^- stretch in in pyroborate units, $BØO_2^{2-}$	[45]
1266–1270; 1352, 1392–1418; 1452–1464	BØ stretch in metaborate unuts, $BØ_2O$ + ν_{as}(B–O–B); B–O–B bridges connect BO_3 units ν(B–O-) stretch in $BØ_2O^-$ units	[45,46]

2.3. Density, Molar Volume, Oxygen Packing Density and Oxygen Molar Volume

Structural information about the glasses obtained was also acquired by measuring the glass density and calculating the molar volume (V_m), oxygen molar volume (V_o) and oxygen packing density (OPD) of the glass system, employing the following relations respectively [53].

$$V_m = \frac{\sum x_i M_i}{\rho_g} \tag{2}$$

$$V_o = V_m \times \left(\frac{1}{\sum x_i n_i}\right) \tag{3}$$

$$OPD = 1000 \times C \times \left(\frac{\rho_g}{M}\right) \tag{4}$$

where x_i is the molar fraction of each oxide component i, M_i the corresponding molecular weight, ρ_g the glass density, n_i is the number of oxygen atoms in each oxide, C is the number of oxygen atoms per formula units, and M is the total molecular weight of the glass compositions. The values obtained are listed in Table 5.

Table 5. Values of physical parameters of glasses $20Li_2O{:}50V_2O_5{:}(30 - x)\ B_2O_3{:}xMoO_3$, x = 10, 20 mol%: density (ρ_g), molar volume (V_m), oxygen molar volume (V_o), oxygen packing density (OPD).

Sample ID	ρ (g/cm^3)	V_m (cm^3/mol)	V_o (cm^3/mol)	OPD (g atom/L)
x = 10	2.964 ± 0.006	39.90	11.08	90.22
x = 20	3.051 ± 0.002	43.48	12.08	82.80

It can be observed that both density and molar volume of the glass possessing higher MoO_3 content at the expense of B_2O_3 content is greater as compared to the glass with lower MoO_3 concentration. In general, density and molar volume are inversely related. However, this anomalous behavior has been reported in other glass systems [54]. In the present system, the increase in glass density can be attributed to the increase in the average molecular mass of the glasses [55]. Since the molar volume (volume occupied by a mass of the glass equal to 1 mole) is significantly influenced by the ionic radii of the incorporated ionic species in the glass, the higher V_m for glass x = 20 as compared with V_m of glass x = 10 is due to the substitution of B^{3+} ions with smaller ionic radii (0.27 Å) by Mo^{6+} ions known to have a higher ionic radius (0.59 Å), resulting in the generation of excess free space, which increases the overall molar volume of this glass. Oxygen molar volume (V_o) and OPD are two parameters that provide information on the packing of the oxygen ions in the glass structure [56]. Low V_o and high OPD are indications of a high degree of network connectivity. It is observed that glass x = 20 is characterized with a greater V_o and reduced OPD values compared with the values of these parameters estimated for the glass x = 10, indicating lower degree of connectivity (increased number of non-bridging oxygens (NBOs)) of the network of the glass having higher MoO_3 content. This assumption is in line with the DSC and IR spectral data revealing the accumulation of BO_3 groups with NBOs (i.e., BO_3^{3-}) in the glass structure of glass x = 20.

2.4. Diffuse Reflectance Spectra (DRS)

UV-vis diffuse spectroscopy is also used to obtained information about the local structure of glasses via ligand to metal charge transfer (LMCT) band position and the corresponding edge energy (E_g) values. Figure 4 represents the diffuse reflectance spectra (a) and the band gap energy E_g (b) and (c) of glasses under investigation.

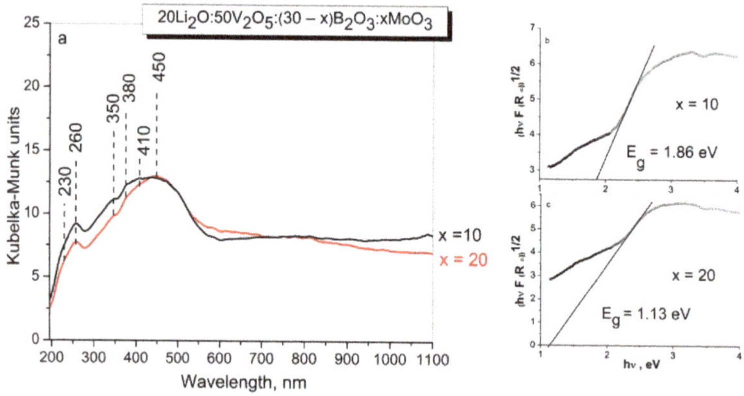

Figure 4. DR-UV-Vis spectra (**a**) and the band gap energy Eg (**b**,**c**) of glasses with the compositions $20Li_2O:50V_2O_5:(30-x)\,B_2O_3:xMoO_3$, x = 10, 20 mol% obtained by melt quenching method.

The diffuse reflectance spectra of the studied glasses consist of bands in the range 200–600 nm, with maxima at 260, 350 nm, due to the ligand–metal charge transfer (LMCT) from oxygen ligands to Mo^{6+} in octahedral coordination and/or V^{5+} ions in tetrahedral coordination, and also bands at 380 nm due to the charge transfer band of the V = O double bond, and bands at 410 and 450 nm due to polymerized VO_6 with octahedral coordination [50,57].

The absence of any absorption in the visible range (600–800 nm) indicates that Mo^{5+} and V^{4+} species are not present in the investigated sample. Assuming the glass obtained to be indirect semiconductors, the plot of the transformed Kubelka–Munk function versus the energy of light (Tauc plot) provides band gap energies, E_g of 1.86 eV for x = 10 and 1.13 eV for x = 20 glasses, respectively. The decrease of E_g values with increase in MoO_3 content is related to the changes in bonding during vitrification, and increase in the number of non-bridging oxygens in the glass structure [58].

2.5. Electrical Properties

The samples with x = 10 and x = 20 which proved to be glasses were heated with a heating rate of 10 K/min to 180 °C, i.e., below the T_g values obtained by DSC, and then the impedance spectra were recorded at different temperatures and frequencies on cooling. After equilibrating the sample at each measurement temperature, the frequency dependence of the impedance modulus and the phase angle was measured. The two samples showed purely resistive behavior at all frequencies in the whole temperature range, i.e., the phase angle was always between 0 and −25°. Thus, the measured samples exhibited behavior which allowed assumption of the impedance modulus at 1 Hz equal to the sample resistance. From this resistance, taking into account the sample thickness and electrode surface, the dc-conductivity σ [S/m] is calculated from Equation (5):

$$\sigma = \frac{1}{\rho} = \frac{1}{R}\frac{d}{S} = \frac{1}{|Z|}\frac{d}{S} \tag{5}$$

where ρ [Ω.m]—sample resistivity, S [m^2]– electrode surface, d [m]—sample thickness, $|Z|$ [Ω]—impedance modulus. The impedance modulus, $|Z|$ [Ω] and phase angle, φ [°] values at 1 Hz, as well as the calculated dc-conductivity σ and the respective activation energies, E_a, assuming, for the glasses x = 10 and x = 20 are given in Table 6 for the different temperatures. In Figure 5, the Arrhenius plots are shown for the glasses x = 10 and x = 20. From the slopes of the Arrhenius plots, the dc-conductivity activation energy

E_a is determined and also given in Table 6. The results from the estimation of the activation energies show that the sample with lower MoO_3 concentration has higher activation energy, which could be attributed to the structural changes occurring in the glass network with increasing molybdenum concentration, i.e., the occurrence of more NBOs as also observed in the IR spectra of the same glass compositions.

Table 6. Data from the electrical measurements: impedance modulus, |Z| and phase angle, φ at 1 Hz, calculated dc-conductivity, σ and the respective activation energies E_a for the samples x = 10 and 20.

x (mol%)	t (°C)	|Z| (Ω)	φ (°)	σ (S/m)	E_a (eV)
20	RT	1.0699×10^3	−0.03	0.0121	0.257 ± 0.001
	40	7.8543×10^2	−0.03	0.0165	
	50	6.4880×10^2	−0.03	0.02	
	60	4.7676×10^2	−0.04	0.0273	
	70	3.5601×10^2	−0.05	0.0365	
	80	2.8176×10^2	−0.05	0.0461	
	90	2.1960×10^2	−0.06	0.0592	
	100	1.7293×10^2	−0.05	0.0751	
	110	1.3925×10^2	−0.05	0.0933	
	120	1.1285×10^2	−0.06	0.115	
	130	9.2586×10^1	−0.06	0.14	
	140	7.7259×10^1	−0.07	0.168	
	150	6.4038×10^1	−0.07	0.203	
	160	5.3605×10^1	−0.08	0.242	
	170	4.5850×10^1	−0.10	0.283	
	180	4.3496×10^1	−0.11	0.299	
10	RT	3.0617×10^4	−0.05	4.16×10^{-4}	0.307 ± 0.002
	40	2.1214×10^4	−0.03	6×10^{-4}	
	50	1.7194×10^4	−0.03	7.41×10^{-4}	
	60	1.1905×10^4	−0.03	0.00107	
	70	8.5299×10^3	−0.03	0.00149	
	80	6.2686×10^3	−0.03	0.00203	
	90	4.5696×10^3	−0/03	0.00279	
	100	3.4185×10^3	−0.04	0.00373	
	110	2.6089×10^3	−0.04	0.00488	
	120	2.0284×10^3	−0.04	0.00628	
	130	1.5909×10^3	−0.05	0.00801	
	140	1.2780×10^3	−0.05	0.00997	
	150	1.0293×10^3	−0.06	0.0124	
	160	8.4183×10^2	−0.06	0.0151	
	170	7.1487×10^2	−0.08	0.0178	
	180	7.3508×10^2	−0.11	0.0173	

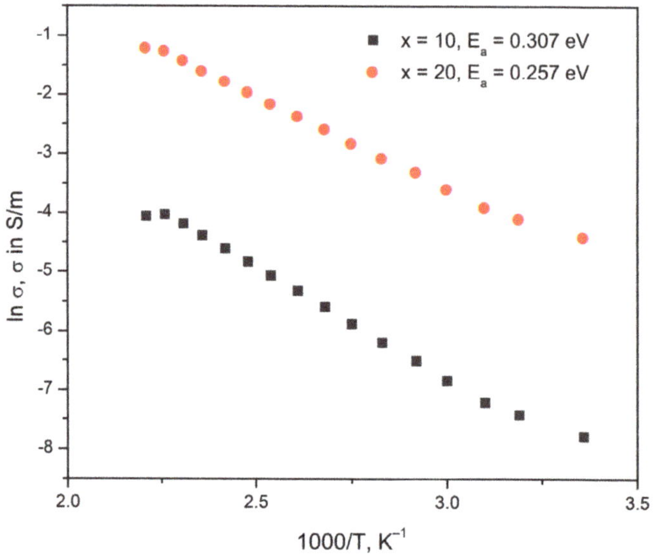

Figure 5. Arrhenius plots of glasses with the compositions $20Li_2O{:}50V_2O_5{:}(30-x)\ B_2O_3{:}xMoO_3$, x = 10, 20 mol%.

Furthermore, the obtained activation energy is the energy of ionic conductivity established by the Li^+ ions under the influence of the applied external electrical field, as the results from the diffuse reflectance spectroscopy reveal that no V^{4+} and Mo^{6+} ions are present in the two studied glasses, which excludes the possibility for the existence of an electronic component of conduction. V_2O_5-containing glasses are known for their electronic conduction established by the phonon-assisted hopping of small polarons [59]; however, the hopping conduction mechanism is only possible to observe if both V^{4+} and V^{5+} ions are present in glass, as for example reported in [22,60]. The same suggestion can be made concerning the hopping of polarons between polyvalent molybdenum ions—as only Mo^{6+} and no Mo^{5+} ions are present in the two investigated glasses, the electronic component of conduction can be excluded.

2.6. All-Solid-State Battery

The electrochemical characterization of the x = 20 glass (glass sample with lower activation energy) obtained was performed by assembling an all-solid-state cell, employing $20Li_2O–50V_2O_5–20MoO_3–10B_2O_3$ glass as a cathode active material. Figure 6a illustrates the impedance characteristics of cells with two different active material contents (C35 and C60) prior to electrochemical testing. The interfacial resistance increases from 223 Ω for the C35 electrode to 308 Ω for the C60 electrode. The C35 electrode contains a higher proportion of the $75Li_2S$-$25P_2O_5$ matrix, which provides better encapsulation and densification of the glass particles, thereby reducing the interfacial resistance between the cathode materials. Figure 6b presents the galvanostatic charge–discharge curves of the C35-based cell at two different current levels. The first cycle, conducted at 0.007 mA (corresponding to 2 mA g^{-1}) within a voltage range of 1.6–4.2 V, delivers a charge capacity of 53.42 mA h g^{-1} and a discharge capacity of 43.31 mA h g^{-1}, with a Coulombic efficiency of 81.07%. In the second cycle, performed at 0.035 mA (10 mA g^{-1}) within a voltage range of 0.9–4.2 V, the cell exhibits a charge capacity of 24.74 mA h g^{-1} and a discharge capacity of 24.87 mA h g^{-1}, achieving a nearly 100% Coulombic efficiency. However, a noticeable increase in the charge voltage and a decrease in the discharge voltage are observed compared to the first cycle. This behavior suggests that under higher current conditions, Li_2S in the cell becomes electrochemically active, leading to the degradation of the active material. Figure 6c displays the galvanostatic charge–discharge profiles of the C60-based cell at various voltages under a constant current of 0.012 mA (2 mA g^{-1}). In the first cycle (2.0–4.2 V), the cell achieves a charge capacity of 8.86 mA h g^{-1}, a discharge capacity of 7.53 mA h g^{-1}, and a Coulombic efficiency of 84.99%. In the second cycle (2.0–4.6 V), the charge and discharge capacities increase to 10.53 and 10.01 mA h g^{-1}, respectively, with a Coulombic efficiency of 95.06%. In the third cycle (2.0–5.0 V), further improvements are observed, with a charge capacity of 15.06 mA h g^{-1}, a discharge capacity of 14.80 mA h g^{-1}, and a Coulombic efficiency of 98.27%. The C60 cell demonstrates stable cycling performance at low current densities, maintaining effective charge–discharge behavior even under high voltage conditions up to 5 V. While the increase in voltage enhances the charge capacity, it also raises the risk of side reactions due to material limitations. To further evaluate the cycling stability, the C60 cell was reassembled and subjected to 50 charge–discharge cycles at 60 °C, within a voltage range of 2.0–4.2 V and at a current of 0.05 mA (8.33 mA g^{-1}). As shown in Figure 6e, the cell exhibits a gradual improvement in capacity over the cycling process. The charge capacity increases from 7.68 to 9.47 mA h g^{-1}, and the discharge capacity rises from 6.88 to 9.09 mA h g^{-1}, while the Coulombic efficiency improves from 89.58% to 95.99%. The higher content of $75Li_2S$-$25P_2O_5$ in the cathode facilitates the effective encapsulation of $20Li_2O$–$50V_2O_5$–$20MoO_3$–$10B_2O_3$ glass particles, which enhances the overall ionic transport capability and reduces the interfacial resistance of the battery. As a result, the electrode exhibits improved ionic conductivity. Therefore, under the same current density, the C35 cell demonstrates superior charge–discharge capacity compared to the C60 cell at equivalent charging voltages. However, as the current increases, the C35 cell rapidly fails. Adverse internal reactions deactivate the active material, leading to a sharp drop in discharge voltage below 2 V.

Under such low discharge voltage conditions, the Li_2S participates in redox reactions, which further compromises the battery's stability and performance. In contrast, the charge–discharge capacity of the C60 cell is constrained by the limited voltage window, making it difficult to achieve higher capacity. Furthermore, increasing the charging voltage poses a greater risk of triggering side reactions. Nevertheless, within a safe voltage range, the C60 cell exhibits excellent cycling stability and maintains reliable electrochemical performance. These findings highlight that optimizing interfacial resistance to achieve higher charge–discharge capacities at lower charging voltages is a crucial direction for the further development of this class of cathode materials. Enhancing ionic conductivity and mitigating parasitic reactions through interfacial engineering will be essential to unlocking the full potential of glass-based cathodes for advanced solid-state batteries.

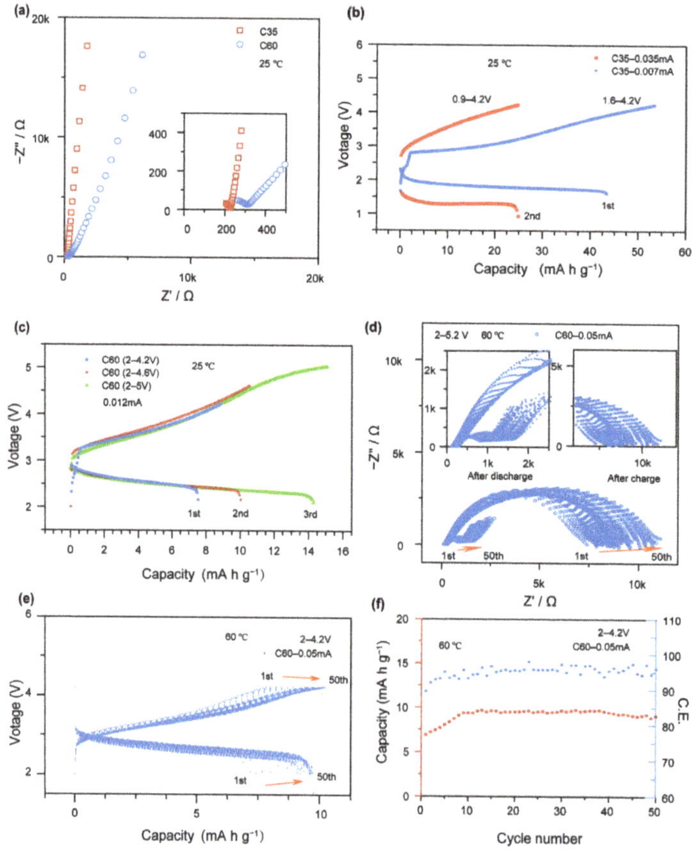

Figure 6. (**a**) Electrochemical impedance spectra of the C35 and C60 cells prior to electrochemical testing; (**b**) galvanostatic charge–discharge curves of the C35 cell at different current densities; (**c**) galvanostatic charge–discharge curves of the C60 cell within different voltage windows; (**d**) electrochemical impedance spectra of the C60 cell measured after charging and after discharging during the 50th cycle within the voltage range of 2.0–4.2 V; (**e**) charge–discharge profiles of the C60 cell over 50 cycles within the voltage range of 2.0–4.2 V; (**f**) capacity and Coulombic efficiency.

3. Materials and Methods

3.1. Sample Preparation

The glass formation ability of two series of compositions with lower and higher V_2O_5 content from the Li_2O-B_2O_3-V_2O_5-MoO_3 system is investigated by applying the melt quenching method. Reagent grade Li_2CO_3, H_3BO_3, V_2O_5 and MoO_3 are used as raw materials. The homogenized batches (5 g) from the first series of compositions with reduced V_2O_5 content (10 mol%) were melted at 850 °C for 15 min in a platinum crucible in air. The melts were cast into graphite molds (cooling rates $\leq 10^1$–10^2 K/s). The homogenized batches (3 g) from the second series of compositions with a higher V_2O_5 concentration (30–50 mol%) were melted at 950 °C for 20 min in a platinum crucible in air. The melts were quenched by pouring and pressing between two copper plates (cooling rates 10^1–10^2 K/s). Figure 7 shows a schematic representation of the preparation steps adopted to obtain the glass and glass–crystalline samples under investigation. The nominal compositions investigated, the experimental conditions applied and the visually observed results are listed in Tables 1 and 2.

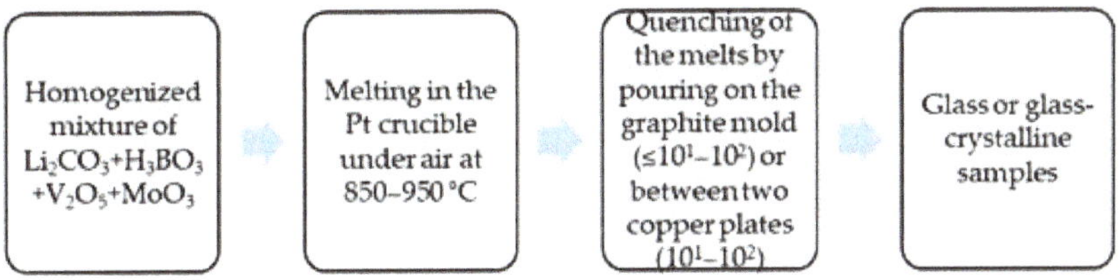

Figure 7. Schematic representation of the preparation steps adopted to obtain the investigated glass and glass–crystalline samples.

3.2. Sample Characterization

The phase formation of the samples was established by X-ray phase analysis using a Bruker D8 advance diffractometer with Cu Kα radiation in the $10 < 2\theta < 60°$ range. The glass transition (T_g) and crystallization peak maximum (T_c) temperatures of the glasses were determined by means of Differential Scanning Calorimetry (DSC) using a Netzsch 404 F3 Pegasus instrument, Germany in the temperature range 25–750 °C at a heating rate of 10 K/min in argon atmosphere. The density of the obtained glasses at room temperature was measured by the Archimedes principle using toluene (ρ = 0.867 g/cm^3) as an immersion liquid on a Mettler Toledo electronic balance of sensitivity 10^{-4} g. The IR spectra of the glasses were measured using the KBr pellet technique on a Nicolet-320 FTIR spectrometer with a resolution of ±4 cm^{-1}, by collecting 64 scans in the range 1600–400 cm^{-1}. A random error in the center of IR bands was found as ±3 cm^{-1}. The optical spectra of powder samples at room temperature were recorded by a spectrometer (Evolution 300 UV-vis Spectrophotometer) employing the integration sphere diffuse reflectance attachment. The spectra of the samples were recorded in the wavelength (λ) range of 190–800 nm with a magnesium oxide reflectance standard used as the baseline. The uncertainty in the observed wavelength is about ±1 nm. The Kubelka–Munk function ($F(R_\infty)$) was calculated from the UV-vis diffuse reflectance spectra. The band gap energy (E_g) for the direct allowed transition (n = 2) was determined by preparing a Tauc plot where the function $(F(R_\infty)h\nu)^n$ was plotted versus hν (incident photon energy). The tangent to the inflection point was determined, and the band gap energy was found as the intersection of the tangent with the horizontal axis, i.e., at $[h\nu F(R_\infty)]^2 = 0$. The electrical properties of selected samples were studied by utilizing measurement of the impedance modulus and the phase angle as a function of frequency and temperature by an impedance analyzer (Zahner IM6, Zahner Elektrik, Kronach, Germany) and two-contacting points measurement. After gold electrodes were thermally evaporated onto the sample base surface, the 1–3 mm thick samples were mounted in the holder, introduced into a resistive furnace with a DC power supply and then connected to the impedance analyzer. The temperature was measured by an Al-Ni-Cr (K-type) thermocouple with an accuracy of 1 °C. The impedance modulus and the phase angle were measured as a function of frequency in the range from 1 Hz to 100 kHz and as a function of temperature from room temperature to 180 °C, i.e., below the T_g. An AC voltage with amplitude 500 mV was applied during the impedance measurements. The accuracy of the impedance modulus measurement is ≤±5% and that of the phase angle ±0.2°. Taking into account the sample geometry, the conductivity as well as several dielectric characteristics of the measured samples were estimated.

3.3. Preparation of Cathode Materials

For the preparation of cathode materials, all sample preparation procedures were carried out in an argon-filled glovebox to prevent moisture and oxygen contamination. Two types of cathode composites were prepared: one consisting of 60 mg of $20Li_2O$-$50V_2O_5$-$20MoO_3$-$10B_2O_3$ glass, 35 mg of $75Li_2S$-$25P_2O_5$ solid electrolyte, and 5 mg of acetylene black (denoted as C60), and the other consisting of 35 mg of $20Li_2O$-$50V_2O_5$-$20MoO_3$-$10B_2O_3$ glass, 60 mg of $75Li_2S$-$25P_2O_5$, and 5 mg of acetylene black (denoted as C35). Each mixture was thoroughly homogenized by dry grinding in a small agate mortar for 10 min to ensure uniform dispersion of the active components.

3.4. Battery Cell Assembly

For Battery Cell Assembly, 100 mg of $75Li_2S$-$25P_2O_5$ solid electrolyte was placed into a PEEK die and pressed under 80 MPa for 2 min to form a dense electrolyte pellet. After removing the upper punch, 10 mg of the pre-mixed cathode composite was evenly spread onto the electrolyte surface and pressed again under 80 MPa for 2 min. The entire assembly was then compressed under 360 MPa for 3 min to enhance interfacial contact and mechanical integrity. For the anode, indium foil with a thickness of 0.1 mm was punched into 8 mm diameter discs and placed on the counter electrode side. The full cell was further pressed under 80 MPa for 2 min to ensure good contact. The assembled cells were then fixed in a customized clamp in preparation for electrochemical testing. All electrochemical measurements were conducted at room temperature using a Princeton Applied Research PMC200 electrochemical workstation. Galvanostatic charge–discharge tests were performed within a voltage window of 0.5–6 V at various current densities ranging from 2 to 10 mA g^{-1}. The voltage limits were adjusted according to the cathode composition and the applied current density in order to evaluate the maximum charge–discharge capacities under different conditions.

4. Conclusions

Novel multicomponent vanadate glasses with the composition $20Li_2O$:$(30 - x)B_2O_3$:$50V_2O_5$:$xMoO_3$, x = 10, 20 mol% were obtained by using the conventional melt quenching method. Several structural units building up the amorphous network, specifically VO_4, VO_5, VO_6, MoO_6, MoO_4, $V(Mo)_2O_8$, BO_3^{3-}, $BØO_2^{2-}$ $BØ_2O^-$ and $BØ_4^-$, were determined by combining IR and DRS spectroscopies. The values of the optical band gap (E_g) of the present glass samples fall within the semiconductor range (1.86 eV and 1.13 eV), which render these glass compositions intriguing for exploration as cathode materials in Li-ion batteries. The prepared glassy material exhibits excellent charge–discharge performance, delivering a reversible capacity of 43.31 mAh g^{-1} at room temperature. Moreover, the material demonstrates remarkable cycling stability, maintaining its capacity without noticeable degradation over 50 cycles at 60 °C, highlighting its potential as an active cathode material for solid-state batteries. However, optimizing interfacial impedance and enhancing the interfacial compatibility between the cathode and solid electrolyte remain significant challenges in all-solid-state systems. This study offers new insights into the application of glassy materials as active cathode components and contributes to the development of all-solid-state lithium metal batteries.

Author Contributions: Conceptualization, M.M.; methodology, M.M., R.H. and N.C.R.-N.; software, R.H. and M.M.; validation, P.V., M.M. and R.H.; formal analysis, X.Y. and M.M.; investigation, X.Y., P.V., M.M., L.A., R.H., M.S. and S.K.; resources, M.M.; data curation, M.M.; writing—original draft preparation, M.M., B.J., R.H. and X.Y.; writing—review and editing, N.C.R.-N. and R.I.; visualization, M.M.; supervision, N.C.R.-N.; project administration, N.C.R.-N.; funding acquisition, N.C.R.-N. All authors have read and agreed to the published version of the manuscript.

Funding: The authors gratefully acknowledge partial financial support from the Elisa project with ref. CNS2023-145494 funded by the Spanish Ministry of Science and Innovation (MCIN/AEI/10.13039/501100011033) and the European Union through the "NextGenerationEU"/PRTR initiative; and 2024ICT207 project from CSIC.

Institutional Review Board Statement: Not applicable.

Informed Consent Statement: Not applicable.

Data Availability Statement: Data are contained within the article.

Conflicts of Interest: The authors declare no conflicts of interest.

References

1. Vermeer, W.; Mouli, G.R.C.; Bauer, P. A Comprehensive Review on the Characteristics and Modeling of Lithium-Ion Battery Aging. *IEEE Trans. Transp. Electrif.* **2022**, *8*, 2205–2232. [CrossRef]
2. Barbosa, J.C.; Gonçalves, R.; Costa, C.M.; Lanceros-Mendez, S. Recent Advances on Materials for Lithium-Ion Batteries. *Energies* **2021**, *14*, 3145. [CrossRef]
3. Afyon, S.; Krumeich, F.; Mensing, C.; Borgschulte, A.; Nesper, R. New High Capacity Cathode Materials for Rechargeable Li-ion Batteries: Vanadate-Borate Glasses. *Sci. Rep.* **2014**, *4*, 7113. [CrossRef]
4. Hassaan, M.Y.; Salem, S.M.; Moustafa, M.G. Study of nanostructure and ionic conductivity of $Li_{1.3}Nb_{0.3}V_{1.7}(PO_4)_3$ glass ceramics used as cathode material for solid batteries. *J. Non-Cryst. Solids* **2014**, *391*, 6–11. [CrossRef]
5. Zhao, E.L.; Zhao, S.X.; Wu, X.; Li, J.W.; Yu, L.Q.; Nan, C.W.; Cao, G. Electrochemical performance of Li_2O-V_2O_5-SiO_2-B_2O_3 glass as cathode material for lithium ion batteries. *J. Materiomics* **2019**, *5*, 663–669. [CrossRef]
6. Gao, Y.; Liu, K.; Zhu, C.; Zhang, X.; Zhang, D. Co-Estimation of State-of-Charge and State-of- Health for Lithium-Ion Batteries Using an Enhanced Electrochemical Model. *IEEE Trans. Ind. Electron.* **2021**, *69*, 2684–2696. [CrossRef]
7. Zhang, W.; Wang, L.; Wang, L.; Liao, C.; Zhang, Y. Joint State-of-Charge and State-of-Available-Power Estimation Based on the Online Parameter Identification of Lithium-Ion Battery Model. *IEEE Trans. Ind. Electron.* **2021**, *69*, 3677–3688. [CrossRef]
8. Wei, Z.; Quan, Z.; Wu, J.; Li, Y.; Pou, J.; Zhong, H. Deep Deterministic Policy Gradient-DRL Enabled Multiphysics-Constrained Fast Charging of Lithium-Ion Battery. *IEEE Trans. Ind. Electron.* **2022**, *69*, 2588–2598. [CrossRef]
9. Ni, Z.; Yang, Y. A Combined Data-Model Method for State-of-Charge Estimation of Lithium-Ion Batteries. *IEEE Trans. Instrum. Meas.* **2022**, *71*, 2503611. [CrossRef]
10. Mustafa, J.; Alqaed, S.; Husain, S.; Jamil, B.; Sharifpur, M.; Cheraghian, G. Effect of Phase Change Materials on Lithium-Ion Plate Batteries. *Batteries* **2023**, *9*, 60. [CrossRef]
11. Wei, Z.; Hu, J.; He, H.; Yu, Y.; Marco, J. Embedded Distributed Temperature Sensing Enabled Multistate Joint Observation of Smart Lithium-Ion Battery. *IEEE Trans. Ind. Electron.* **2023**, *70*, 555–565. [CrossRef]
12. Kittaneh, O. On the Theory of the Arrhenius-Normal Model with Applications to the Life Distribution of Lithium-Ion Batteries. *Batteries* **2023**, *9*, 55. [CrossRef]
13. Shui, Z.Y.; Li, X.H.; Feng, Y.; Wang, B.C.; Wang, Y. Combining Reduced-Order Model with Data-Driven Model for Parameter Estimation of Lithium-Ion Battery. *IEEE Trans. Ind. Electron.* **2023**, *70*, 1521–1531. [CrossRef]
14. Krishnamoorthy, U.; Ayyavu, P.G.; Panchal, H.; Shanmugam, D.; Balasubramani, S.; Al-rubaie, A.J.; Al-khaykan, A.; Oza, A.D.; Hembrom, S.; Patel, T.; et al. Efficient Battery Models for Performance Studies-Lithium Ion and Nickel Metal Hydride Battery. *Batteries* **2023**, *9*, 52. [CrossRef]
15. Delmas, C.; Cognac-Auradou, H.; Cocciantelli, J.M.; Menetrier, M.; Doumerc, J.P. The $Li_xV_2O_5$ system: An overview of the structure modifications induced by the lithium intercalation. *Solid State Ionics.* **1994**, *69*, 257–264. [CrossRef]
16. Attos, O.; Massot, M.; Mavi, H.S.; Julien, C. Spectroscopic investigations of the network structure in borovanadate glasses. *MRS Online Proc. Libr.* **1997**, *455*, 477–482. [CrossRef]
17. Bih, L.; El Omari, M.; Réau, J.M.; Haddad, M.; Boudlich, D.; Yacoubi, A.; Nadiri, A. Electronic and ionic conductivity of glasses inside the Li_2O–MoO_3–P_2O_5 system. *Solid State Ion.* **2000**, *132*, 71–85. [CrossRef]
18. Bih, L.; Nadiri, A.; El Omari, M.; Yacoubi, A.; Haddad, M. FTIR, EPR and X-ray investigation of mixed valence molybdenum phosphate A_2O–$(MoO_3)_2$–P_2O_5 (A=Li,Na) glasses. *Phys. Chem. Glas.* **2002**, *43*, 153–157.
19. Jozwiak, P.; Garbarczyk, J.E. Mixed electronic–ionic conductivity in the glasses of the Li_2O–V_2O_5–P_2O_5 system. *Solid State Ion.* **2005**, *176*, 2163–2166. [CrossRef]
20. Al Hajry, A.; Al-Shahrani, A.; El-Desoky, M.M. Structural and other physical properties of barium vanadate glasses. *Mater. Chem. Phys.* **2006**, *95*, 300–306. [CrossRef]
21. Bih, L.; Abbas, L.; Nadiri, A.; Amraoui, Y.E.; Mezzane, D.; Khemakhem, H. DC and AC conductivities of the yLi_2O–(1-y)[0.35(MoO3)2–0.65(P2O5)] glasses. *M. J. Condenced Mater.* **2006**, *7*, 70–73.
22. Rao, L.S.; Reddy, M.S.; Reddy, M.R.; Veeraiah, N. Dielectric dispersion in Li_2O–MoO_3–B_2O_3 glass system doped with V_2O_5. *J. Alloys Compd.* **2008**, *464*, 472–482. [CrossRef]
23. Takahashi, H.; Karasawa, T.; Sakuma, T.; Garbarczyk, J.E. Electrical conduction in the vitreous and crystallized Li_2O–V_2O_5–P_2O_5 system. *Solid State Ion.* **2010**, *81*, 27–32. [CrossRef]

24. Barczyński, R.J.; Król, P.; Murawski, L. AC and DC conductivities in V_2O_5–P_2O_5 glasses containing alkaline ions. *J. Non-Cryst. Solids* **2010**, *356*, 1965–1967. [CrossRef]
25. Gowda, V.C.; Chethana, B.K.; Reddy, C.N. Ion transport studies in lithium phospho-molybdate glasses containing Cl^- ion. *Mat. Sci. Eng. B-Adv.* **2013**, *178*, 826–833. [CrossRef]
26. Saetova, N.S.; Raskovalov, A.A.; Antonov, B.D.; Yaroslavtseva, T.V.; Reznitskikh, O.G.; Zabolotskaya, E.V.; Kadyrova, N.I.; Telyatnikova, A.A. Conductivity and spectroscopic studies of Li_2O-V_2O_5-B_2O_3 glasses. *Ionics* **2018**, *24*, 1929–1938. [CrossRef]
27. Kindle, M.; Kmiec, S.; d'Anciães, I.; Silva, A.; Eckert, H.; Martin, S.W.; Song, M.K.; McCloy, J.S. Structural properties of alumina-doped lithium borovanadate glasses and glass-ceramics. *J. Non-Cryst. Solids* **2019**, *521*, 119551. [CrossRef]
28. Saetova, N.S.; Raskovalov, A.A.; Antonov, B.D.; Denisova, T.A.; Zhuravlev, N.A. Structural features of Li_2O–V_2O_5–B_2O_3 glasses: Experiment and molecular dynamics simulation. *J. Non-Cryst. Solids* **2020**, *545*, 120253. [CrossRef]
29. Banagar, A.V.; Kumar, M.P.; Nagaraja, N. Effect of Mixed Transition Metal Ions in B_2O_3-V_2O_5-MoO_3 Glass System. *J. Electron. Mater.* **2020**, *49*, 7370–7378. [CrossRef]
30. Ori, G.; Montorsi, M.; Pedone, A.; Siligardi, C. Insight into the structure of vanadium containing glasses: A molecular dynamics study. *J. Non-Cryst. Solids* **2011**, *357*, 2571–2579. [CrossRef]
31. Boora, M.; Malik, S.; Kumar, V.; Bala, M.; Arora, S.; Rohilla, S.; Kumar, A.; Dalal, J. Investigation of structural and impedance spectroscopic properties of borate glasses with high Li^+ concentration. *Solid State Ion.* **2021**, *368*, 115704. [CrossRef]
32. Ukpong, A.M. Controlling the crystallization of lithium borovanadate phases in an oxide glass composite using the CALPHAD approach. *Mater. Today Proc.* **2021**, *38*, 1059–1070. [CrossRef]
33. Yang, X.; Muñoz, F.; Vargas, P.; Palomar, T.; Rosero-Navarro, N.C. Electrochemical Stability and Ionic Conductivity of AlF_3 Containing Lithium Borate Glasses: Fluorine Effect, Strength or Weakness? *Battery Energy* **2025**, *4*, e70007. [CrossRef]
34. Rosero-Navarro, N.C.; Watanabe, H.; Jalem, R.; Mena, M.; Yang, X.; Sugio, S.; Ito, H.; Tateyama, Y.; Miura, A.; Tadanaga, K. Microstructure-controlled Li ion conductive oxide–based ceramic solid electrolytes supporting high current densities. *Electrochim. Acta* **2025**, *528*, 146233. [CrossRef]
35. Warren, Z.; Seghers, M.; Boissière, P.; Aparicio, M.; Rosero-Navarro, N.C. Effect of Temperature during Open Vessel Sintering on Li_6PS_5Cl Solid-State Electrolyte Powders for Use in Cathode Composite Electrodes. *ACS Electrochem.* **2025**, *1*, 1093–1100. [CrossRef]
36. Peng, L.; Lei, T.; Cheng, S.; Xie, J. Synergistic fluorine-nitrogen interfaces enabling stable high-voltage sulfide-based all-solid-state lithium metal batteries. *Nano Res.* **2025**, *18*, 94907614. [CrossRef]
37. Guo, S.; Li, W.; Wu, X.; Guo, X.; Gong, Z.; Wang, E.; Wang, R.; Luo, J.; Mi, L.; Kang, J.; et al. Functional Separator Induced Interface Potential Uniform Reformation Enabling Dendrite-Free Metal Batteries. *Adv. Funct. Mater.* **2025**. [CrossRef]
38. Miura, A.; Rosero-Navarro, N.C.; Sakuda, A.; Tadanaga, K.; Phuc, N.H.H.; Matsuda, A.; Machida, N.; Hayashi, A.; Tatsumisago, M. Liquid-phase syntheses of sulfide electrolytes for all-solid-state lithium battery. *Nat. Rev. Chem.* **2019**, *3*, 189–198. [CrossRef]
39. Men, M.; Wu, J.; Liu, G.; Zhang, J.; Zhang, N.; Yao, X. Sulfide solid electrolyte synthesized by liquid phase approach and application in all-solid-state lithium batteries. *Acta Phys.-Chim. Sin.* **2025**, *41*, 100004. [CrossRef]
40. Warren, Z.; Rosero-Navarro, N.C. Solution-Based Suspension Synthesis of Li_2S–P_2S_5 Glass-Ceramic Systems as Solid-State Electrolytes: A Brief Review of Current Research. *ACS Omega* **2024**, *9*, 31228–31236. [CrossRef]
41. International Centre for Diffraction Data. Pennsylvania, USA 2025. Available online: www.ICDD.com.
42. Hamad, A.H.; Marzouk, M.A.; ElBatal, H.A. The Effect of Bi_2O_3 on Optical, FTIR and Thermal Properties of SrO-B_2O_3 glasses. *Silicon* **2016**, *8*, 121–131. [CrossRef]
43. Soliman, A.A.; Kashif, I. Copper oxide content dependence of crystallization behavior, glass forming ability, glass stability and fragility of lithium borate glasses. *Phys. B* **2010**, *405*, 247–253. [CrossRef]
44. Zhu, L.; Xu, T.F.; Nie, Q.H.; Shen, X. Spectral properties and thermal stability of erbium TeO_2-WO_3-La_2O_3 glass. *J. Inorg. Mater.* **2006**, *21*, 351–356.
45. Varsamis, C.P.; Makris, E.N.; Valvi, C.; Kamitsos, E.I. Short-range structure, the role of bismuth and property-structure correlations in bismuth borate glasses. *Phys. Chem. Chem. Phys.* **2021**, *23*, 10006–10020. [CrossRef] [PubMed]
46. Milanova, M.; Aleksandrov, L.; Yordanova, A.; Iordanova, R.; Tagiara, N.S.; Herrmann, A.; Gao, G.; Wondraczek, L.; Kamitsos, E.I. Structural and luminescence behavior of Eu^{3+} ions in ZnO-B_2O_3-WO_3 glasses. *J. Non-Cryst. Solids* **2023**, *600*, 122006. [CrossRef]
47. Iordanova, R.; Milanova, M.; Aleksandrov, L.; Khanna, A. Structural study of glasses in the system B_2O_3-Bi_2O_3-La_2O_3-WO_3. *J. Non-Cryst. Solids* **2018**, *481*, 254–259. [CrossRef]
48. Markova-Velichkova, M.; Iordanova, R.; Dimitriev, Y. Glass formation in the V_2O_5-MoO_3-ZnO system. *Phys. Status Solidi C* **2011**, *8*, 3159–3162. [CrossRef]
49. Khan, S.; Singhm, K. Structual, optical, thermal and conducting properties of $V_{2-x}Li_xO_{5-\delta}$ ($0.15 \leq x \leq 0.30$) system. *Sci. Rep.* **2020**, *10*, 1089.
50. Bachvarova-Nedelcheva, A.; Iordanova, R.; Kostov, K.L.; Ganev, V.; Yordanov, S.; Dimitriev, Y. Synthesis and structural chatacterization of a glass in the Ag_2O-SeO_2-MoO_3 system. *J. Non-Cryst. Solids* **2018**, *481*, 138–147. [CrossRef]

51. Milanova, M.; Iordanova, R.; Tatsumisago, M.; Hayashi, A.; Tzvetkov, P.; Nihtianova, D.; Markov, P.; Dimitriev, Y. Soft mechanochemical synthesis and electrochemical behavior of $LiVMoO_6$ for all-solid-state lithium batteries. *J. Mater. Sci.* **2016**, *51*, 3574–3584. [CrossRef]
52. Milanova, M.; Iordanova, R.; Aleksandrov, L.; Hassan, M.; Dimitriev, Y. Glass formation and structure of glasses in the $ZnO—Bi_2O_3—WO_3—MoO_3$ system. *J. Non-Cryst. Solids* **2011**, *357*, 2713–2718. [CrossRef]
53. Ray, N.H. Composition-properties relationship in Inorganic Oxide Glasses. *J. Non-Cryst. Solids* **1974**, *15*, 423–434. [CrossRef]
54. Saddeek, Y.B.; Azooz, M.A.; Saddek, A.B. Ultrasonic investigations of some bismuth borate glasses doped with Al_2O_3. *Bull. Mater. Sci.* **2015**, *38*, 241–246. [CrossRef]
55. Aryal, P.; Kesavulu, C.R.; Kim, H.J.; Lee, S.W.; Kang, S.J.; Kaewkhao, J.; Chanthima, N.; Damdee, B. Optical and luminescence characteristics of Eu^{3+}-doped B_2O_3: SiO_2: Y_2O_3: CaO glasses for visible red laser and scintillation material applications. *J. Rare Earths* **2018**, *36*, 482–491. [CrossRef]
56. Villegas, M.A.; Fernández Navarro, J.M. Physical and structural properties of glasses in the $TeO_2–TiO_2–Nb_2O_5$ system. *J. Eur. Ceram. Soc.* **2007**, *27*, 2715–2723. [CrossRef]
57. Centi, G.; Perathoner, S.; Trifiro, F.; Aboukals, A.; Aissi, C.F.; Guelton, M. Physicochemiacl Characterizationof V-Silicates. *J. Phus. Chem.* **1992**, *96*, 2617–2629. [CrossRef]
58. Aleksandrov, L.; Iordanova, R.; Dimitriev, Y.; Georgiev, N.; Komatsu, T. Eu^{3+} doped $1La_2O_3$:$2WO_3$:$1B_2O_3$ glass and glass–ceramic. *Opt. Mater.* **2014**, *36*, 1366–1372. [CrossRef]
59. Margha, F.H.; El-Bassyouni, G.T.; Turky, G.M. Enhancing the electrical conductivity of vanadate glass system (Fe_2O_3, B_2O_3, V_2O_5) via doping with sodium or strontium cations. *Ceram. Int.* **2018**, *45*, 11838–11843. [CrossRef]
60. Rao, L.S. AC conductivity and polarization phenomenon of Li_2O-MoO_3-B_2O_3:V_2O_5 glasses. *J. Alloys Compd.* **2019**, *787*, 1280–1289.

inorganics

MDPI

Article

Controlled Synthesis of Alkali Metal Hydroxide Particles via Solvothermal Processing

Chiara Tuccio [1,2], Francesco Armetta [1,3,*], Delia Francesca Chillura Martino [1], Ramūnas Skaudžius [4] and Maria Luisa Saladino [1,3,*]

1 Department of Biological, Chemical and Pharmaceutical Sciences and Technologies (STEBICEF), University of Palermo, Viale delle Scienze, Bld.17, I-90128 Palermo, Italy; chiara.tuccio@unipa.it (C.T.); delia.chilluramartino@unipa.it (D.F.C.M.)

2 Department of Cultures and Societies, University of Palermo, Viale delle Scienze, Bld.15, I-90128 Palermo, Italy

3 Consiglio Nazionale Delle Ricerche, Istituto per i Processi Chimico Fisici, CNR-IPCF-Messina, Viale Ferdinando Stagno d'Alcontres 37, I-98158 Messina, Italy

4 Institute of Chemistry, Faculty of Chemistry and Geosciences, Vilnius University, Naugarduko 24, LT-03225 Vilnius, Lithuania; ramunas.skaudzius@chgf.vu.lt

* Correspondence: francesco.armetta01@unipa.it (F.A.); marialuisa.saladino@unipa.it (M.L.S.)

Abstract: This study presents a solvothermal approach starting from micron-sized hydroxide precursors, which combines features of top-down size reduction and bottom-up recrystallization, leading to nanoscale hydroxide particles. The method is based on autoclave treatment at a moderate temperature (180 °C) and a pressure of 8 bar, using different mixtures of water and isopropanol. The hydroxide precursors, used in micrometric form without surfactants or additives, were converted into nanoscale particles through a one-pot, one-step process. The nanomaterials obtained were characterized using X-ray diffraction (XRD), Fourier transform infrared spectroscopy (FTIR), optical microscopy (MO), scanning electron microscopy (SEM), transmission electron microscopy (TEM) and Brunauer–Emmett–Teller (BET) surface area analysis to assess their structural, morphological and textural characteristics. The results show that solvent composition and precursor concentration strongly influence the crystalline phase, particle morphology, dispersion stability and surface area. Well-defined acicular and fibrous morphologies were obtained for $Ba(OH)_2$ and $Sr(OH)_2$, while $Mg(OH)_2$ formed spherical and hexagonal structures, respectively. Of all the conditions tested, the 75:25 water/isopropanol ratio produced the most stable systems. This work provides a method to produce alkaline earth hydroxide nanoparticles with tunable properties.

Keywords: solvothermal synthesis; hydroxide nanoparticles; barium; magnesium; strontium; top-down approach; autoclave; particle morphology; needles

1. Introduction

The controlled production of nanomaterials has become a central topic in materials chemistry [1,2], due to the ability to finely modulate particle size, morphology, crystallinity, and surface properties. Nanoparticles, particularly those composed of alkaline earth metal hydroxides such as barium, magnesium, and strontium, are increasingly employed in a wide range of applications [3–5], from catalysis and environmental remediation to biomedicine, advanced materials, and more recently, cultural heritage conservation.

Synthesis strategies for nanoparticles generally fall into two main categories: bottom-up and top-down approaches [6,7]. The bottom-up (chemical) method, also referred to as "molecular nanotechnology," involves the growth of nanostructures through the assembly of atoms or molecules, allowing for controlled development from molecular precursors [8]. The top-down (physical) approach, by contrast, achieves nanoscale dimensions by reducing bulk materials into smaller particles, using mechanical, chemical, or other forms of energy [9].

Despite the growing interest in alkaline earth metal hydroxides, currently available synthesis methods often present several limitations, including operational complexity, the use of toxic solvents, or the need for stabilizing agents.

For instance, magnesium hydroxide nanoparticles can be obtained via simple hydrothermal synthesis using magnesium salts and urea, yielding flower-like structures of stacked nanosheets [10]. Alternatively, reverse microemulsion methods enable precise control over particle size, producing monodisperse and hydrophobic nanoparticles ideal for nonpolar media [11]. Or they can be obtained by homogeneous chemical precipitation from aqueous solutions containing magnesium salts (chloride, nitrate, sulfate, perchlorate) and sodium hydroxide as precipitating agents [12].

The synthesis of barium hydroxide nanoparticles is limited by their high solubility in water, which makes the bottom-up approach difficult. An effective alternative is a top-down method, in which micrometer-sized crystals of $Ba(OH)_2$ are ground in short-chain alcohols (1- or 2-propanol) under high temperature and pressure, resulting in particles of about 100 nm [13].

Strontium hydroxide synthesis also suffers from limitations related to water solubility. However, homogeneous-phase protocols at low temperatures have been described, employing simple commercial salts ($Sr(NO_3)_2$ and NaOH) under controlled conditions. This process allows for the production of nearly spherical $Sr(OH)_2$ particles with sizes around 30 nm. Although these approaches have led to promising results, many still require extended purification phases and complex equipment or are difficult to scale up [14].

Among the most promising top-down techniques is hydrothermal synthesis [15], which enables the direct formation of crystalline phases in aqueous or mixed-solvent media under an elevated temperature and pressure. Hydrothermal or solvothermal treatments in autoclaves are particularly attractive due to their ability to generate high-purity, crystalline materials without intermediate calcination or sintering steps. Moreover, the use of green solvents such as water and short-chain alcohols (e.g., ethanol, isopropanol) reduces toxicity and improves particle dispersion in the final colloidal systems.

This work presents a one-pot, one-step solvothermal route for the efficient and reproducible synthesis of barium, magnesium, and strontium hydroxide nanoparticles. The novelty of this work lies in combining a mixed-solvent system (water/isopropanol) with a solvothermal process starting from bulk precursors, enabling nanoparticle formation without surfactants or complex purification steps. The methodology relies solely on simple and commercially available precursors combined with non-toxic solvents (water and isopropanol) in varying proportions. Isopropanol and 1-propanol (the latter for comparison testing) were chosen as solvents because they have already been used successfully in previous solvothermal syntheses of alkaline earth metal hydroxides [13]. The synthesis is carried out in sealed autoclaves at moderate temperatures (up to 180 °C), without the addition of surfactants, chelating agents, or other auxiliaries.

In response to current challenges in sustainable nanomaterial production, we report a new method for synthesizing alkaline earth metal hydroxide nanoparticles. This approach is environmentally friendly and economically viable, as it avoids the use of hazardous chemicals and complex purification steps. The influence of synthesis parameters—such as precursor concentration and solvent ratio—on particle morphology and size was systematically investigated. The resulting materials were characterized using X-ray diffraction (XRD),

Fourier-transform infrared spectroscopy (FTIR), optical microscopy, scanning electron microscopy (SEM), transmission electron microscopy (TEM), and nitrogen gas adsorption.

2. Results and Discussion

A summary of the obtained particles in different conditions is shown in Table 1.

Table 1. Synthesis conditions and obtained products at 180 °C, 8 hPa, 24 h.

Label	Type	Solvent	Concentration (mol/L)	Aspect	Particles	Composition
CT1	$Ba(OH)_2$	water	0.099	particles in suspension	powder	$Ba(OH)_2$ and $BaCO_3$
CT2	$Ba(OH)_2$	isopropanol	0.099	colloidal solution	powder	$Ba(OH)_2$
CT3	$Ba(OH)_2$	1-propanol	0.099	precipitate	powder	$Ba(OH)_2$
CT4	$Ba(OH)_2$	water/isopropanol = 25:75	0.025	unstable milky colloidal solution	powder and crystals	$Ba(OH)_2$
CT5	$Ba(OH)_2$	water/isopropanol = 75:25	0.070	stable colloidal solution for 1 h	crystals	$Ba(OH)_2$
CT6	$Ba(OH)_2$	water/isopropanol = 50:50	0.050	unstable colloidal solution	powder	$Ba(OH)_2$
CT7	$Ba(OH)_2$	water in presence nitrogen	0.099	particles in suspension	powder	$Ba(OH)_2$ and $BaCO_3$
SR1	$Sr(OH)_2$	water/isopropanol = 75:25	0.148	stable milky colloidal solution for 1 h	crystals	$Sr(OH)_2$
MG1	$Mg(OH)_2$	water/isopropanol = 75:25	0.017	stable colloidal solution for 1 h	powder	$Mg(OH)_2$

Regarding the synthesis of $Ba(OH)_2$ particles, their composition and stability are affected by the used solvent. In all experiments, the total solvent volume was kept constant at 50 mL, while only the solvent type and the percentage ratio were varied. The product obtained by using water as a solvent was a mixture of hydroxides and carbonates, also insufflating nitrogen. The product obtained by using 1-propanol as a solvent was hydroxide, but the dispersion was not stable. The particles obtained by using a mixture of water and isopropanol gave the best results in terms of composition and stability. In detail, the most stable dispersion was obtained with a 75% water and 25% isopropanol mixture. Based on these promising results for $Ba(OH)_2$, the same synthetic methodology, solvent types, and solvent ratio were applied to the preparation of $Sr(OH)_2$ and $Mg(OH)_2$. All dispersions showed good stability under the same conditions. The stability of the dispersions was first assessed qualitatively through visual observation over time, monitoring the appearance of turbidity or sedimentation. Under these conditions, the systems remained macroscopically stable for approximately one hour at room temperature, with the best performance observed for the 75:25 water/isopropanol mixture. Although qualitative, these observations were reproducible across multiple preparations, confirming the enhanced colloidal stability achieved under these solvent conditions. To further evaluate dispersion stability, turbidimetric measurements were carried out. The absorbance of the dispersions remained unchanged for 60 min, indicating stability within this time range. Moreover, after gentle agitation, the dispersions appeared homogeneous, and the Tyndall effect was clearly visible when illuminated with a red laser, further confirming their colloidal nature (inset in Figure 1).

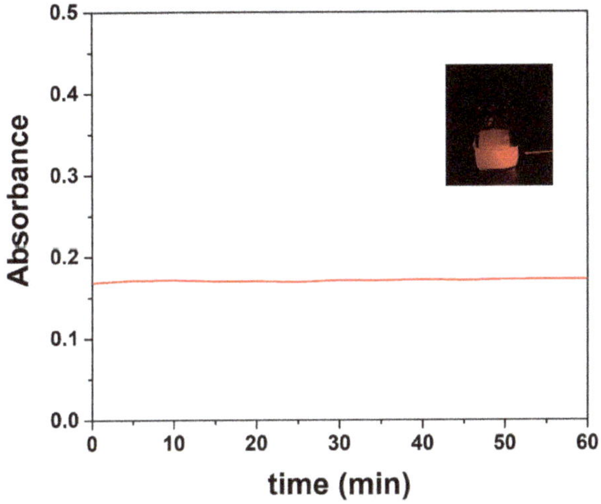

Figure 1. Absorbance of the dispersion monitored for 60 min. Tyndall effect observed in the $Sr(OH)_2$ dispersion under red laser illumination. The arrow indicates the laser beam path through the dispersion.

For the above findings, here we report the characterization of the three samples prepared in 75% water and 25% isopropanol mixture ($Mg(OH)_2$ = MG1, $Sr(OH)_2$ = SR1, $Ba(OH)_2$ = CT5).

The XRD patterns of the powders show the crystalline phases formed under autoclave synthesis conditions (Figure 2). MG1 exhibits characteristic peaks corresponding to the brucite phase (magnesium hydroxide, $Mg(OH)_2$ ref code 01-074-2220) with a hexagonal crystal system. SR1 is identified as strontium hydroxide octahydrate ($Sr(OH)_2 \cdot 8H_2O$ ref code 00-027-1438), with a tetragonal crystal system, where the (202) peak appears more intense and broader. Similarly, CT5 corresponds to barium hydroxide octahydrate ($Ba(OH)_2 \cdot 8H_2O$ ref code 01-076-0916), crystallized in a monoclinic system. The autoclave synthesis promotes the formation of hydrated hydroxide phases for strontium and barium, as the controlled temperature and pressure conditions favored water incorporation into their crystalline structures. In the case of magnesium, the small ionic radio of the cation likely mitigates the incorporation of water. The FTIR spectra of the synthesized compounds for different metals exhibit three similar profiles. All spectra show a broad absorption band in the range around 3000 cm^{-1}, assigned to the O–H stretching vibration of hydroxyl groups. The absence of characteristic carbonate bands indicates that no CO_3^{2-} species were formed during the solvothermal treatment, in agreement with the XRD results that confirmed the exclusive presence of hydroxide phases. In the low-wavenumber region (<600 cm^{-1}), bands corresponding to M–O and M–OH lattice vibrations (where M = Ba, Sr, or Mg) are observed, further confirming the formation of crystalline hydroxide structures.

According to XRD patterns, the absence of carbonate bands indicates that the synthesis in the above solvent in the autoclave favored the stabilization of hydroxide phases without carbonate formation.

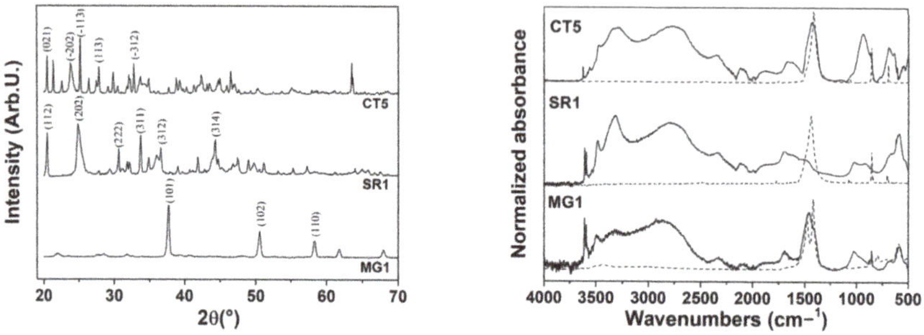

Figure 2. (**left**) XRD patterns and (**right**) FTIR spectra of $Mg(OH)_2$, $Sr(OH)_2$ and $Ba(OH)_2$ compounds obtained by the hydrothermal synthesis ($Mg(OH)_2$ = MG1, $Sr(OH)_2$ = SR1, $Ba(OH)_2$ = CT5).

In addition, the acicular crystals typical of barium hydroxide and strontium hydroxide were observed in SEM images (Figure 3) in the dispersions prepared with 75% water and 25% isopropanol and a homogeneous particle distribution. On the contrary, for magnesium hydroxide (MG1 sample), no characteristic morphology was visible macroscopically; further details were observed under the optical microscope. The barium hydroxide (CT5) and strontium hydroxide (SR1) samples showed particles with both acicular and cubic shape with nanometric dimensions. These cubic crystals suggest that controlled crystallization is favored by the solvent mixtures. The magnesium hydroxide (MG1) particles show homogeneous and spherical distribution.

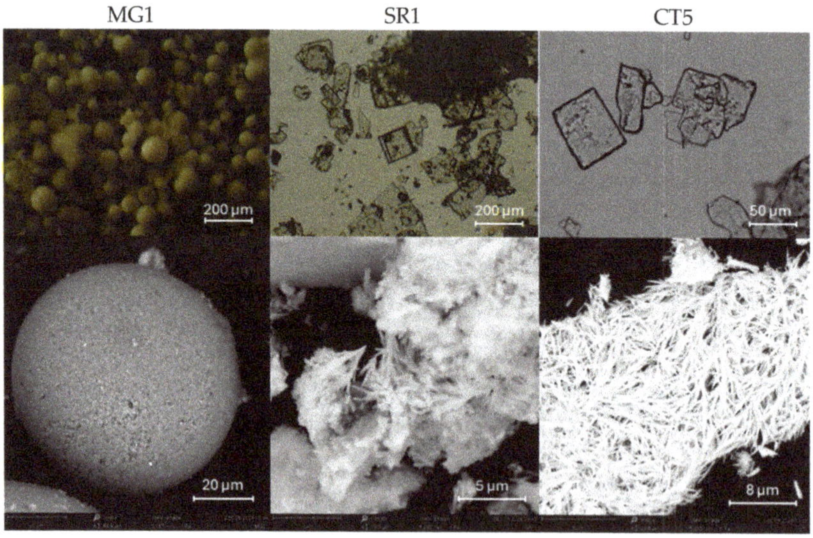

Figure 3. (**up**) Images obtained under the optical microscope (OM)OM; (**down**) SEM micrographs of MG1, SR1 and CT5 samples.

SEM observations confirmed the similarities found for the barium hydroxide (CT5) and strontium hydroxide (SR1) samples, which are characterized by a fibrous structure. A perfectly delineated spherical shape is observed for the magnesium hydroxide (MG1) samples, whose particle size distribution is reported in Figure 4. From the graph, it can be seen that the particles range in size from around 60 µm to 290 µm.

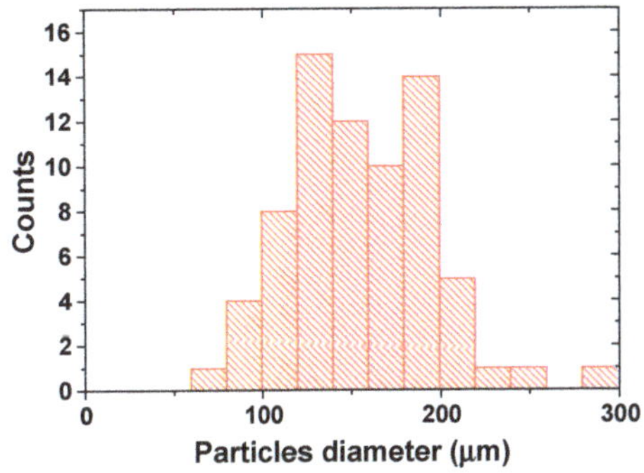

Figure 4. Particle size distribution for $Mg(OH)_2$ particles, obtained by OM images.

The TEM micrographs (Figure 5) of the CT5 sample show three different morphologies of barium hydroxide particles. Type 2 and 3 structures show a marked acicular morphology, which is typical for barium hydroxide: in type 2 structures, well-defined needles arranged in a radial pattern are observed, while in type 3 structures, bundles of flexible nanocrystals are formed. Type 1, on the other hand, shows disordered aggregates of nanosheets, indicative of a less oriented growth. The observed morphological variety reflects the influence of the synthesis conditions on the structural development of the material.

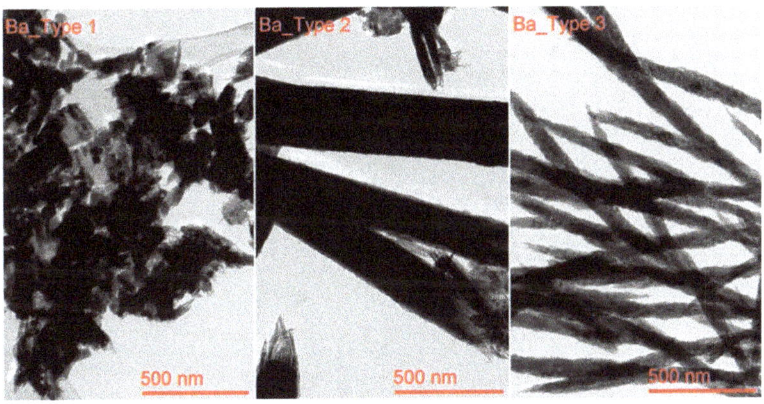

Figure 5. TEM micrographs of CT5 sample.

Micrographs analysis (Table 2) supports the morphological distinctions among the three barium hydroxide particle types. Type 2 exhibits the highest average length (~163 nm) and area (~121 nm^2), consistent with the formation of well-defined, radially arranged needle-like crystals. The large standard deviation in length and area also indicates significant dimensional variability, likely due to differences in needle growth or partial overlap. Type 3 structures, although shorter (~53 nm), are characterized by higher pixel intensities, reflecting denser or thicker nanocrystals arranged in bundled configurations. In contrast, Type 1 shows the smallest dimensions (mean length ~50 nm, area ~51 nm^2) and a more moderate intensity range, in line with its morphology of disordered nanosheets. These quantitative differences confirm that the synthesis conditions result in markedly distinct crystallization pathways, influencing not only the shape but also the size distribution and density of the resulting particles.

TEM micrographs (Figure 6-up) of MG1 sample show magnesium hydroxide particles. The sample exhibits a relatively homogeneous morphology, consisting mainly of overlapping lamellar structures accompanied by well-defined hexagonal particles. The coexistence of plates and hexagonal shapes suggests controlled growth along preferential crystallographic planes, influenced by the hydrothermal conditions of synthesis.

Table 2. Analysis of particle morphology from TEM micrographs of CT5 sample.

			CT5 Type 1			
Label	Area	Mean	Min	Max	Angle	Lenght
Mean	50.967	74.845	58.235	118.942	−53.685	49.98
SD	14.485	37.161	32.813	56.004	82.66	14.538
Min	30	18.565	14.91	30	−165.5	29
Max	77	160.855	144.093	241.136	175.236	76.485
			CT5 Type 2			
Label	Area	Mean	Min	Max	Angle	Lenght
Mean	121.088	48.872	29.815	156.531	−67.419	162.673
SD	70.634	22.027	18.928	61.880	56.562	95.338
Min	22.488	16.214	5.054	61.000	−145.491	29.624
Max	217.750	105.665	77.000	255.000	14.931	293.412
			CT5 Type 3			
Label	Area	Mean	Min	Max	Angle	Lenght
Mean	39.491	115.408	89.863	168.329	−85.771	52.615
SD	13.88	28.427	28.341	40.109	54.933	18.747
Min	18.1	60.057	43.927	92	−175.236	23.884
Max	67.464	164.805	144.796	254.667	20.323	90.68

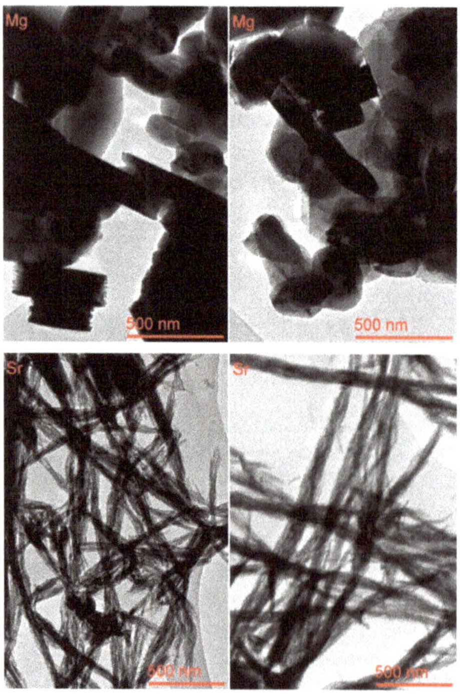

Figure 6. TEM micrographs of MG1 (**up**) and SR1 samples (**down**).

The images of micrographs (Table 3) confirm the relatively uniform size and shape distribution of magnesium hydroxide particles, despite local morphological variations observable in different regions of the same sample. The measured mean particle area is approximately 200–225 nm^2, with average lengths ranging from 139 to 157 nm. The standard deviations, particularly in the second image region, indicate moderate polydispersity, likely due to overlapping or aggregation effects. Notably, the pixel intensity is lower in the second region, suggesting a local decrease in thickness or electron density, possibly associated with differences in the stacking or orientation of lamellar structures. Overall, the data support the interpretation of a homogeneous material composed of plate-like and hexagonal particles with a controlled crystallization pattern.

Table 3. Analysis of particle morphology from TEM micrographs of MG1 sample.

			MG1 (image left)			
Label	Area	Mean	Min	Max	Angle	Lenght
Mean	199.556	59.086	45.028	116.717	−100.87	139.199
SD	51.374	29.015	23.948	59.915	36.37	36.257
Min	96.787	27.325	23.961	42.713	−170.074	66.981
Max	360.936	158.205	135.354	255	−25.074	252.955
			MG1 (image right)			
Label	Area	Mean	Min	Max	Angle	Lenght
Mean	225.702	40.944	27.241	81.298	−85.167	157.486
SD	107.575	25.117	22.143	46.628	41.144	75.625
Min	118.968	5.009	0	14	−147.171	82.958
Max	594.838	107.997	91.71	215	−3.521	416.796

The TEM micrographs (Figure 6-down) of the SR1 sample show strontium hydroxide particles. The sample shows a homogeneous morphology characterized by thin, elongated acicular structures intertwined to form fibrillar bundles and networks. This arrangement suggests marked anisotropic growth typical of strontium hydroxide, favored by the hydrothermal synthesis conditions. Micrograph analysis (Table 4) supports the morphological homogeneity observed in the TEM micrographs. Both regions of the $Sr(OH)_2$ sample exhibit similar characteristics, with elongated particles forming fibrous networks. The average particle length increases from ~66 nm in the left region to ~91 nm in the right, accompanied by a wider dispersion (SD ~32 nm), indicating variability in bundle thickness and possible overlap. The higher mean area in the second region (~166 nm^2) also suggests denser or more bundled nanofibers. Pixel intensity values are consistent across both regions, with moderate variation, reflecting relatively uniform density or thickness among the fibrils. The narrow angular distributions further support the aligned, interwoven structure of the acicular aggregates, typical of anisotropic crystal growth under hydrothermal conditions. The morphological differences observed among $Ba(OH)_2$, $Sr(OH)_2$, and $Mg(OH)_2$ particles are mainly related to the different chemical natures and solubilities of the cations, which affect the nucleation and growth processes during the solvothermal treatment. The solvent composition also plays a role, as the mixed water/isopropanol system can influence surface energy and crystal growth direction, leading to the formation of distinct shapes such as acicular, plate-like, or spherical particles.

Table 4. Analysis of particle morphology from TEM images of SR1 sample.

			SR1 (image left)			
Label	Area	Mean	Min	Max	Angle	Lenght
Mean	120.638	101.344	69.482	164.1	−89.693	65.584
SD	38.842	31.937	25.643	43.278	70.48	21.576
Min	67.738	17.917	8	31	−173.157	35.559
Max	232.244	141.83	108.086	218.638	15.642	128.071
			SR1 (image right)			
Label	Area	Mean	Min	Max	Angle	Lenght
Mean	165.904	94.001	69.293	155.544	−100.689	90.622
SD	57.631	30.005	25.698	52.173	41.601	32.11
Min	103.22	14.756	7	44	−163.301	56.109
Max	322.562	155.239	123.333	232	−26.565	177.867

Nitrogen adsorption–desorption measurements provided insight into the specific surface area and porosity of the hydroxide samples (Table 5). Among the three samples, MG1 exhibited the highest BET surface area, with a value of 14.55 ± 0.03 m^2/g, indicative of a well-developed porous structure. SR1 presented an intermediate surface area of 5.18 ± 0.09 m^2/g, while CT5 showed a markedly lower value of 2.52 ± 0.06 m^2/g, suggesting a more compact morphology or greater degree of particle aggregation. The average pore diameters also varied considerably: CT5 featured the largest pores (14.24 nm), followed by SR1 (11.35 nm) and MG1 (9.85 nm). These values suggest that MG1 possesses a more finely structured mesoporous framework, which correlates well with its lamellar morphology observed in TEM images. Conversely, the low surface area yet larger pore size in CT5 may be attributed to the formation of dense or aggregated needle-like structures. SR1's intermediate values in both parameters likely reflect its fibrillar morphology with moderate porosity and structural openness.

Table 5. BET analysis results: specific surface area and average pore diameter.

Sample	**BET Surface Area, m^2/g**	**Average Pore Size, nm**
SR1	5.178 ± 0.09	11.3
MG1	14.55 ± 0.03	9.8
CT5	2.52 ± 0.06	14.2

These results highlight how optimized synthesis conditions, particularly the use of a water/isopropanol solvent system and moderate solvothermal parameters, favor the controlled growth of crystalline hydroxide structures and the formation of stable, well-defined nanoparticles. The strong correlation between particle morphology (as observed by TEM) and surface area/porosity (from BET analysis) confirms the reliability and reproducibility of the proposed one-pot method. In general, different morphologies of hydroxide particles arise from the combined influence of thermodynamic and kinetic factors during nucleation and growth. In particular, the solvent composition, precursor concentration, temperature, and intrinsic solubility of each hydroxide play crucial roles. The solvent (water/alcohol ratio) affects the dielectric constant and polarity of the medium, modifying ion solvation and the supersaturation level, which in turn govern nucleation and crystal growth rates. Moreover, the different solubilities of $Ba(OH)_2$, $Sr(OH)_2$, and $Mg(OH)_2$ lead to distinct precipitation kinetics: the less soluble hydroxides tend to nucleate rapidly, producing smaller and less ordered particles, while the more soluble species undergo slower growth and recrystallization, forming larger or more defined morphologies. In addition, alcohol molecules may selectively adsorb onto specific crystal faces, influencing anisotropic growth and further contributing to morphological diversity.

The final stages involved filtering and drying the resulting solution to obtain the particles in powder form. The particles were dispersed (Figure 7) in different solvents in order to obtain a dispersed system that is stable over time. The particles obtained in powder form were redispersed in water, isopropanol, and a water/isopropanol mixture (75:25), with the aim of evaluating the best dispersion system. To improve the stability and homogeneity of the system, the dispersions were sonicated in an ultrasonic bath for 30 min.

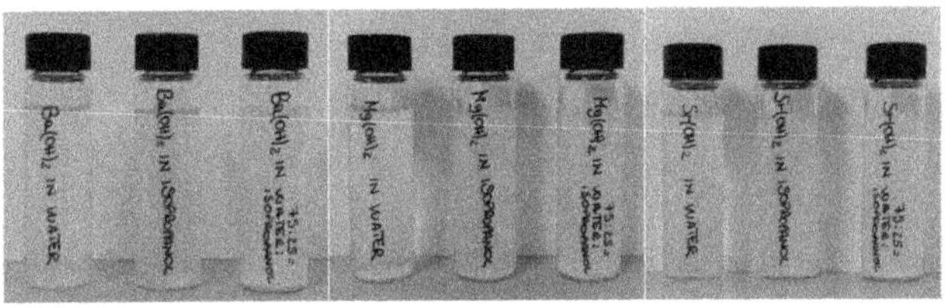

Figure 7. Redispersion of hydroxide in water, isopropanol, and 75:25 water/isopropanol mixture.

The results showed that the dispersions remained stable for about 1 h, with the best performance observed in the water/isopropanol mixture (75:25) for the same percentages used for hydrothermal synthesis. Compared to previously documented approaches, which were often characterized by bottom-up, multi-step synthesis and the use of multiple precursors, the methodology developed in this work is distinguished by the use of a top-down approach, simplifying the process and making it more efficient and sustainable. The use of isopropanol has also been explored in other studies, but an important optimization is reported here due to the reduction in the amount of alcohol used. This choice not only reduces the overall toxicity of the process but also allows for lower production costs.

3. Materials and Methods

3.1. Materials

All the chemical reagents and solvents used in this work were supplied by Sigma Aldrich: technical-grade barium hydroxide ($Ba(OH)_2$), technical-grade ~95% strontium hydroxide ($Sr(OH)_2$), ~94% magnesium hydroxide ($Mg(OH)_2$) from BioUltra at ≥99.0% (KT), isopropyl alcohol at ≥99.7%, FCC, FG, and 1-propanol. Deionized water was prepared with an ionized water purification system (Demineraliser model minidem, CTS conservation) and was used in all procedures.

3.2. *Synthesis of Barium, Strontium, and Magnesium Hydroxide Nanoparticles*

Barium, strontium, and magnesium hydroxide were synthesized using the solvothermal method using $Ba(OH)_2$, Sr $(OH)_2$, and $Mg(OH)_2$ at micrometer sizes as reagents and different solvents. The concentrations of barium hydroxide and strontium hydroxide were set according to their solubility limit in water at 25 °C, with values of 17 g/L and 18 g/L, respectively. Since the solubility of magnesium hydroxide in water is very low (0.009 g/L) at 25 °C, 1 g of magnesium hydroxide was used in 50 mL of solution. The synthesis was carried out in an autoclave provided by Parr Instrument Company (Moline, IL, USA) using a Model 4838 Reactor controller coupled to Model 4793 general-purpose pressure vessels with four valves. The synthesis scheme is reported in Figure 8.

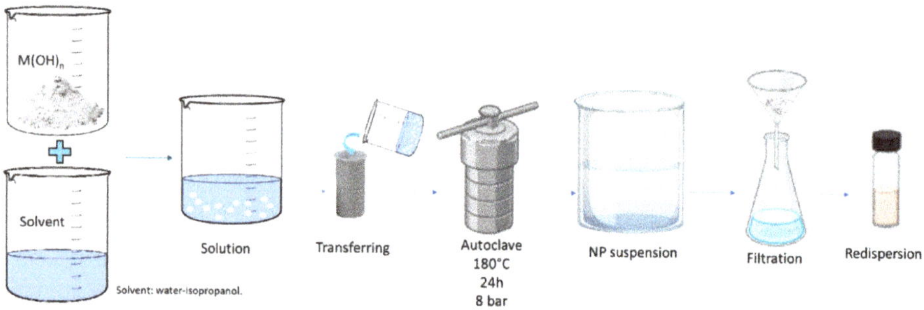

Figure 8. Synthesis scheme of hydroxides.

The preparation of the synthesis was carried out through the following phases:

1. Initial crushing: The micrometer-sized hydroxide particle grains were placed in a mortar and ground. This mechanical crushing process made it possible to reduce the particle size, increase their exposed surface area, and improve efficiency in subsequent steps.
2. Solvent dispersions: The ground particles were then dispersed in the solvents.
3. Autoclave treatment: The suspension of particles was transferred to the autoclave in a sealed container that can withstand high temperatures and pressures. The autoclave was heated at 180 °C and 8 bar for 24 h.
4. Evaluation of the stability of the system: Once the process was completed, a suspension of nanoparticles dispersed in the solvent was obtained. The resulting mixtures were thus sonicated in an ultrasonic bath for 30 min, collected via filtration, and dried at 60 °C overnight.
5. The nanoparticles were redispersed in the proper solvent.

3.3. Characterization

The *XRD patterns* were acquired by a Philips PW 1050/39 powder diffractometer in Bragg–Brentano geometry, which uses Cu Kα radiation (λ = 1.54056 Å) as the X-ray source. The X-ray tube is powered by a generator that works at a power of 40 KV and a current of 30 mA. XRD patterns were acquired in the range of 20° to 70° (2θ) with a step of 0.05° and an acquisition time for each step of 5 s. Identification of mineralogical phases was performed using X'pert HighScore® software.

The *FT-IR spectra* were acquired with a Bruker VERTEX 70V FTIR spectrophotometer with a Platinum ATR accessory (diamond crystal) in the spectral range of 4000–400 cm^{-1}, using a step of 2 cm^{-1} and averaging over 120 scans at 2 hPa. The data analysis was performed using OPUS 7.5® software.

The *optical microscope images* were captured using an Optika optical microscope equipped with an integrated 5-megapixel digital camera, ensuring high-resolution imaging and accurate documentation. The microscope featured objective lenses with magnification levels of 4×, 10×, 40×, and 100×, allowing for detailed analysis across multiple scales.

For the morphological and dimensional analysis of the synthesized nanoparticles, a Phenom Pro X *scanning electron microscope (SEM)* (Thermo Fisher Scientific, Waltham, MA, USA) was used. The SEM was operated at accelerating voltages of 15 kV, allowing for magnifications between 500× and 13,500×. The instrument was equipped with an Energy Dispersive X-ray Spectroscopy (EDS) detector supplied by Alfates, which allowed for compositional analysis. Samples were imaged digitally using transmission electron microscopy (TEM). Images and EDX spectra were acquired using a Tecnai G2 F20 X-TWIN (FEI, Eindhoven, The Netherlands, 2011) microscope with a Schottky-type field emission electron source, a high-angle annular dark field (HAADF) detector, a single- and double-tilted sample holder, and an 11MPix ORIUS SC1000B (Gatan) CCD camera. Samples were mounted directly on a holey carbon-coated copper grid (Agar Scientific Ltd., Stansted, UK).

The crystalline structure was determined with a *Transmission Electron Microscope* (TEM) working in diffraction mode, while for micrographs, an objective aperture was additionally inserted for enhanced contrast.

The surface areas of the produced materials were examined by conducting *nitrogen gas adsorption* via the BET (Brunauer–Emmet–Teller) method using a surface area and porosity analyzer, TriStar II 3020 (Micromeritics, USA). Prior to these measurements, each sample was degassed for 1.5 h at 120 °C under nitrogen flow using the FlowPrep 060 sample degas system (Micromeritics, USA).

4. Conclusions

The hydrothermal synthesis performed in an autoclave, using a solvent mixture of 75% water and 25% isopropanol at 180 °C for 24 h, has proven to be a highly effective and reproducible method for producing stable and well-dispersed nanoparticles of barium, strontium, and magnesium hydroxides. The optimized solvent composition favored uniform suspensions with limited particle aggregation. This one-pot method yielded nanoparticles with distinct morphologies. Acicular and fibrous morphologies were obtained for $Ba(OH)_2$ and $Sr(OH)_2$, while $Mg(OH)_2$ formed spheric and hexagonal structures. The simplicity and environmental compatibility of this method make it a promising strategy for the sustainable production of functional nanomaterials, with potential applications in advanced technological fields.

Author Contributions: M.L.S. and D.F.C.M.: Conceptualization and methodology, writing—review and editing. C.T.: Synthesis, XRD patterns and FT-IR spectra, writing. F.A.: SEM and MO investigation. R.S.: TEM and gas adsorption. D.F.C.M.: Funding. All authors have read and agreed to the published version of the manuscript.

Funding: This research work is supported by the MUR-PNRR SAMOTHRACE Project (ECS00000022). F.A. thanks MIUR for the Project PON Ricerca e Innovazione 2014–2020—Avviso DD 407/2018 "AIM Attrazione e Mobilità Internazionale" (AIM1808223).

Institutional Review Board Statement: Not applicable.

Informed Consent Statement: Not applicable.

Data Availability Statement: The datasets generated and/or analyzed during the current study are available from the corresponding author on reasonable request.

Acknowledgments: SEM investigations were performed at the ATEN-Center of University of Palermo.

Conflicts of Interest: The authors declare no conflicts of interest.

References

1. Gleiter, H. Nanostructured Materials: Basic Concepts and Microstructure. *Acta Mater.* **2000**, *48*, 1–29. [CrossRef]
2. Rao, C.N.R.; Müller, A.; Cheetham, A.K. *The Chemistry of Nanomaterials: Synthesis, Properties and Applications*; John Wiley & Sons, Ltd.: Weinheim, Germany, 2004; pp. 1–20. ISBN 978-3-527 60247-6.
3. Mohajerani, A.; Burnett, L.; Smith, J.V.; Kurmus, H.; Milas, J.; Arulrajah, A.; Horpibulsuk, S.; Abdul Kadir, A. Nanoparticles in Construction Materials and Other Applications, and Implications of Nanoparticle Use. *Materials* **2019**, *12*, 3052. [CrossRef]
4. Fathi-Achachelouei, M.; Knopf-Marques, H.; Ribeiro da Silva, C.E.; Barthès, J.; Bat, E.; Tezcaner, A.; Vrana, N.E. Use of Nanoparticles in Tissue Engineering and Regenerative Medicine. *Front. Bioeng. Biotechnol.* **2019**, *7*, 113. [CrossRef]
5. David, M.E.; Ion, R.-M.; Grigorescu, R.M.; Iancu, L.; Andrei, E.R. Nanomaterials Used in Conservation and Restoration of Cultural Heritage: An Up-to-Date Overview. *Materials* **2020**, *13*, 2064. [CrossRef]

6. Abid, N.; Khan, A.M.; Shujait, S.; Chaudhary, K.; Ikram, M.; Imran, M.; Haider, J.; Khan, M.; Khan, Q.; Maqbool, M. Synthesis of Nanomaterials Using Various Top-down and Bottom-up Approaches, Influencing Factors, Advantages, and Disadvantages: A Review. *Adv. Colloid Interface Sci.* **2022**, *300*, 102597. [CrossRef]
7. Sabatier, P.A. Top-Down and Bottom-Up Approaches to Implementation Research: A Critical Analysis and Suggested Synthesis. *J. Public Policy* **1986**, *6*, 21–48. [CrossRef]
8. Jasti, R.; Bertozzi, C.R. Progress and Challenges for the Bottom-up Synthesis of Carbon Nanotubes with Discrete Chirality. *Chem. Phys. Lett.* **2010**, *494*, 1–7. [CrossRef] [PubMed]
9. Amendola, V.; Meneghetti, M.; Granozzi, G.; Agnoli, S.; Polizzi, S.; Riello, P.; Boscaini, A.; Anselmi, C.; Fracasso, G.; Colombatti, M.; et al. Top-down Synthesis of Multifunctional Iron Oxide Nanoparticles for Macrophage Labelling and Manipulation. *J. Mater. Chem.* **2011**, *21*, 3803–3813. [CrossRef]
10. Balducci, G.; Diaz, L.B.; Gregory, D.H. Recent Progress in the Synthesis of Nanostructured Magnesium Hydroxide. *CrystEngComm* **2017**, *19*, 6067–6084. [CrossRef]
11. Wu, J.; Yan, H.; Zhang, X.; Wei, L.; Liu, X.; Xu, B. Magnesium Hydroxide Nanoparticles Synthesized in Water-in-Oil Microemulsions. *J. Colloid Interface Sci.* **2008**, *324*, 167–171. [CrossRef]
12. Giorgi, R.; Bozzi, C.; Dei, L.; Gabbiani, C.; Ninham, B.W.; Baglioni, P. Nanoparticles of $Mg(OH)_2$: Synthesis and Application to Paper Conservation. *Langmuir* **2005**, *21*, 8495–8501. [CrossRef]
13. Giorgi, R.; Ambrosi, M.; Toccafondi, N.; Baglioni, P. Nanoparticles for Cultural Heritage Conservation: Calcium and Barium Hydroxide Nanoparticles for Wall Painting Consolidation. *Chem.–A Eur. J.* **2010**, *16*, 9374–9382. [CrossRef]
14. Ciliberto, E.; Condorelli, G.G.; La Delfa, S.; Viscuso, E. Nanoparticles of Sr(OH)2: Synthesis in Homogeneous Phase at Low Temperature and Application for Cultural Heritage Artefacts. *Appl. Phys. A* **2008**, *92*, 137–141. [CrossRef]
15. Gan, X.Y.; Jayatissa, A.H.; Yu, Z.; Chen, X.; Li, M. Hydrothermal Synthesis of Nanomaterials. *J. Nanomater.* **2020**, *2020*, 8917013. [CrossRef]

inorganics

MDPI

Communication

SiO_2 Electret Formation via Stamp-Assisted Capacitive Coupling: A Chemophysical Surface Functionalisation

Edoardo Chini [1], Denis Gentili [1], Andrea Liscio [2] and Massimiliano Cavallini [1,*]

[1] Istituto per lo Studio dei Materiali Nanostrutturati (ISMN)-Consiglio Nazionale Delle Ricerche (CNR), Via P. Gobetti 101, 40129 Bologna, Italy

[2] Rome Unit, Istituto Per la Microelettronica e Microsistemi (IMM)-Consiglio Nazionale Delle Ricerche (CNR), Via Del Fosso Del Cavaliere 100, 00133 Roma, Italy

* Correspondence: massimiliano.cavallini@cnr.it

Abstract

This work introduces a new method for creating patterned SiO_2 electrets using stamp-assisted capacitive coupling (SACC), enabling surface functionalisation without direct electrode contact. SACC applies an alternating current through capacitive coupling between a conductive stamp and an insulating substrate in high-humidity conditions, forming a nano-electrochemical cell that drives localised reactions. Using thermally grown SiO_2 films, we achieve submicrometre patterning with minimal topographical impact but significant electronic alterations. Characterisation via Kelvin Probe Force Microscopy and Electric Force Microscopy confirms the formation of charged regions replicating the stamp pattern, with adjustable surface potential shifts up to -1.7 V and charge densities reaching 300 $nC \cdot cm^{-2}$. The process can be scaled to areas of 1 cm^2 and is compatible with conventional laboratory equipment, offering a high-throughput alternative to scanning-probe lithography. SACC combines simplicity, accuracy, and scalability, opening new opportunities for patterned electret production and functional surface engineering.

Keywords: electrets; functional patterning; SiO_2; electrochemical lithography; surface functionalization

1. Introduction

SiO_2 is one of the most widely used dielectric materials in modern technology, serving both passive and active roles. Beyond its established use as an insulating layer in silicon-based electronics, SiO_2 can acquire active functionalities via surface functionalization. These include charge trapping [1–3], optical sensing [4], enhancing charge transport in organic semiconductors [5], regenerable resistive switching [6] and controlled material growth or assembly [7]. These capabilities extend its relevance from microelectronics to emerging fields including organic electronics [5], memories [6] and surface sensing [4].

Herein, we demonstrate the formation of SiO_2 electrets through a novel approach that enables precise submicrometre patterning.

Electrets, known for over a century, are often considered the electrostatic analogues of permanent magnets [8]. Their ability to produce persistent electric fields has enabled a broad spectrum of applications [9,10], including microphones [11], nanoxerography, [12] magnetic [13] and pressure [14] photo sensors [15], actuators, [16] energy harvesting systems [17], soft electronics, [18] and silicon-based photovoltaic technologies [10,19]. Many of these applications rely on precise spatial control of charge localisation. Common inorganic electrets include SiO_2 [1,3,19,20], Si_3N_4 [21], and transition metal oxides [22], while

 https://doi.org/10.3390/inorganics14010021

organic electrets are typically polymers or molecular compounds [18]. However, despite their technological importance, electret materials are poorly studied, and the diversity of materials and methods for fabricating electrets remains limited. Conventional approaches rely on strong electric fields to induce dipole alignment or structural changes, such as local amorphisation, reduction, or doping [20]. Spatial control is usually achieved using techniques like scanning probe lithography [22] or electrified microcontact printing [18].

In this communication, we introduce a preliminary study on a novel strategy for the formation of SiO_2 electrets, introducing stamp-assisted capacitive coupling. This method, which uses the experimental setup of parallel electrochemical nanolithography, enables precise submicrometer patterning without direct electrode contact by applying alternating current via capacitive coupling between an electrified probe and a conductive thin film on an insulating substrate. The resulting chemophysical surface functionalisation was characterised using Kelvin Probe Force Microscopy [23] and Electric Force Microscopy [24], confirming the successful creation of patterned electret regions. This approach opens new avenues for scalable, high-resolution electret fabrication and broadens the functional versatility of SiO_2 in advanced electronic and sensing applications.

2. Results and Discussion

2.1. Process

To functionalise SiO_2 films, we used stamp-assisted capacitive coupling (SACC). SACC was originally introduced as a scanning-probe-based technique, called Electrode-Free Anodic Oxidation Nanolithography [25], for electrochemically assisted etching of graphene deposited on insulating substrates. SACC operates by applying a high-frequency alternating current (>10 kHz) via capacitive coupling between a conductive probe and the substrate, triggering electrochemical reactions at the probe-surface interface. This setup removes the need for direct electrical contact, enabling localised processing of insulating materials, although it does not achieve the same level of control as traditional electrochemical nanolithography (e.g., precise local oxidation or reduction) [26,27].

In this study, SACC was adapted for parallel processing to enable large-area patterning. Instead of an AFM probe, a conductive stamp composed of parallel lines was used. Under high-humidity conditions (~95% RH), a nanometric water meniscus naturally forms between the stamp protrusions and the SiO_2 surface, creating a confined nano- cell. When an AC bias is applied through the conductive stamp, localised capacitive coupling occurs only beneath the protrusions, reproducing the stamp geometry on the substrate [25]. Despite SACC not reaching the spatial resolution of scanning probe-based methods, nor enabling controllable neutralisation or redistribution of charges, beyond spatial control, SACC allows fine-tuning of local effects by adjusting the applied bias, process duration, or repeating the treatment in targeted regions. Figure 1 shows the scheme of the process.

To demonstrate SACC's effectiveness for electret fabrication, experiments were conducted on thermally grown SiO_2 films deposited on highly doped silicon substrates, commonly used in organic electronics. SiO_2 is a well-known model electret material. The treatment involved applying an AC bias of 15 V at 50 kHz for varying durations under 95% relative humidity. The stamp used was a polymer replica of a blank compact disc coated with a 100 nm gold layer (Figure 2a). The stamp consisted of parallel lines with a thickness of 220 nm and a spacing of 1.5 μm.

For short treatments (<1 h), SACC leaves the SiO_2 morphology essentially unaltered. Prolonged exposure (>1 h) induces only minor roughness changes (RMS < 0.2 nm), within the range of instrumental uncertainty.

https://doi.org/10.3390/inorganics14010021

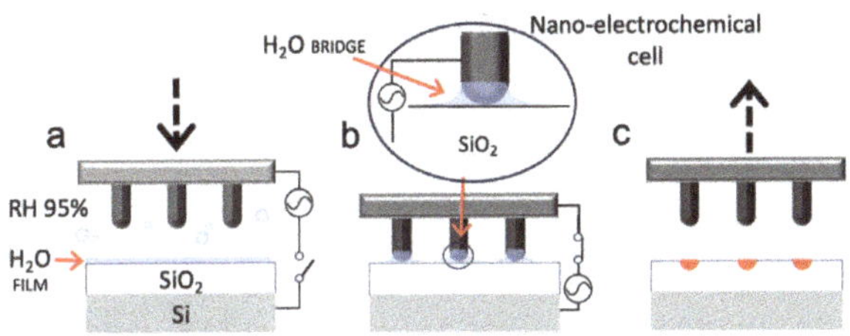

Figure 1. Scheme of stamp-assisted capacitive coupling (SACC). (**a**) When the stamp is placed in contact with the surface (black arrow), a thin water layer is formed in highly humid conditions (Relative Humidity > 95%). (**b**) Stamp motifs create a meniscus through capillary force upon contact with the surface, forming a two-electrode nano-electrochemical cell. By applying an AC bias voltage, electrochemical reactions take place within this nano-cell. (**c**) The electrochemical reaction occurs exclusively beneath the stamp protrusions, resulting in the patterning of surface structures that replicate the stamp relief features.

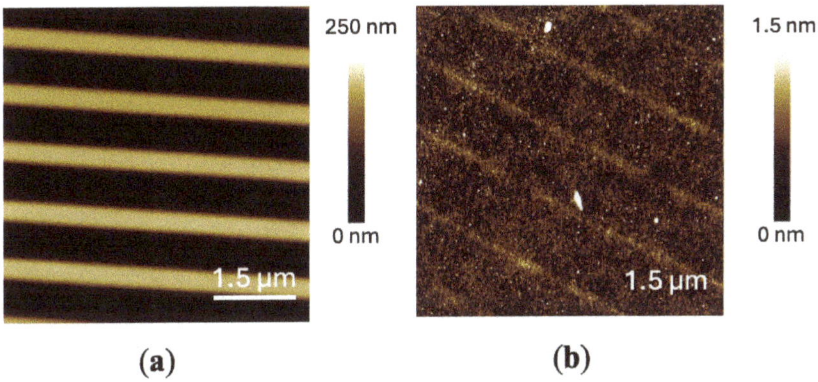

Figure 2. (**a**) AFM topographic image of the stamp (the colour bar indicates the z scale. (**b**) AFM topographic image of the SiO_2 surface treated by stamp-assisted capacitive coupling, (SACC) for 1 h.

X-ray photoelectron spectroscopy and Raman spectroscopy do not show any chemical composition of the SiO_2 surface after processing. Therefore, to assess the impact of SACC on the electronic properties of SiO_2, we used Electric Force Microscopy (EFM) and Kelvin Probe Force Microscopy (KPFM).

In SACC, the minimum achievable pattern size depends on the features of the stamp. In principle, resolutions down to 10 nm are possible when using scanning probe techniques [25]. With our current stamp-based approach, we attained a resolution of approximately 100 nm. However, below 200 nm, it becomes challenging to assess the homogeneity of the effects caused by the treatment. At this stage, we consider 100 nm to be the practical resolution limit for SACC.

2.2. Electric Force Microscopy

EFM is highly sensitive to nanoscale variations in surface charge distribution, whereas KPFM maps surface potential (SP), providing insight into changes in work function and local charge states.

EFM operates in a two-pass mode, detecting electrostatic forces between the oscillating probe and the sample [24]. These forces generate a gradient that induces a phase shift in the cantilever oscillation. For small oscillation amplitudes, the interaction can be decomposed into capacitive forces (quadratically dependent on the applied EFM voltage, (V_{EFM}) and

https://doi.org/10.3390/inorganics14010021

Coulombic forces, linearly dependent on VEFM due to static charges or multipoles on the surface [28]:

$$\Delta\phi \propto F' \approx \frac{1}{2}C'_{s-t}V_{EFM}^{2} + E_s C''_{s-t}V_{EFM} \quad (1)$$

where $\Delta\phi$ is the Phase shift, F′ is the force gradient $C'_{s\text{-}t}$ and $C''_{s\text{-}t}$ are the first and second derivative capacitances with respect to the tip–sample distance, respectively, and E_S is the electric field caused by static charge.

Although SACC did not produce significant morphological changes, EFM imaging revealed clear contrasts replicating the stamp pattern. Figure 3a shows a representative EFM phase image of a patterned SiO_2 film. Voltage-dependent EFM measurements (Figure 3b) reveal that pristine SiO_2 responds with a purely quadratic phase–bias dependence, characteristic of neutral dielectrics. In contrast, SACC-treated regions show a pronounced linear component and a negative shift of the phase maximum, demonstrating the presence of trapped negative charges.

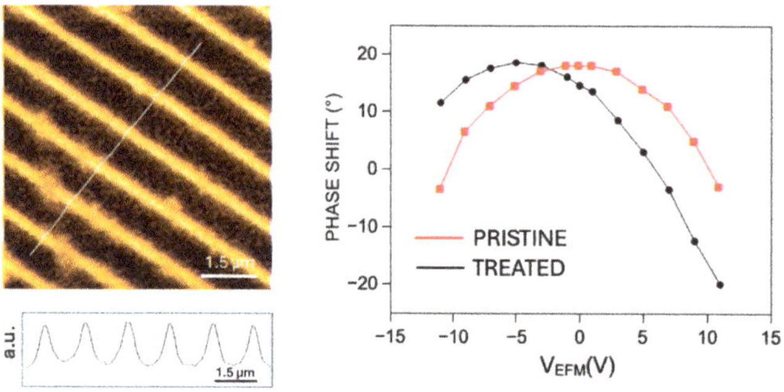

(a) (b)

Figure 3. (a) Electric Force Microscopy (phase shift) image of the SiO2 surface after 30 min of SACC treatment (bright colour indicates a higher signal), along with its line profile (white lines). (b) Voltage-dependent EFM measurements comparing a pristine sample (red curve) and an SACC-treated sample (30-min treatment).

2.3. *Kelvin Probe Force Microscopy*

KPFM measurements further demonstrated a decrease in surface potential (SP) within patterned regions (Figure 4a). The SP contrast increases approximately linearly with treatment duration up to 60 min, up to −1.7 V (Figure 4b), beyond which local film degradation or increased roughness affects measurement accuracy. The trapped charge density (σ) was estimated using a parallel-plate capacitor model:

$$\Delta SP = \frac{\sigma h_{\text{eff}}}{\varepsilon_0 \varepsilon_r} \quad (2)$$

where ΔSP is the surface potential contrast, h_{eff} is the effective dielectric thickness of the SiO_2 layer (~200 nm), and $\varepsilon_r \approx 3$ is its relative permittivity. Using these parameters, extended SACC treatments yield surface-change densities up to $\sigma \approx 300$ nC·cm^{-2}, fully consistent with reported values for SiO_2-based electrets. This model provides a reliable first-order estimate because the tip radius ($\approx$20 nm) is significantly larger than the patterned features, ensuring that the local electrostatic interaction can be approximated as a macroscopic capacitive geometry.

https://doi.org/10.3390/inorganics14010021

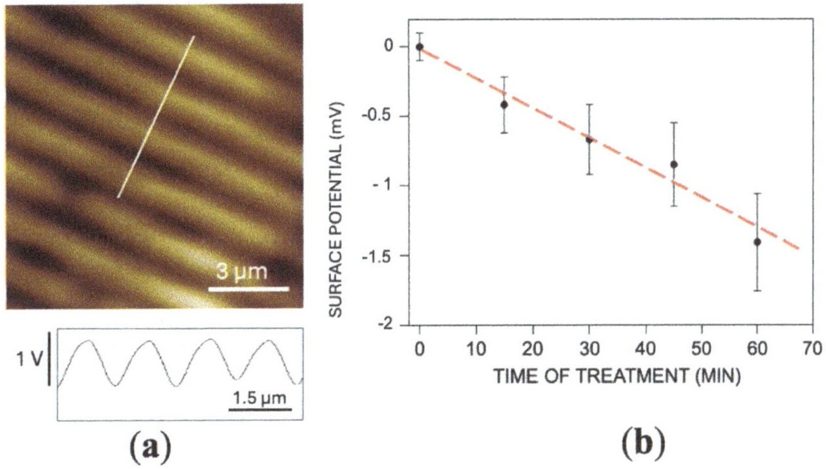

Figure 4. (**a**) Surface potential map of the SiO_2 surface after 30 min of treatment using SACC (bright colour indicates a higher signal), along with the corresponding line profile (white lines). (**b**) Changes in surface potential over time for different SACC treatment durations. The dashed red line is a guide for the eyes.

After 90 days, the surface potential decreased by approximately 45% under ambient conditions (air, relative humidity 50–75%) and under light exposure, regardless of the initial charge. This retention time is more than enough for most electrect applications.

Finally, SACC processing can be repeated within the same region to increase charge density, expand coverage, or create complex patterns (Figure 5).

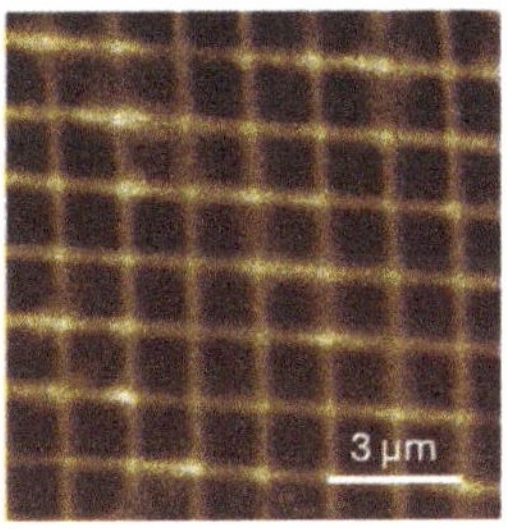

Figure 5. Electric Force Microscopy (phase shift) image of the SiO_2 surface after 30 min of Electrode-Free Electrochemical Nanolithography treatment. The sample was obtained by rotating the sample 90° after the first process application.

3. Materials and Method

Electrochemical apparatus. The SACC process was carried out using a homemade apparatus described in detail in references [26,27]. A commercial hygrometer (Thermopro) monitors relative humidity.

Stamp Fabrication. Elastomeric stamps were produced using polydimethylsiloxane (PDMS, Sylgard 184, Dow Corning) by a replica molding process. The polymer was cured at 70 °C for 1 h, after which the solidified replica was carefully detached from the master template and rinsed with absolute ethanol to eliminate any residual uncured polymer. Additionally, the metallic layer from standard blank compact discs was employed directly as a stamping surface. The resulting pattern features an array of parallel lines with a periodicity of approximately 1.6 µm, an apex width of 200 nm, and a depth of ~220 nm (Figure 2a).

https://doi.org/10.3390/inorganics14010021

Scanning Probe Microscopy (KPFM, EFM). SPM characterisations were performed using a Bruker Multimode 8 microscope. Measurements were carried out in air at ambient conditions.

4. Conclusions

We demonstrated a parallel implementation of stamp-assisted capacitive coupling that can produce patterned electrets over areas up to 1 cm^2 using standard laboratory equipment. The method maintains a slightly lower precision than scanning-probe techniques but enables high-throughput processing, which is essential for technological applications. Applied to thermally grown SiO_2, SACC achieves high-quality charge patterning and chemical surface functionalisation, highlighting its potential for integration in advanced electronic and sensing devices. Because the mechanism relies on capacitive coupling rather than direct electrical contact, the method is compatible with a wide range of inorganic and organic substrates. SACC therefore offers a versatile and scalable route to engineered electrets and functionalised surfaces. Expected developments include the use of various materials, such as carbon-based materials [29], polymers and low-dimensional semiconductors [30], for sensors [31–34] and optoelectronic applications.

Author Contributions: Conceptualization M.C., methodology D.G. and A.L.; investigation E.C. All authors have read and agreed to the published version of the manuscript.

Funding: This research was funded by European Union–Next Generation EU from the Italian Ministry of University and Research, Project PRIN 2022SRHPF2 "Molecular assisted atom vacancies arrangement to modulate magnetism in 2D transition metal dichalcogenides" (MAVAM).

Data Availability Statement: Experimental data are available upon reasonable request to the corresponding author.

Conflicts of Interest: The authors declare no conflicts of interest.

References

1. Liang, F.; Li, H.Y.; Wang, Y.; Kuang, S.Y.; Fan, Y.J.; Wang, Z.L.; Zhu, G. Charge Distribution and Stability of SiO_2 Nanoarray Electret. *Chemnanomat* **2020**, *6*, 212–217. [CrossRef]
2. Ma, X.L.; Zhao, D.; Xue, M.Q.; Wang, H.; Cao, T.B. Selective Discharge of Electrostatic Charges on Electrets Using a Patterned Hydrogel Stamp. *Angew. Chem. Int. Ed.* **2010**, *49*, 5537–5540. [CrossRef]
3. Zhang, Y.H.; Zhao, D.; Tan, X.X.; Cao, T.B.; Zhang, X. AFM Force Mapping for Characterizing Patterns of Electrostatic Charges on SiO_2 Electrets. *Langmuir* **2010**, *26*, 11958–11962. [CrossRef]
4. Melucci, M.; Zambianchi, M.; Favaretto, L.; Palermo, V.; Treossi, E.; Montalti, M.; Bonacchi, S.; Cavallini, M. Multicolor, large-area fluorescence sensing through oligothiophene-self-assembled monolayers. *Chem. Commun.* **2011**, *47*, 1689–1691. [CrossRef] [PubMed]
5. Gentili, D.; Sonar, P.; Liscio, F.; Cramer, T.; Ferlauto, L.; Leonardi, F.; Milita, S.; Dodabalapur, A.; Cavallini, M. Logic-gate devices based on printed polymer semiconducting nanostripes. *Nano Lett.* **2013**, *13*, 3643–3647. [CrossRef]
6. Cavallini, M.; Hemmatian, Z.; Riminucci, A.; Prezioso, M.; Morandi, V.; Murgia, M. Regenerable Resistive Switching in Silicon Oxide Based Nanojunctions. *Adv. Mater.* **2012**, *24*, 1197–1201. [CrossRef]
7. Leclère, P.; Surin, M.; Lazzaroni, R.; Kilbinger, A.F.M.; Henze, O.; Jonkheijm, P.; Biscarini, F.; Cavallini, M.; Feast, W.J.; Meijer, E.W.; et al. Surface-controlled self-assembly of chiral sexithiophenes. *J. Mater. Chem.* **2004**, *14*, 1959–1963. [CrossRef]
8. Heaviside, O. Electromagnetic induction and its propagation. *Electrician* **1885**, *15*, 230–231.
9. Gerhard, R. Dielectric materials for electro-active (electret) and/or electro-passive (insulation) applications. In Proceedings of the 2nd International Conference on Electrical Materials and Power Equipment, Guangzhou, China, 7–10 April 2019; pp. 91–96.
10. Suzuki, Y. Recent progress in MEMS electret generator for energy harvesting. *IEEJ Trans. Electr. Electron. Eng.* **2011**, *6*, 101–111. [CrossRef]
11. Kamogashira, T.; Yamasoba, T.; Kikuta, S.; Kondo, K. A Sleep Sensor Made with Electret Condenser Microphones. *Clocks Sleep* **2025**, *7*, 28. [CrossRef] [PubMed]
12. Jacobs, H.O.; Campbell, S.A.; Steward, M.G. Approaching nanoxerography: The use of electrostatic forces to position nanoparticles with 100 nm scale resolution. *Adv. Mater.* **2002**, *14*, 1553–1557. [CrossRef]

https://doi.org/10.3390/inorganics14010021

13. Zimoch, L.; Schröder, S.; Elzenheimer, E.; Kaps, S.; Strunskus, T.; Faupel, F.; Höft, M.; Adelung, R. Electret integrated magnetic field sensor based on magnetostrictive polymer composite with nT resolution. *Sci. Rep.* **2025**, *15*, 1561. [CrossRef]
14. Yasuda, T.; Komine, R.; Nojiri, R.; Takabe, Y.; Nara, K.; Kaneko, T.; Horigome, S.; Takeda, Y.; Wang, Y.; Kawaguchi, S.; et al. Ultra-Rapidly Responsive Electret-Based Flexible Pressure Sensor via Functional Polymeric Nanoparticle Synthesis. *Adv. Funct. Mater.* **2024**, *34*, 2402064. [CrossRef]
15. Hu, L.; Li, X.; Guo, X.Y.; Xu, M.X.; Shi, Y.Q.; Herve, N.B.; Xiang, R.; Zhang, Q. Electret Modulation Strategy to Enhance the Photosensitivity Performance of Two-Dimensional Molybdenum Sulfide. *ACS Appl. Mater. Interfaces* **2023**, *15*, 59704–59713. [CrossRef]
16. Xu, S.; Guan, X.; Bian, K.; Zhu, Q.; Dai, N.; Zhao, X.; Qiu, Y.; Zheng, S.; Dong, Y.; Zhong, J.; et al. Electret actuators enabling dual functions of optical ranging and audio feedback to elevate non-contact human-machine interactions. *Nano Energy* **2024**, *125*, 109553. [CrossRef]
17. Mescheder, U.; Müller, B.; Baborie, S.; Urbanovic, P. Properties of SiO_2 electret films charged by ion implantation for MEMS-based energy harvesting systems. *J. Micromech. Microeng.* **2009**, *19*, 094003. [CrossRef]
18. Guo, Z.F.; Patil, Y.; Shinohara, A.; Nagura, K.; Yoshida, M.; Nakanishi, T. Organic molecular and polymeric electrets toward soft electronics. *Mol. Syst. Des. Eng.* **2022**, *7*, 537–552. [CrossRef]
19. Bonilla, R.S.; Wilshaw, P.R. Potassium ions in SiO_2: Electrets for silicon surface passivation. *J. Phys. D Appl. Phys.* **2017**, *51*, 025101. [CrossRef]
20. Minami, T.; Utsubo, T.; Yamatani, T.; Miyata, T.; Ohbayashi, Y. SiO_2 electret thin films prepared by various deposition methods. *Thin Solid Films* **2003**, *426*, 47–52. [CrossRef]
21. Juberi, A.M.; Saha, P.C.; Faruqe, O.; Park, C. Partial Discharge Characteristics of SiO_2/Si_3N_4 Electret Incorporated AlN Substrates. *IEEE Trans. Transp. Electrif.* **2024**, *10*, 7796–7803. [CrossRef]
22. Kim, Y.J.; Yang, C.H. Electret formation in transition metal oxides by electrochemical amorphization. *NPG Asia Mater.* **2020**, *12*, 1. [CrossRef]
23. Zahmatkeshsaredorahi, A.; Millan-Solsona, R.; Jakob, D.S.; Collins, L.; Xu, X.G. Kelvin probe force microscopy under ambient conditions. *Nat. Rev. Methods Prim.* **2025**, *5*, 53. [CrossRef]
24. Albonetti, C.; Chiodini, S.; Annibale, P.; Stoliar, P.; Martinez, R.V.; Garcia, R.; Biscarini, F. Quantitative phase-mode electrostatic force microscopy on silicon oxide nanostructures. *J. Microsc.* **2020**, *280*, 252–269. [CrossRef]
25. Li, H.Y.; Ying, Z.; Lyu, B.S.; Deng, A.L.; Wang, L.L.; Taniguchi, T.; Watanabe, K.; Shi, Z.W. Electrode-Free Anodic Oxidation Nanolithography of Low-Dimensional Materials. *Nano Lett.* **2018**, *18*, 8011–8015. [CrossRef] [PubMed]
26. Gentili, D.; Calabrese, G.; Lunedei, E.; Borgatti, F.; Mirshokraee, S.A.; Benekou, V.; Tseberlidis, G.; Mezzi, A.; Liscio, F.; Candini, A.; et al. Tuning Electronic and Functional Properties in Defected MoS_2 Films by Surface Patterning of Sulphur Atomic Vacancies. *Small Methods* **2025**, *9*, 2401486. [CrossRef] [PubMed]
27. Gentili, D.; Chini, E.; Cavallini, M. Generation and Tuning of Semiconductor Electronic and Functional Properties through Electrochemical Patterning. *Acc. Mater. Res.* **2025**, *6*, 1094–1104. [CrossRef]
28. Qi, G.C.; Yan, H.; Guan, L.; Yang, Y.L.; Qiu, X.H.; Wang, C.; Li, Y.B.; Jiang, Y.P. Characteristic capacitance in an electric force microscope determined by using sample surface bias effect. *J. Appl. Phys.* **2008**, *103*, 114311. [CrossRef]
29. Lui, C.H.; Liu, L.; Mak, K.F.; Flynn, G.W.; Heinz, T.F. Ultraflat graphene. *Nature* **2009**, *462*, 339–341. [CrossRef] [PubMed]
30. Cavallini, M.; Gentili, D. Atomic Vacancies in Transition Metal Dichalcogenides: Properties, Fabrication, and Limits. *Chempluschem* **2022**, *87*, e202100562. [CrossRef]
31. Du, Y.; Shen, P.H.; Liu, H.F.; Zhang, Z.W.; Yang, F.Y.; Chu, D.P.; Ren, T.L.; Wang, Z.L.; Wei, D. Meta-structured electret heterointerface for resilient and adaptive tele-perception in embodied intelligence. *Matter* **2025**, *8*, 102363. [CrossRef]
32. Gao, H.; Martin, L.S.; Hughes, L.B.; Leitao, N.T.; Put, P.; Zhou, H.; Koyluoglu, N.U.; Meynell, S.A.; Jayich, A.C.B.; Park, H.; et al. Signal amplification in a solid-state sensor through asymmetric many-body echo. *Nature* **2025**, *646*, 68–73. [CrossRef] [PubMed]
33. Kelly, M.T.; Chun, J.K.M.; Bocarsly, A.B. A silicon sensor for SO_2. *Nature* **1996**, *382*, 214–215. [CrossRef]
34. Du, Y.; Shen, P.; Liu, H.; Zhang, Y.; Jia, L.; Pu, X.; Yang, F.; Ren, T.; Chu, D.; Wang, Z.; et al. Multi-receptor skin with highly sensitive tele-perception somatosensory. *Sci. Adv.* **2024**, *10*, eadp8681. [CrossRef] [PubMed]

https://doi.org/10.3390/inorganics14010021

inorganics

MDPI

Article

A Comprehensive Study of the Optical, Structural, and Morphological Properties of Chemically Deposited ZnO Thin Films

Sayra Guadalupe Ruvalcaba-Manzo [1,*], Rafael Ramírez-Bon [2,*], Ramón Ochoa-Landín [3] and Santos Jesús Castillo [1]

[1] Departamento de Investigación en Física, Universidad de Sonora, Hermosillo 83000, Sonora, Mexico; santos.castillo@unison.mx

[2] Centro de Investigación y de Estudios Avanzados del Instituto Politécnico Nacional, Unidad Querétaro, Libramiento Norponiente 2000, Fraccionamiento Real de Juriquilla, Querétaro 76230, Querétaro, Mexico

[3] Departamento de Física, Universidad de Sonora, Hermosillo 83000, Sonora, Mexico; ramon.ochoa@unison.mx

* Correspondence: sayra.manzo@unison.mx (S.G.R.-M.); rrbon@cinvestav.mx (R.R.-B.)

Abstract: Zinc oxide (ZnO) is a wide bandgap semiconductor with optoelectronic and photocatalytic properties, which depend on its optical, structural, and morphological characteristics. In this study, we synthesized ZnO thin films by chemical bath deposition (CBD) and then thermally annealed them at 400 °C and 600 °C to evaluate the effect of thermal treatments. We characterized their structural, optical, morphological, and chemical properties using X-ray diffraction (XRD), UV–Vis spectroscopy, scanning electron microscopy (SEM), and X-ray photoelectron spectroscopy (XPS). The optical bandgap values were 3.20 eV for the as-grown thin films, and 3.23 eV and 3.21 eV after annealing at 400 °C and 600 °C, respectively. SEM micrographs revealed a change from elongated agglomerates in the as-grown thin films to uniform flower-like structures after annealing at 600 °C. XPS analysis confirmed ZnO formation in all samples, and we detected residual precursor species only in the as-grown thin films, which were completely removed by annealing at 600 °C. These results demonstrate that the CBD synthesis of ZnO can tune its optical and morphological properties through thermal annealing, making it suitable for optoelectronic, sensing, and photocatalytic applications.

Keywords: thin films; semiconductors; zinc oxide

1. Introduction

Zinc oxide (ZnO) is an n-type semiconductor with a direct bandgap from 3.20 to 3.40 eV [1,2], making it an attractive material for a wide variety of optoelectronic applications, including light-emitting diodes (LEDs), lasers, and photodetectors, due to its high emission efficiency and thermal stability, which allow it to operate at elevated temperatures [3–6]. Furthermore, ZnO exhibits piezoelectric properties, allowing its use in gas sensors and biosensors [7,8]. Moreover, its broad applicability also extends to renewable energy technologies implemented in solar cells [9] and as a photocatalyst for the degradation of environmental contaminants [10–13]. One challenge commonly reported for ZnO thin films is the aging effect caused by oxygen adsorption at the active surface on the thin film [14–16]. This phenomenon can cause rapid variations in both optical and electrical properties depending on ambient humidity and exposure time [15]. In practical device applications, such effects are typically mitigated through surface passivation strategies such as encapsulation [14,15] or protective coatings [16], which ensure the long-term stability of ZnO-based devices.

Structurally, ZnO is an interesting material that can crystallize in the wurtzite, zincblende, and rocksalt phases. The hexagonal wurtzite phase, being the most thermodynamically stable under ambient conditions, is the most common [17,18]. The literature reports preferred orientations along the (100), (002), and (101) crystalline planes. However, the preferential orientation is not a fixed characteristic of ZnO. The growth parameters, such as temperature and precursor concentration, strongly influence it. For example, the (002) orientation typically dominates at temperatures above 140 °C, whereas the (101) orientation becomes more predominant at lower synthesis temperatures [18].

A variety of synthesis routes have been reported for ZnO thin films, including thermal evaporation [19], spray pyrolysis [8,13], sol–gel [12,20], SILAR (Successive Ionic Layer Adsorption and Reaction) [21,22], spin coating [23], chemical bath deposition (CBD) [5,7,24,25], and sonochemical synthesis [6,10]. However, CBD shows up for its simplicity, low cost, scalability, and ability to produce uniform thin films without the need for complex equipment or high-vacuum environments. Also, CBD offers facile control over sample thickness, stoichiometry, and crystal growth by adjusting parameters such as pH, precursor concentration, and deposition time, qualities that make it attractive for both fundamental research and industrial applications.

In this work, ZnO thin films were synthesized by chemical bath deposition (CBD) using zinc acetate as the metal precursor. We investigated their structural, optical, chemical, and morphological properties using UV–vis diffuse reflectance spectroscopy, X-ray diffraction (XRD), scanning electron microscopy (SEM), and X-ray photoelectron spectroscopy (XPS). Unlike many previous CBD-based studies, this work highlights how annealing not only removes hydroxide-related residuals but also induces morphological reorganization, transforming elongated agglomerates into compact flower-like structures with higher symmetry and uniformity after thermal annealing. These morphological changes, attributed to ZnO complete crystallization and enhanced atomic mobility at higher temperatures, directly impact light scattering, surface reactivity, and carrier transport, which are key factors that govern the performance of ZnO thin films in optoelectronic, sensing, and photocatalytic applications [3–13,26,27]. In the present study, we avoided encapsulation to isolate and analyze the intrinsic effects of thermal annealing on the structure, composition, and morphology of ZnO growth by CBD. However, it is important to note that in device fabrication, encapsulation remains a common strategy to enhance long-term stability. The novelty of this study lies in demonstrating the combined role of precursor selection and thermal annealing conditions in enhancing the structural crystallinity, chemical purity, and surface morphology stability of ZnO samples.

2. Materials and Methods

2.1. Materials

This study synthesized ZnO thin films using zinc acetate dihydrate ($Zn(CH_3COO)_2{\cdot}2H_2O$, purity 99%, CAS No. 5970-45-6) as the zinc precursor. Triethanolamine (TEA, CAS No. 102-71-6) was employed as the complexing agent, while ammonium hydroxide (NH_4OH, 29%, CAS No. 1336-21-6) was used as the hydroxide ion source. All reagents, except NH_4OH, were dissolved in deionized water to prepare homogeneous aqueous solutions and were used without further purification.

2.2. Synthesis Procedure

The chemical bath deposition (CBD) method promotes nucleation and thin-film growth on glass substrates by chemical reaction of a metal ion source, a complexing agent, and a hydroxide ion source.

The synthesis of ZnO thin films involves adding 12 mL of 0.5 M zinc acetate solution and 3 mL of 29% ammonium hydroxide. The ammonium hydroxide ensures a strongly alkaline medium (pH~11), generating hydroxide ions (OH^-) and promoting the precipitation of $Zn(OH)_2$ as an intermediate phase. A high pH drove the hydrolysis of Zn^{2+} ions and prevented uncontrolled precipitation, favoring the subsequent dehydration of $Zn(OH)_2$ into ZnO. Subsequently, the addition of 5 mL of 1 M triethanolamine (TEA) into the solution as a complexing agent regulated molecular assembly and promoted controlled film growth.

All reagents were dissolved in 40 mL of deionized water, initiating deposition on glass substrates at 70 °C for two hours. Following this stage, the substrate was removed from the chemical bath and placed into a newly prepared solution for an additional hour at 70 °C under identical conditions. Although we did not monitor the pH continuously during the process, the initial pH measurement confirmed alkaline conditions throughout the thin film deposition.

The proposed reaction mechanism consists of the following steps:

1. Dissociation of zinc acetate

$$Zn(CH_3COO)_2{\cdot}2H_2O \rightarrow Zn^{2+} + 2CH_3COO^- + 2H_2O, \quad (1)$$

2. Dissociation of ammonium hydroxide

$$NH_4OH \leftrightarrow NH_3 + H_2O \leftrightarrow (NH_4)^+ + OH^-, \quad (2)$$

3. Formation of $Zn(OH)_2$

$$Zn^{2+} + 2OH^- \rightarrow Zn(OH)_2, \quad (3)$$

$$Zn(OH)_2 + 2NH_4OH \rightarrow [Zn(OH)_4]^{2-} + 2NH_4^+, \quad (4)$$

$$[Zn(OH)_4]^{2-} + 2NH_4^+ \rightarrow Zn(OH)_2 + 2NH_3 + 2H_2O, \quad (5)$$

4. Formation of the zinc–TEA complex ion to avoid precipitation and obtain ZnO thin films.

$$Zn^{2+} + C_6H_{15}NO_3 \rightarrow [Zn(C_6H_{15}NO_3)_2]^{2+}, \quad (6)$$

$$[Zn(C_6H_{15}NO_3)_2]^{2+} + 2OH^- \rightarrow Zn(OH)_2 + 2C_6H_{15}NO_3, \quad (7)$$

$$Zn(OH)_2 \rightarrow ZnO + H_2O, \quad (8)$$

Although the reaction involves the dehydration of $Zn(OH)_2$ into ZnO and H_2O, XPS and XRD analyses revealed residual precursor species in the as-grown thin films, which indicates that secondary phases or impurities can coexist under low-temperature CBD conditions. The presence of hydroxide and intermediate species (e.g., $Zn(OH)_2$ and TEA-derived organics) remained in the as-deposited films because chemical bath deposition occurs below the boiling point of water. To eliminate these residues and enhance crystallization of the wurtzite ZnO phase, thermal annealing at 400 °C and 600 °C was applied to the samples. After synthesis, all ZnO samples were dried and stored under ambient laboratory conditions, with a relative humidity of approximately 35–45%. Post-deposition thermal annealing was then applied in each case for two hours, followed by natural cooling to ambient temperature. The resulting thickness of the ZnO films was 5.37 ± 0.11 mm for the as-deposited films, and 4.50 ± 0.30 mm and 4.23 ± 0.21 mm for the samples annealed at 400 and 600 °C. Table A1. Shows the order of addition, molar concentrations, and volumes of the reagents used in ZnO synthesis.

2.3. Characterization

The structural, optical, chemical, and morphological properties of the ZnO thin films were analyzed using various characterization techniques. Optical properties were measured using a Lambda 19 PERKIN ELMER double-beam UV–Vis spectrophotometer (PerkinElmer Inc., Waltham, MA, USA), equipped with an integrating sphere, which is ideal for measuring diffuse reflectance spectra of optically opaque samples in the wavelength range of 300 to 800 nm. X-ray diffraction (XRD) measurements were performed with a Rigaku Geigerflex D/Max-B diffractometer (Rigaku Corporation, Tokyo, Japan), employing a CuKα radiation source (λ = 1.5406 Å) and a graphite monochromator, operated at 40 kV and 20 mA. The chemical composition and surface analysis of the ZnO thin films were performed using a Φ Phi 5100 X-ray Photoelectron Spectrometer (LVAC-PHI, Chigasaki, Kanagawa, Japan; distributed by ULVAC-PHI, Inc., Chanhassen, MN, USA). This system operates under ultra-high vacuum (UHV) conditions, with a base pressure of 2.1×10^{-9} Torr and is equipped with a magnesium X-ray source (Perkin Elmer), operating in the energy range of 0 to 1100 eV. Prior to measurement, argon ion sputtering cleaned the surface of ZnO thin films under UHV conditions. However, the presence of carbon and oxygen peaks appeared due to their adsorption onto the sample surface from the ambient environment. The thickness of the thin films was measured by optical profilometry using a Bruker Contour GT-i optical profilometer (Bruker Nano Surfaces Division, Tucson, AZ, USA). Finally, morphology was analyzed using a Thermo Scientific Phenom ProX scanning electron microscope (Thermo Fisher Scientific, Eindhoven, The Netherlands), which offers an electronic magnification range from 80× to 130,000×, a spatial resolution of $\leq$10 nm, and operates at selectable accelerating voltages of 5 kV, 10 kV, and 15 kV.

The characterization sequence was as follows: diffuse reflectance spectroscopy (UV–Vis) was performed immediately after drying, XRD and SEM analyses were obtained within the following 48 h, and XPS characterization within the next 48 h. This procedure ensured consistent exposure and minimized variability due to environmental factors. By applying an identical protocol for all samples, variations in the results could be attributed to synthesis and thermal annealing conditions rather than environmental exposure.

3. Results and Discussion

The optical properties of the thin films were investigated using the diffuse reflectance technique, from which absorption, reflectance, and transmittance spectra were obtained. In addition, X-ray diffraction (XRD) was employed to determine the stoichiometry and crystalline structure of the ZnO thin films, and the diffraction patterns were compared with standard crystallographic databases (JCPDS). SEM provided a detailed analysis of the surface morphology, revealing changes induced by thermal annealing. Furthermore, XPS analysis was used to identify the oxidation states of the chemical elements present in the ZnO films, as well as to detect residual compounds associated with impurities or secondary phases formed during the synthesis process.

3.1. Optical Properties via Diffuse Reflectance

The spectrum obtained through the diffuse reflectance technique corresponds to the reflectance of the material as a function of wavelength. From these data, the Kubelka–Munk function and transmittance can be obtained, with the Kubelka–Munk function as an approximation of the absorption coefficient of optically opaque samples [28].

The normalized optical spectra of ZnO thin films in the as-grown state and after thermal annealing at 400 °C and 600 °C are presented in Figure 1. Figure 1a shows the absorption spectra, where negligible absorption is observed above ~420 nm, while a sharp absorption edge appears at shorter wavelengths. The position of the edge shifts with annealing temperature, occurring at ~410 nm for the as-grown film, ~402 nm after annealing at 400 °C, and ~400 nm after annealing at 600 °C. Figure 1b displays the transmission spectra, where all films exhibit high transparency (>60%) in the visible range with a steep absorption edge between ~350 and 400 nm. The as-grown sample shows the highest transmittance, while the 600 °C annealed film exhibits the lowest transmittance, consistent with the reflectance increase shown in Figure 1c. The reflectance spectra reveal a sharp rise in the near-UV region associated with the fundamental absorption edge of ZnO. The thin film as grown presents the lowest reflectance across the visible range, whereas annealing at 400 °C leads to a moderate increase, and annealing at 600 °C results in the highest reflectance, particularly above ~450 nm, suggesting changes in morphology and defect concentration [29–33].

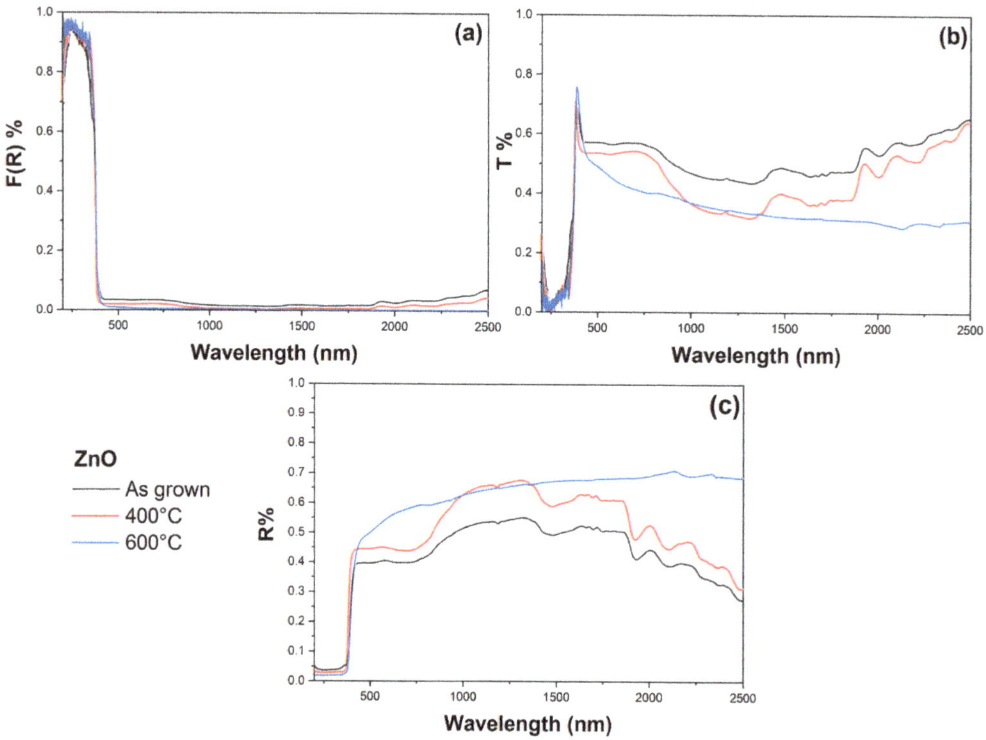

Figure 1. Normalized optical spectra of ZnO samples: (**a**) absorption spectra, (**b**) transmission spectra, and (**c**) reflection spectra for as-grown, thermally annealed films at 400 °C and 600 °C.

Moreover, to determine the optical band gap, the data were analyzed using both the Tauc and derivative methods, which provide an estimate of the energy necessary for electronic transitions between the valence and conduction bands. In the Tauc method, the linear portion of the $(\alpha h\upsilon)^n$ vs. $h\upsilon$ plot is extrapolated to its intersection with the energy axis, yielding the optical band gap value. In contrast, the derivative method involves plotting $\frac{d(\ln\alpha)}{d(h\upsilon)}$ vs. $h\upsilon$, where the maximum of the band corresponds to the band gap energy.

Figure 2 shows the results obtained for ZnO thin films: as-grown, thermal annealed at 400 °C, and annealed at 600 °C. Figure 2a–c correspond to the derivative method, with the maximum value indicating the band gap value for as-grown (3.20 eV), annealed at 400 °C (3.23 eV), and annealed at 600 °C (3.21 eV) ZnO thin films. Figure 2d presents the Tauc analysis $[F(R)h\upsilon]^2$ vs. $h\upsilon$ from which the band gap energies were determined as 3.21 eV,

3.23 eV, and 3.20 eV for the same respective conditions, in agreement with the observed in the absorption edge positions. Minor differences between the visual estimation from the absorption edge and the calculated E_g values can be attributed to band tailing and defect-related states within the gap, as well as spectral broadening effects that smooth the absorption onset. The results obtained are consistent with literature values of band gap for ZnO thin films, which report a direct band gap between 3.20 and 3.40 eV [3,34–37].

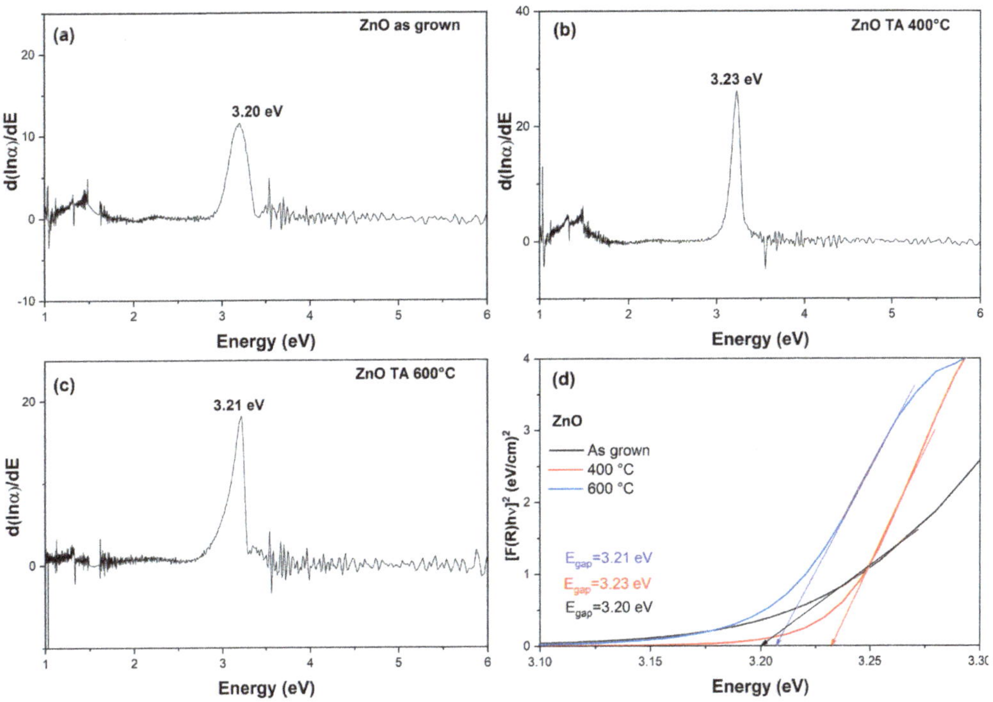

Figure 2. Direct band gap of ZnO thin films determined by (**a**–**c**) the first derivative method and (**d**) the Tauc Method.

Considering the experimental precision of the Tauc method (±0.1 eV), the band gap values of 3.20–3.23 eV can be regarded as independent of thermal annealing temperature. The slightly lower value of 3.20 eV lies at the lower limit of the reported range for ZnO. It is most likely associated with localized states within the band gap, originating from intrinsic defects such as oxygen vacancies [31,32], zinc interstitials [31,32], or other structural imperfections [33]. Although the optical band gap does not change significantly with thermal annealing, complementary XRD and SEM results demonstrate that thermal treatment improves the quality of the ZnO thin films.

3.2. Structural Properties

Figure 3 shows the X-ray diffraction (XRD) patterns of the ZnO thin films. The horizontal axis corresponds to the diffraction angle 2θ (in degrees), while the vertical axis represents the intensity in arbitrary units. The diffraction peaks are well defined and indexed to the (100), (002), (101), (102), (110), (103), (200), (112), and (201) crystallographic planes, with a preferred orientation along the (101) plane [24,37]. Comparison with the crystallographic reference card JCPDS 36-1451 confirms that all samples correspond to a polycrystalline hexagonal ZnO structure [24,34,37], with peak positions falling within the experimental tolerance of ±0.2° in 2θ [24,34]. For the as-grown and 400 °C thin films, some peaks appear slightly shifted relative to the JCPDS 36-1451 card. These shifts are related to lattice strain and residual hydroxyl-related species formed during low-temperature CBD growth, consistent with the XPS evidence of –OH groups adsorbed on the surface. These features do not indicate the presence of a separate crystalline $Zn(OH)_2$ phase but rather point to stress and defect incorporation in the wurtzite ZnO lattice. In contrast, the 600 °C annealed films exhibit well-defined peaks that align closely with the JCPDS 36-1451 reference card, reflecting improved phase purity and crystallinity. The higher intensity of these peaks confirms the beneficial effect of thermal treatment, which enhances atomic mobility and eliminates hydroxyl-related defects.

$$2d\sin\Theta = n\lambda, \tag{9}$$

Using Wulff-Bragg's law (Cu Kα, λ = 1.5406 Å, n = 1), the interplanar distances, d, were calculated in Table 1 for all diffraction peaks. The experimental d values match those of wurtzite ZnO (JCPDS 36-1451) for the diffraction peaks (100), (002), (101), (102), (110), (103), (200), (112), and (201) planes. In addition, the relative intensity and sharpness of the peaks (i.e., their sensitivity) provide evidence of the influence of thermal annealing. For the as-grown and 400 °C thin films, the diffraction peaks are of lower intensity, indicating the presence of lattice strain and structural defects [38]. In contrast, the increase in the intensity of diffraction peaks in the sample annealed at 600 °C is attributed to improved crystallinity and structural quality.

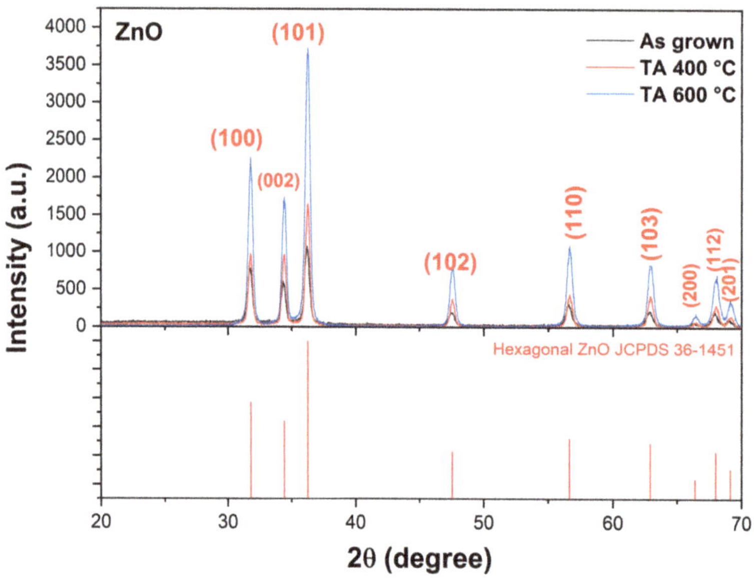

Figure 3. XRD pattern of ZnO thin films compared with the standard crystallographic references of ZnO (JCPDS 36-1451).

Table 1. Experimental diffraction data (2θ, interplanar distances d, and relative intensities) of ZnO thin films as-grown and thermal annealed at 400 °C and 600 °C, compared with reference JCPDS for ZnO (36-1451, $P6_3mc$).

ZnO	Experimental Peak Position 2θ (°)	d_{hkl} (Å)	Intensity (a.u.)	hkl	[a] d_{hkl} (Å)
As grown	31.74	2.8169	784	(1 0 0)	2.8143
	34.34	2.6093	604	(0 0 2)	2.6033
	36.22	2.4781	1085	(1 0 1)	2.4759
	47.46	1.9141	197	(1 0 2)	1.9111
	56.5	1.6274	311	(1 1 0)	1.6247
	62.84	1.4776	224	(1 0 3)	1.4771
	66.3	1.4087	69	(2 0 0)	1.4072
	67.9	1.3793	205	(1 1 2)	1.3782
	68.98	1.3603	107	(2 0 1)	1.3583
TA 400 °C	31.74	2.8169	973	(1 0 0)	2.8143
	34.44	2.602	967	(0 0 2)	2.6033
	36.26	2.4755	1648	(1 0 1)	2.4759
	47.54	1.9111	375	(1 0 2)	1.9111
	56.62	1.6243	445	(1 1 0)	1.6247
	62.88	1.4768	427	(1 0 3)	1.4771
	66.48	1.4053	78	(2 0 0)	1.4072
	67.98	1.3779	300	(1 1 2)	1.3782
	69.12	1.3579	145	(2 0 1)	1.3583
TA 600 °C	31.77	2.8143	2268	(1 0 0)	2.8143
	34.42	2.6034	1739	(0 0 2)	2.6033
	36.26	2.4755	3735	(1 0 1)	2.4759
	47.56	1.9103	820	(1 0 2)	1.9111
	56.62	1.6243	1097	(1 1 0)	1.6247
	62.88	1.4768	838	(1 0 3)	1.4771
	66.44	1.406	189	(2 0 0)	1.4072
	68.06	1.3765	702	(1 1 2)	1.3782
	69.08	1.3585	360	(2 0 1)	1.3583

TA: thermal annealing, [a]: interplanar distances for ZnO obtained from JCPDS 36-1451.

Table 2 summarizes the lattice parameters *a* and c, their uncertainties, and the calculated c/a ratios for the ZnO thin films synthesized as-grown and after thermal annealing at 400 °C and 600 °C. For a hexagonal system, d is related to the Miller indices (h k l) and the lattice constants a and c by:

$$\frac{1}{d^2} = \frac{4}{3} \cdot \frac{h^2 + hk + k^2}{a^2} + \frac{l^2}{c^2}. \tag{10}$$

The experimental lattice parameters confirm the wurtzite structure in all samples. For reference, the crystallographic card JCPDS 36-1451 reports lattice constants of a = 3.24982 Å and c = 5.20661 Å, corresponding to a c/a ratio of 1.602. The experimental values obtained for the as-grown films differ slightly from reference lattice constants, which suggests the presence of residual strain and hydroxide-related incorporation during CBD growth at low temperature. Thermal annealing at 400 °C partially relaxes this strain, while annealing at 600 °C yields lattice parameters almost identical to the reference values reported, confirming improved crystallinity, reduced defect density, and phase purity.

Table 2. Lattice parameters (*a*, *c*), *c*/*a* ratios, and structural phase of as-grown ZnO thin films, annealed at 400 °C and 600 °C, obtained from experimental XRD data.

ZnO Sample [1]	a (Å) $\pm$ σ_a	c (Å) $\pm$ σ_c	c/a	Structural Phase
As grown	3.220 ± 0.016	5.199 ± 0.045	1.614	ZnO (strained)
TA 400 °C	3.247 ± 0.001	5.205 ± 0.003	1.603	ZnO (strained, improved crystallinity)
TA 600 °C	3.247 ± 0.002	5.202 ± 0.005	1.602	ZnO (wurtzite)

[1] Lattice parameters were refined from multiple diffraction peaks using Wulff–Bragg's law and the hexagonal lattice equation. Uncertainties ($\pm\sigma_a$, $\pm\sigma_c$) were obtained by least-squares fitting.

3.3. Morphological Properties

According to the literature, a variety of morphologies have been reported for ZnO depending on the synthesis method and processing conditions. Similar morphologies of ZnO thin films obtained have been observed, including hexagonal or tubular nanorods [5,39,40], plate-like [41,42], granular-shaped [40,42], as well as flower-like [43–48], depending on the thermal treatment and synthesis parameters [32,39–41,43,44]. Figure 4 shows scanning electron microscopy (SEM) images of the ZnO thin films. Figure 4a corresponds to surface micrographs acquired at a working scale of 1 μm. The images reveal clusters of elongated microcrystals with approximate sizes of 5 μm, growing in random directions. Figure 4b micrograph shows a reorganization of the surface morphology compared to the as-grown samples. A more homogeneous distribution of the microcrystal can be observed, which appears to be more densely packed and better adhered to the substrate after the thermal treatment, i.e., well-defined flower-like structures are visible, indicating improved order and compaction. This transformation is associated with the removal of hydroxide-related residues and the enhancement of atomic rearrangements, which promote the growth of larger, more stable ZnO crystals [35,41].

Figure 4. SEM micrographs of ZnO thin films (**a**) as grown, (**b**) with thermal annealing of 400 °C, and (**c**) with thermal annealing of 600 °C.

Figure 4c reveals a notable evolution in the morphology of ZnO samples subjected to thermal annealing at 600 °C. The flower-like structures appear more defined and have greater uniformity in shape and arrangement. The improvement in definition and uniformity is associated with the complete conversion to the wurtzite ZnO phase and an increase in crystallinity. The observed structures suggest increased atomic mobility during the thermal treatment, favoring the formation of crystals with structural quality. Such distinctive flower-like morphologies are particularly attractive due to their large surface area and structural tunability, which broaden the scope of potential applications. In the biomedical field, ZnO flower-like morphology is promising for biosensing [43], antibacterial coatings [43,45], and tissue engineering [43], while in optoelectronic and sensing devices, their unique surface features can improve light scattering, UV shielding [46,47], photocatalytic activity [44,48], and gas detection [49–51].

3.4. Compositional Properties

The low-resolution X-ray photoelectron spectroscopy (XPS) analysis, also known as a survey scan, was performed for the ZnO thin film in the binding energy range from 1100 to 0 eV, as shown in Figure 5. The presence of zinc (Zn), carbon (C), and oxygen (O) was detected. However, carbon and oxygen are considered atmospheric contaminants adsorbed onto the sample surface, and therefore also contribute to the O 1s and C 1s spectral regions.

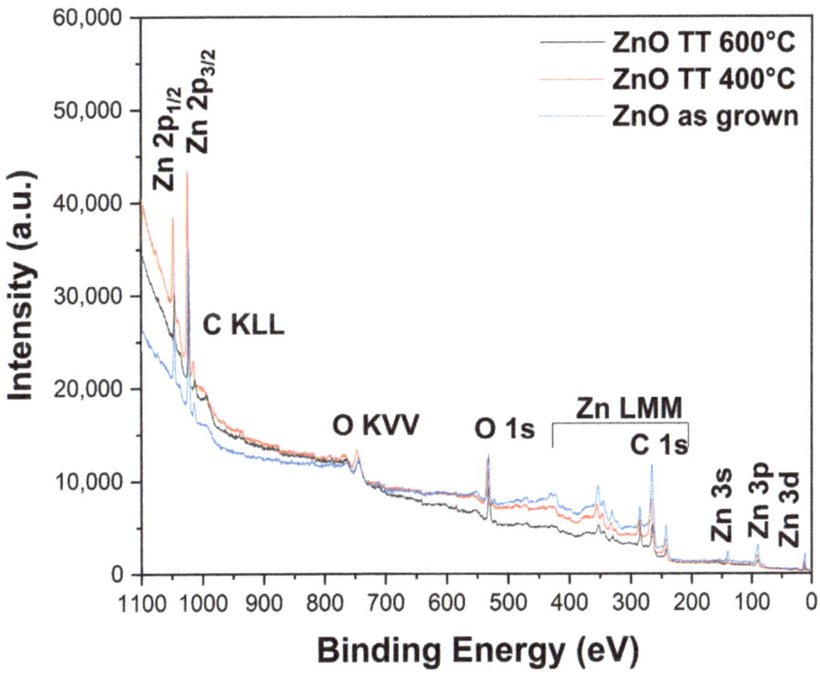

Figure 5. Survey XPS spectrum of ZnO thin films.

To obtain more precise information on the chemical states of the elements, a high-resolution XPS analysis was performed. In this technique, the photoelectron detector is focused on specific binding energy regions associated with the region of the main elements of interest, allowing precise identification of their oxidation states and chemical environments. The high-resolution spectra of the C 1s, O 1s, and Zn 2p states for the ZnO thin films in the as-grown state and after thermal annealing (TA) at 400 °C and 600 °C are presented in Figure 6.

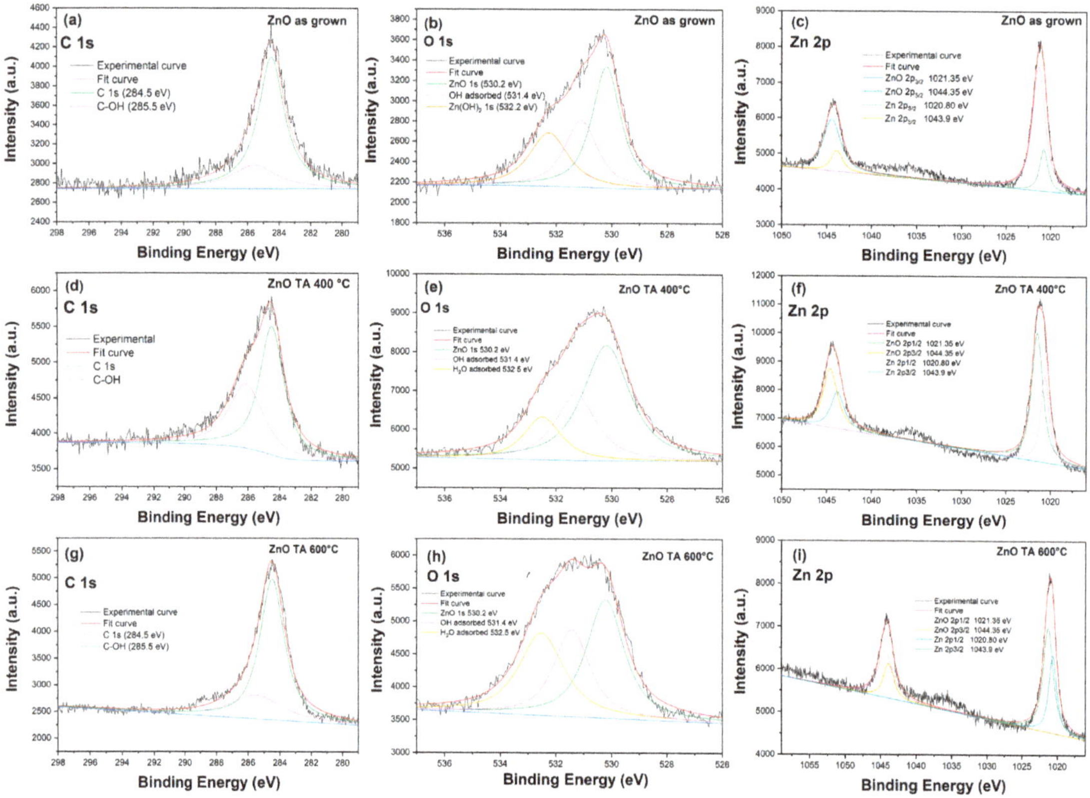

Figure 6. High-resolution XPS spectra of ZnO thin films as grown, annealed at 400 °C and annealed at 600 °C, for the elements (**a**,**d**,**g**) carbon; (**b**,**e**,**h**) oxygen; and (**c**,**f**,**i**) zinc.

In the C 1s region, illustrated in Figure 6a,d,g, the spectra were deconvoluted into two main components: a peak at 284.5 eV, attributed to adventitious carbon contamination, and another at 285.5 eV, associated with C–OH [52]. The O 1s state, shown in Figure 6b,e,h, reveals three oxygen species: attributed to the ZnO formation at ~530.2 eV, OH^- species adsorbed on the surface at ~531.4 eV, and molecular H_2O adsorbed at ~532.5 eV [53–55]. Changes in the relative intensities of the OH^- and H_2O peaks with increasing annealing temperature indicate modifications in surface chemistry and in the degree of hydroxylation or hydration of the ZnO thin films. In the Zn 2p region, presented in Figure 6c,f,i, the spectra exhibit the Zn $2p_3/_2$ and Zn $2p_1/_2$ spin–orbit doublet, located at ~1021.35 eV and ~1044.35 eV, respectively, with an experimental spin–orbit splitting of 23.0 eV for all cases [56–58]. This doublet splitting is consistent with Zn^{2+} in ZnO. The binding energies and the spin–orbit separation after thermal annealing confirm the Zn^{2+} oxidation state, although variations in peak intensities may reflect changes in surface composition or defect density.

4. Conclusions

ZnO thin films were successfully synthesized on glass substrates by chemical bath deposition (CBD) under alkaline conditions adjusted with NH_4OH to maintain a high pH environment for the ZnO formation. Optical, structural, morphological and chemical characterizations confirmed the formation of the hexagonal wurtzite phase of ZnO. XRD analysis revealed that the as-grown thin films exhibit wurtzite ZnO diffraction peaks slightly shifted with respect to the JCPDS reference, attributed to lattice strain and defect incorporation during low-temperature growth. After thermal annealing at 400 °C, diffraction peaks became more intense, indicating a reduction in the strain stress and improved crystallinity, while the sample annealed at 600 °C exhibits a wurtzite ZnO structure with higher intensities of diffraction peaks, confirming crystalline quality.

Tauc plot and derivative methods yielded direct band gap values of 3.20 eV, 3.23 eV, and 3.21 eV for the as-grown ZnO thin films, thermally annealed at 400 °C and at 600 °C, respectively, in agreement with reported literature values. The band gap values remain nearly constant, but the absorption edge sharpness is improved with thermal annealing, which we attribute to reduced defect density and relaxation of residual strain in the ZnO lattice.

SEM micrographs revealed changes in the morphology of the ZnO thin films. The as-grown thin films exhibited elongated agglomerates oriented in random directions, while thermal annealing promoted their reorganization into flower-like structures. At 400 °C, the agglomerates became more densely packed and firmly adhered to the substrate, and at 600 °C, well-defined, highly symmetrical formations were obtained, consistent with enhanced atomic mobility.

XPS confirmed ZnO formation in all samples and detected residual precursor species only in the as-grown thin films. These residues were removed entirely upon thermal annealing. It is important to clarify that the detection of OH species by XPS does not necessarily imply the formation of a crystalline $Zn(OH)_2$ phase. XPS is highly surface-sensitive and can detect adsorbed hydroxyl groups or residual precursor-related species, whereas XRD requires long-range crystallographic periodicity to produce diffraction peaks. Therefore, hydroxyl groups clearly observed in the O 1s spectra do not translate into $Zn(OH)_2$ diffraction peaks in the diffractogram. The main contribution is attributed to wurtzite ZnO, and the possible presence of surface hydroxyl residues is discussed by XPS results.

The thickness of the ZnO films was 5.37 ± 0.11 µm for the as-grown sample and decreased slightly to 4.50 ± 0.30 µm and 4.23 ± 0.21 µm after annealing at 400 °C and 600 °C, respectively, consistent with densification and morphological compaction during thermal treatment.

These results emphasize that while the band gap is stable mainly across thermal annealing conditions, improvements in crystallinity, defect reduction, and morphological reorganization are decisive for enhancing ZnO thin films for optoelectronic, sensing, and photocatalytic applications.

Author Contributions: S.G.R.-M.: conceptualization, investigation, methodology, writing—original draft preparation, funding acquisition, formal analysis; R.R.-B.: validation, resources, supervision, formal analysis, writing—review and editing; R.O.-L.: formal analysis, methodology, investigation; S.J.C.: validation, visualization, supervision. All authors have read and agreed to the published version of the manuscript.

Funding: This research was funded by the SECIHTI Postdoctoral Fellowship Program, Mexico.

Institutional Review Board Statement: Not applicable.

Informed Consent Statement: Not applicable.

Data Availability Statement: The original contributions presented in this study are included in the article. Further inquiries can be directed to the corresponding authors.

Acknowledgments: The authors would like to express their sincere gratitude to Guillermo Suárez Campos, from the Solid-State Laboratory of DIFUS at the Universidad de Sonora, for his valuable support and guidance in the use of the PerkinElmer UV/VIS/NIR Lambda 19 spectrophotometer. We also extend our thanks to the staff of the XPS Laboratory at Universidad de Sonora for their assistance in the XPS characterization of the samples. Special thanks are also extended to the Geochemistry and Scanning Electron Microscopy Laboratory of the Department of Geology, under the direction of Diana Meza Figueroa and Belem González Grijalva, for their valuable contributions in carrying out the characterizations required for this research.

Conflicts of Interest: The authors declare no conflicts of interest.

Abbreviations

The following abbreviations are used in this manuscript:

ZnO	Zinc oxide
CBD	Chemical bath deposition
XRD	X-ray Diffraction
SEM	Scanning electron microscopy
XPS	X-ray photoelectron spectroscopy
UHV	Ultra-high vacuum
TA	Thermal annealing

Appendix A

Appendix A.1

Table A1 summarizes the reagents used for the chemical bath deposition (CBD) of ZnO thin films. The table specifies the chemical precursors, their molar concentrations, and the volumes employed in the synthesis process.

Table A1. Precursors, concentrations, and volumes used for the chemical bath deposition of ZnO thin films.

Reagent	Molar Concentration	Volume
$Zn(CH_3COO)_2$	0.5 M	12 mL
NH_4OH (29%)	-	3 mL
$C_6H_{15}NO_3$	1 M	5 mL
Deionized water	-	40 mL
$Zn(CH_3COO)_2$	0.5 M	12 mL

Appendix A.2

Unit-cell parameters of wurtzite ZnO (space group $P6_3mc$) were refined from multiple Bragg reflections using (10). For each sample, experimental spacings d_i (Cu Kα, $\lambda = 1.5406$ Å) were obtained from 2θ and indexed against JCPDS 36-1451.

A linear least-squares fit of $y_i = 1/d_i^2$ versus $\left[X_{1,i} = \frac{4}{3}(h^2 + hk + k^2), X_i = l^2\right]$ yielded A and B, from which $a = \sqrt{4/(3A)}$ and $c = 1/\sqrt{B}$. Uncertainties were derived from the covariance of the fit and propagated to a and c. The refined parameters and c/a ratios are summarized in Table 2 and discussed in Section 3.2.

References

1. Pawar, V.; Jha, P.K.; Panda, S.K.; Jha, P.A.; Singh, P. Band-gap engineering in ZnO thin films: A combined experimental and theoretical study. *Phys. Rev. Appl.* **2018**, *9*, 054001. [CrossRef]
2. Wisz, G.; Virt, I.; Sagan, P.; Potera, P.; Yavorskyi, R. Structural, optical and electrical properties of zinc oxide layers produced by pulsed laser deposition method. *Nanoscale Res. Lett.* **2017**, *12*, 253. [CrossRef]

3. Pauporte, T.; Lincot, D. Electrodeposition of semiconductors for optoelectronic devices: Results on zinc oxide. *Electrochim. Acta* **2000**, *45*, 3345–3353. [CrossRef]
4. Kaawash, N.M.; Halge, D.I.; Narwade, V.N.; Alegaonkar, P.S.; Bogle, K.A. Rationally designed state-of-the-art approach for enhancing the ultraviolet photon detection performance of ZnO thin film. *Physica B* **2023**, *654*, 414624.
5. Lokhande, C.D.; Gondkar, P.M.; Mane, R.S.; Shinde, V.R.; Han, S.-H. CBD grown ZnO-based gas sensors and dye-sensitized solar cells. *J. Alloys Compd.* **2009**, *475*, 304–311. [CrossRef]
6. Hammed, N.A.; Aziz, A.; Usman, A.I.; Qaeed, M. The sonochemical synthesis of vertically aligned ZnO nanorods and their UV photodetection properties: Effect of ZnO buffer layer. *Ultrason. Sonochem.* **2019**, *50*, 172–181. [CrossRef] [PubMed]
7. Abubakar, S.; Paiman, S.; Tan, T.S.; Rahman, S.F.A. Piezoelectric properties of zinc oxide nanostructure synthesized via chemical bath solution. In Proceedings of the IEEE International Conference on Sensors and Nanotechnology, Penang, Malaysia, 24–25 July 2019; pp. 1–4.
8. Ramos-Serrano, J.; Alcántara-Iniesta, S.; Acosta-Osorno, M.; Calixto, M.E. Growth of highly c-axis oriented ZnO thin films by spray pyrolysis for piezoelectric applications. *Mater. Sci. Semicond. Process.* **2022**, *150*, 106911. [CrossRef]
9. Hodes, G. Semiconductor and ceramic nanoparticle films deposited by chemical bath deposition. *Phys. Chem. Chem. Phys.* **2007**, *9*, 2181–2196. [CrossRef]
10. Yu, H.; Fan, H.; Wang, X.; Wang, J.; Cheng, P. Synthesis of flower-like ZnO nanostructures by sonochemical route and their photocatalytic activity. *Optik* **2015**, *126*, 4397–4400. [CrossRef]
11. Lenka, S.; Badamali, S.K. Nanostructured ZnO as an efficient heterogeneous photocatalyst towards degradation of lignin under visible light irradiation. *Mol. Catal.* **2023**, *545*, 113255. [CrossRef]
12. Thein, M.T.; Pung, S.-Y.; Aziz, A.; Itoh, M. The role of ammonia hydroxide in the formation of ZnO hexagonal nanodisks using sol–gel technique and their photocatalytic study. *J. Exp. Nanosci.* **2014**, *9*, 1078–1080.
13. Bizarro, M. High photocatalytic activity of ZnO and ZnO:Al nanostructured films deposited by spray pyrolysis. *Appl. Catal. B Environ.* **2010**, *96*, 601–608.
14. Rai, K.B.; Yadav, R.P. Tailoring of ZnO thin films: Effect of number of coating and sample ageing. *Int. J. Math. Phys.* **2023**, *14*, 95–102. [CrossRef]
15. Rodriguez-Davila, R.A.; Chapman, R.A.; Shamsi, Z.H.; Castillo, S.J.; Young, C.D.; Quevedo-Lopez, M.A. Low temperature, highly stable ZnO thin-film transistors. *Microelectron. Eng.* **2023**, *279*, 112063.
16. Zhang, W.; Chen, X.Y.; Xu, Z.; Wu, L.; Yang, Y.; Tsang, S.; Chen, S. The Positive Aging Effect of ZnO Nanoparticles Induced by Surface Stabilization. *J. Phys. Chem. Lett.* **2020**, *11*, 5863–5870. [CrossRef]
17. Singh, P.; Kumar, A.; Kaushal, A.; Kaur, D.; Pandey, A.; Goyal, R. In situ high temperature XRD studies of ZnO nanopowder prepared via cost effective ultrasonic mist chemical vapour deposition. *Bull. Mater. Sci.* **2008**, *31*, 573–577. [CrossRef]
18. Cheng, Y.; Yuan, K.; Chen, M. ZnO thin films prepared by atomic layer deposition at various temperatures from 100 to 180 °C with three-pulsed precursors in every growth cycle. *J. Alloys Compd.* **2016**, *685*, 329–335.
19. Utlu, G. Structural investigation of ZnO thin films obtained by annealing after thermal evaporation. *Sakarya Univ. J. Sci.* **2019**, *23*, 1034–1042.
20. Saleh, W.; Saeed, N.; Twej, W.; Alwan, M. Synthesis Sol-Gel Derived Highly Transparent ZnO Thin Films for Optoelectronic Applications. *Adv. Mater. Phys. Chem.* **2021**, *2*, 11–16.
21. Gonugade, M.D.; Powar, S.B.; Salokhe, B.S.; Lokhande, C.D.; Kim, J.H.; Gurav, K.V. SILAR deposited nanocrystalline ZnO films as LPG sensor. *Mater. Today* **2021**, *43*, 2668–2672. [CrossRef]
22. Pawar, J.R.; Dudhamal, M.E.; Chaudhari, V.A.; Bogle, K.; Joshi, R. SILAR synthesized ZnO thin films. *Adv. Mater. Res.* **2022**, *1169*, 87–91. [CrossRef]
23. Popielarski, P.; Mosinska, L.; Bala, W.; Paprocki, K.; Zorenko, Y.; Zorenko, T.; Sypniewska, M. Persistent photoconductivity in ZnO thin films grown on Si substrate by spin coating method. *Opt. Mater.* **2019**, *97*, 109343. [CrossRef]
24. Fu, Y.; Chen, J. Characteristics of zinc oxide film prepared by chemical bath deposition method. *Key Eng. Mater.* **2014**, *609–610*, 739–743. [CrossRef]
25. Kahraman, S.; Bayansal, F.; Çetinkara, H.A.; Çakmak, H.M.; Güder, H.S. Characterization of CBD grown ZnO films with high c-axis orientation. *Mater. Chem. Phys.* **2012**, *134*, 1036–1041.
26. Lee, G. Relationship between crystal structure and photoluminescence properties of ZnO films formed by oxidation of metallic Zn. *Electron. Mater. Lett.* **2010**, *6*, 109–112. [CrossRef]
27. Jiang, K.-B.; Huang, W.-Q.; Song, T.-T.; Wu, P.-X.; Wang, W.-F.; Chen, Q.-S.; Wang, M.-S.; Guo, G.-C. Photobreeding heterojunction on semiconductor materials for enhanced photocatalysis. *Adv. Funct. Mater.* **2023**, *33*, 2305609. [CrossRef]
28. Alcaraz de la Osa, R.; Iparragirre, I.; Ortiz, D.; Saiz, J.M. The extended Kubelka–Munk theory, its application to spectroscopy. *ChemTexts* **2020**, *6*, 2. [CrossRef]
29. Anjum, A.; Ahmed, R.; Umar, Z.; Azzam, S.; Hussain, T.; Sarwar, M.; Baig, M. Structure and defects-related optical properties of highly (002)-oriented zinc oxide thin films. *Phys. B Condens. Matter* **2022**, *644*, 414195.

30. Kaushik, V.; Bhardwaj, K.; Kumar, D.; Kumar, M.; Sharma, S. Effect of various processing parameters on the properties of ZnO thin films. *Hybrid Adv.* **2024**, *7*, 100295. [CrossRef]
31. Clark, S.; Robertson, J.; Lany, S.; Zunger, A. Intrinsic defects in ZnO calculated by screened exchange and hybrid density functionals. *Phys. Rev. B* **2010**, *81*, 115311. [CrossRef]
32. Kamarulzaman, N.; Kasim, M.; Rusdi, R. Band gap narrowing and widening of ZnO nanostructures and doped materials. *Nanoscale Res. Lett.* **2015**, *10*, 1034. [CrossRef]
33. Oba, F.; Nishitani, S.; Adachi, H.; Tanaka, I.; Kohyama, M.; Tanaka, S. Ab initio study of symmetric tilt boundaries in ZnO. *Phys. Rev. B.* **2001**, *63*, 045410.
34. Foo, K.L.; Kashif, M.; Hashim, U.; Liu, W. Effect of different solvents on the structural and optical properties of zinc oxide thin films for optoelectronic applications. *Ceram. Int.* **2014**, *40*, 753–761. [CrossRef]
35. Shaikh, S.K.; Inamdar, S.I.; Ganbavle, V.V.; Rajpure, K.Y. Chemical bath deposited ZnO thin film based UV photoconductive detector. *J. Alloys Compd.* **2016**, *664*, 242–249. [CrossRef]
36. Arab, L.; Amri, A.; Meftah, A.; Latif, A.; Tibermacine, T.; Sengouga, N. Effect of the annealing process on the properties of ZnO thin films prepared by the sol–gel method. *Chem. Phys. Impact* **2023**, *7*, 100124.
37. Al-Gaashani, R.; Radiman, S.; Daud, A.R.; Tabet, N.; Al-Douri, Y. XPS and optical studies of different morphologies of ZnO nanostructures prepared by microwave methods. *Ceram. Int.* **2013**, *39*, 2283–2292. [CrossRef]
38. Cai, R.; Yang, D.; Wu, J.; Zhang, L.; Wu, C.; Chen, X.; Wang, Y.; Wan, S.; Hou, F.; Yan, Q.; et al. Fabrication of ultrathin $Zn(OH)_2$ nanosheets as drug carriers. *Nano Res.* **2016**, *9*, 1780–1790. [CrossRef]
39. Shaban, M.; Zayed, M.; Hamdy, H. Nanostructured ZnO thin films for self-cleaning applications. *RSC Adv.* **2017**, *7*, 617–631. [CrossRef]
40. Tangade, H.; Pusawale, S.; Shirguppikar, S. Synthesis and characterization of ZnO thin films deposited by chemical route. *Mater. Today Proc.* **2020**, *33*, 5147–5149. [CrossRef]
41. Szczęsny, R.; Scigala, A.; Derkowska-Zielinska, B.; Skowroński, L.; Cassagne, C.; Boudebs, G.; Viter, R.; Szłyk, E. Synthesis, optical, and morphological studies of ZnO powders and thin films fabricated by wet chemical methods. *Materials* **2020**, *13*, 2559. [CrossRef]
42. Thool, G.; Singh, A.; Singh, R.; Gupta, A.; Susan, M. Facile synthesis of flat crystal ZnO thin films by solution growth method: A micro-structural investigation. *J. Saudi Chem. Soc.* **2014**, *18*, 712–721. [CrossRef]
43. Krishna, S.; Jakmunee, J.; Mishra, Y.; Prakash, J. ZnO based 0–3D diverse nano-architectures, films and coatings for biomedical applications. *J. Mater. Chem. B* **2024**, *12*, 2950–2984. [CrossRef]
44. Ghamsari, S.; Alamdari, S.; Han, W.; Park, H. Impact of nanostructured thin ZnO film in ultraviolet protection. *Int. J. Nanomed.* **2016**, *12*, 207–216. [CrossRef]
45. Patil, V.; Vanalakar, S.; Vhanalakar, S.; Kamble, A.; Dongale, T.; Kurhe, D.; Kamble, P.; Patil, S.; Shendage, S.; Patil, P.; et al. Chemically synthesized hierarchical flower-like ZnO microstructures. *Z. Phys. Chem.* **2018**, *233*, 1183–1200.
46. Bao, Y.; Lu, G.; Feng, C.; Zhang, J.; Liu, C.; Simion, D. Hollow flower-like ZnO: Synthesis, growth mechanism and application in polyacrylate. *Adv. Powder Technol.* **2020**, *31*, 1975–1984. [CrossRef]
47. Kushwaha, S.; Bahadur, L. Studies of structural and morphological characteristics of flower-like ZnO thin film and its application as photovoltaic material. *Optik* **2013**, *124*, 5696–5701.
48. Ikhmayies, S. Synthesis of flower-like ZnO micro/nano structures by the spray pyrolysis technique. *JOM* **2020**, *72*, 621–627. [CrossRef]
49. Mane, S.; Nimbalkar, A.; Go, J.; Patil, N.; Dhasade, S.; Thombare, J.; Burungale, A.; Shin, J. NO_2 sensing properties of 3D flower-like ZnO nanostructure decorated with thin porous petals synthesized using a simple sol–gel drop-casting method. *Appl. Phys. A* **2021**, *127*, 13.
50. Bhat, P.; Naveen Kumar, S.K.; Nagaraju, P. Fabrication of ultrasensitive hexagonal disc structured ZnO thin film sensor to trace nitric oxide. *J. Asian Ceram. Soc.* **2020**, *9*, 96–105. [CrossRef]
51. Rai, P.; Raj, S.; Ko, K.; Park, K.; Yu, Y. Synthesis of flower-like ZnO microstructures for gas sensor applications. *Sens. Actuators B Chem.* **2013**, *178*, 107–112.
52. Morais, A.; Alves, J.P.C.; Lima, F.A.S.; Lira-Cantu, M.; Nogueira, A.F. Enhanced photovoltaic performance of inverted hybrid bulk-heterojunction solar cells using TiO_2/reduced graphene oxide films as electron transport layers. *J. Photon. Energy* **2015**, *5*, 057408.
53. Mun, B.S.; Liu, Z.; Motin, M.A.; Roy, P.C.; Kim, C.M. In situ observation of H_2 dissociation on the ZnO (0001) surface under high pressure of hydrogen using ambient-pressure XPS. *Int. J. Hydrogen Energy* **2018**, *43*, 8655–8661. [CrossRef]
54. Lu, Y.; Lin, Y.; Wang, D.; Wang, L.; Xie, T.; Jiang, T. A high performance cobalt-doped ZnO visible light photocatalyst and its photogenerated charge transfer properties. *Nano Res.* **2011**, *4*, 1144–1152. [CrossRef]
55. Chen, T.; Liu, S.-Y.; Xie, Q.; Detavernier, C.; Meirhaeghe, R.L.; Qu, X.-P. The effects of deposition temperature and ambient on the physical and electrical performance of DC-sputtered n-ZnO/p-Si heterojunction. *Appl. Phys. A* **2010**, *98*, 357–365.

56. Zhang, H.; Li, W.; Qin, G.; Fang, L.; Ruan, H.; Tan, M.; Wu, F.; Kong, C. Effect of surface carbon contamination on the chemical states of N-doped ZnO thin films. *Appl. Phys. A* **2018**, *124*, 1565. [CrossRef]
57. Claros, M.; Setka, M.; Jimenez, Y.P.; Vallejos, S. AACVD synthesis and characterization of iron and copper oxides modified ZnO structured films. *Nanomaterials* **2020**, *10*, 471. [CrossRef]
58. Lai, Y.; Wang, Y.; Cheng, S.; Yu, J. Defects and resistive switching of zinc oxide nanorods with copper addition grown by hydrothermal method. *J. Electron. Mater.* **2014**, *43*, 2676–2682. [CrossRef]

inorganics

MDPI

Review

Research Progress on the Preparation, Modification, and Applications of g-C_3N_4 in Photocatalysis and Piezoelectric Photocatalysis

Mengyang Li [1,†], Liuqing Yang [1,†], Yizhe Song [1], Hongru Hou [1], Yujie Fang [2], Yucheng Liu [3], Lihao Xie [1] and Dingze Lu [1,*]

1 School of Science, Xi'an Polytechnic University, No. 19 of Jinhua South Road, Beilin District, Xi'an 710048, China; luckymeetyou420@gmail.com (M.L.); yangliuqingbo@163.com (L.Y.); 42308060110@stu.xpu.edu.cn (Y.S.); 42308060204@stu.xpu.edu.cn (H.H.); lhxie1999@163.com (L.X.)
2 Frontiers Science Center for Flexible Electronics, Xi'an Institute of Flexible Electronics (IFE), Northwestern Polytechnical University, Xi'an 710072, China; fangyujie@mail.nwpu.edu.cn
3 Engineering Training Center, Xi'an Polytechnic University, Xi'an 710048, China; liuyucheng@xpu.edu.cn
* Correspondence: ludingze@whu.edu.cn
† These authors contributed equally to this work.

Abstract: The metal-free polymeric semiconductor graphitic carbon nitride (g-C_3N_4) has emerged as a promising material for photocatalytic applications due to its responsiveness to visible light, adjustable electronic structure, and stability. This review systematically summarizes recent advances in preparation strategies, including thermal polycondensation, solvothermal synthesis, and template methods. Additionally, it discusses modification approaches such as heterojunction construction, elemental doping, defect engineering, morphology control, and cocatalyst loading. Furthermore, it explores the diverse applications of g-C_3N_4-based materials in photocatalysis, including hydrogen (H_2) evolution, carbon dioxide (CO_2) reduction, pollutant degradation, and the emerging field of piezoelectric photocatalysis. Particular attention is given to g-C_3N_4 composites that are rationally designed to enhance charge separation and light utilization. Additionally, the synergistic mechanism of photo–piezocatalysis is examined, wherein a mechanically induced piezoelectric field facilitates carrier separation and surface reactions. Despite significant advancements, challenges persist, including limited visible-light absorption, scalability issues, and uncertainties in the multi-field coupling mechanisms. The aim of this review is to provide guidelines for future research that may lead to the development of high-performance and energy-efficient catalytic systems in the context of environmental and energy applications.

Keywords: g-C_3N_4; photocatalysis; piezoelectric photocatalysis; modification; energy conversion

1. Introduction

With the rapid development of human society and the acceleration of modernization, the impact of environmental pollution and energy shortage is deepening [1]. Since the Second Industrial Revolution, the rapid development of the global economy has been accompanied by excessive fossil fuel combustion, uncontrolled industrial wastewater discharge, and the misuse of antibiotics. These actions, akin to an invisible sword, have gradually eroded Earth, which is essential for human survival, triggering a series of severe energy crises and environmental pollution issues. The intensification of the greenhouse effect, the accumulation of heavy metals in soil, the eutrophication of aquatic systems, and other phenomena serve as silent warnings that we are confronting unprecedented ecological challenges. More critically, these pollutants continue to accumulate within the food chain and, even at very low concentrations, can inflict significant harm on the human body, thereby threatening human life and health. For nearly a century, the greenhouse effect has remained a major environmental issue afflicting humanity [2]. Nowadays, China's "dual-carbon" goal has been elevated to the level of national strategy and has become an important part of building a strong modern socialist country [3,4]. Among all gases confirmed to contribute to the greenhouse effect, carbon dioxide (CO_2) is the predominant emitter, responsible for approximately 60% of total greenhouse gas emissions [5]. The effective and efficient utilization of solar energy can significantly decrease the reliance on fossil fuels, thereby contributing to the achievement of the "dual carbon" goals [6]. Zou (2021) summarized international practices for achieving carbon neutrality, categorizing them into four main aspects: carbon substitution, carbon emission reduction, carbon sequestration, and carbon recycling [7]. The escalating energy crisis and environmental pollution underscore the urgency of exploring clean renewable energy carriers and reducing excessive dependence on traditional fossil fuels [8]. In light of this challenge, there is an urgent need to actively develop and utilize alternative clean energy sources. The primary objectives are to reduce carbon emissions, optimize the energy structure, and achieve resource recycling.

In the natural environment, solar energy, waste heat, and mechanical vibrations of water bodies represent significant potential energy sources. Through the implementation of efficient collection and conversion technologies, these forms of energy can be transformed into usable power. This approach not only reduces our dependence on fossil fuels but also addresses the pressing issue of environmental pollution [9–11]. Photocatalysis is a promising strategy for reducing dependence on fossil fuels and addressing environmental pollution issues [12]. The efficient conversion of clean solar energy to chemical energy through semiconductor photocatalytic technology is regarded as a viable technical approach to mitigate energy shortages and environmental pollution [13]. Solar-driven photocatalysis can effectively eliminate environmental pollutants by generating photogenerated electron–hole pairs and triggering a series of redox reactions. However, its practical application is significantly hindered by low photocatalytic activity, primarily due to the high recombination rate of photogenerated carriers [14,15]. The construction of internal electric fields in photocatalytic materials provides a driving force for charge separation, reduces the probability of recombination between photogenerated electron–hole pairs, and enhances the rates of reduction and oxidation reactions on their surfaces [16]. Although various methods have been employed to construct internal electric fields, including the design of heterojunctions, the effects of conventional internal electric fields are often short-lived due to the extreme saturation of external charges and internal photogenerated carriers [17,18]. Therefore, developing a strategy for the sustainable provision of carrier drives is essential. Notably, water bodies contain an abundance of fluctuating mechanical energies and converting these mechanical energies into a cyclic and continuous internal electric field through the piezoelectric effect can assist in the continuous separation of photogenerated charges. The polluted water environment mainly includes organic pollutants and heavy metal ions [19]. Organic pollutants primarily include bisphenol A (BPA), p-chlorophenol (4-CP), rhodamine B (RhB), and methylene blue (MB). Heavy metal ions predominantly consist of hexavalent chromium (Cr^{6+}), mercury ions (Hg^{2+}), trivalent arsenic (As^{3+}), and certain radioactive elements such as uranium (U) and thorium (Th). The mechanical and catalytic environments provided by fluctuations in the aquatic ecosystem, along with natural light, facilitate the piezoelectric photocatalysis process, which enables the harmless degradation of organic pollutants and the reduction of heavy metal ions. Consequently, the integration of the piezoelectric effect and photocatalytic effect presents a pioneering solution for the sustainable degradation of pollutants in aquatic environments (Figure 1). This strategy harnesses renewable energy sources, including solar energy, fluctuating mechanical energy, and thermal energy, resulting in reaction products that do not generate secondary pollution. This approach aligns with the scientific objectives of achieving a green cycle and low carbon emissions, thereby possessing significant research and application value.

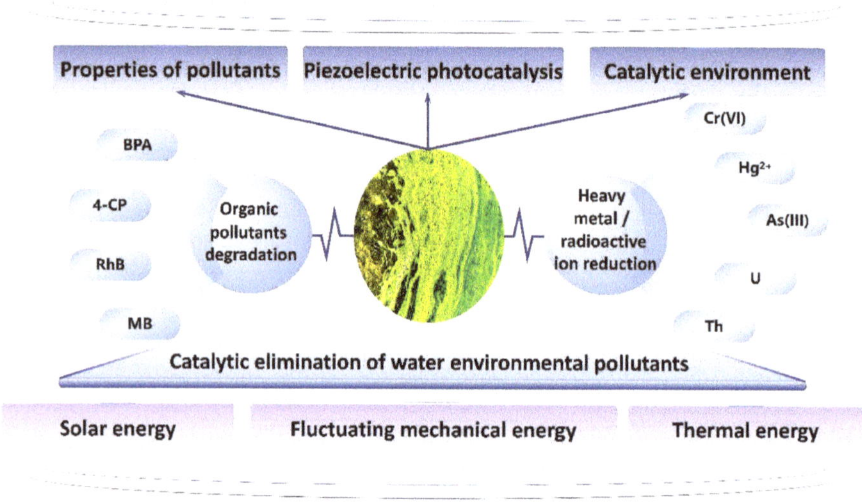

Figure 1. Piezoelectric–photocatalytic solution to water pollution.

The sensitivity and response of photocatalytic outputs from single materials possessing photocatalytic properties are significantly below the actual demand. Therefore, optimizing the design of semiconductor photocatalytic materials, selecting appropriate piezoelectric materials, and integrating the properties of both semiconductors and piezoelectric materials are essential for enhancing carrier transfer efficiency. Such enhancements can improve the piezoelectric output capacity, bringing it closer to actual energy demands. Consequently, the design and development of a nano-piezoelectric photocatalyst that integrates a photosensitive semiconductor with a piezoelectric semiconductor to broaden its application scope is a topic of considerable contemporary interest.

Graphitic carbon nitride (g-C_3N_4) is a novel non-metallic polymer semiconductor material characterized by its unique two-dimensional layered structure and conjugated system. Its appropriate band gap of approximately 2.7 eV endows it with a significant ability to respond to visible light. Furthermore, g-C_3N_4 exhibits remarkable physical and chemical stability, along with distinctive electronic properties, which render it a promising candidate in the field of photocatalysis. Consequently, it has found widespread applications in energy conversion, environmental remediation, and catalysis [20]. In 2009, Wang et al. [21] were the first to apply graphitic carbon nitride (g-C_3N_4) in photocatalysis, demonstrating its capability to catalyze the decomposition of water under visible light irradiation, resulting in the production of H_2 and O_2. Ong et al. [22] synthesized g-C_3N_4 through thermal polymerization using various nitrogen-containing organic compounds. Numerous scholars have conducted continuous research on g-C_3N_4, revealing that this material possesses unique advantages in the field of photocatalysis. g-C_3N_4 can be synthesized from various nitrogen-rich precursors, such as melamine, urea, and dicyandiamide, through a straightforward thermal polymerization method. These raw materials are readily available and suitable for large-scale production [23]. As a metal-free catalyst, g-C_3N_4 is devoid of toxic heavy metal elements and does not generate secondary pollution during its application. It adheres to the principles of green chemistry, and its excellent biocompatibility further enhances its potential applications in biomedical and environmental fields [20]. Compared to traditional wide-band-gap semiconductors like TiO_2 (3.2 eV), g-C_3N_4 features a narrower band gap of approximately 2.7 eV, enabling more effective utilization of visible light for photocatalytic reactions [24]. The optical absorption range, charge separation efficiency, and surface reactivity of g-C_3N_4 can be further optimized through strategies such as nanostructure design, element doping, and heterostructure construction [25]. These advantages position g-C_3N_4 materials as having significant potential for application in environmental remediation, particularly in water treatment and air purification. They can effectively degrade and remove a wide range of pollutants, including organic dyes, antibiotics, heavy metal ions, gaseous pollutants, pesticides, endocrine disruptors, and other organic compounds [20].

However, g-C_3N_4 also presents several limitations, including a small specific surface area, a narrow light response range, and the propensity for rapid recombination of photogenerated electrons and holes, which restrict its photocatalytic performance [26]. The primary reason for the deficiencies in the photocatalytic activity of g-C_3N_4 is its amorphous nature, which leads to a significant number of carrier complexation centers on its surface. Although numerous researchers have successfully synthesized high-crystallinity g-C_3N_4 photocatalysts that exhibit superior capabilities for photogenerated carrier separation, these high-crystalline g-C_3N_4 catalysts often lack reactive sites due to the elimination of many non-condensed amino groups. Consequently, this limitation restricts their efficiency in utilizing photogenerated carriers. To enhance the photocatalytic performance of g-C_3N_4, it is essential to optimize its structure by creating additional reactive sites. However, improvements in photocatalytic performance remain constrained by the modification of a single material, as the inherent issue of a high carrier complexation rate necessitates the development of more advanced structures. Techniques such as semiconductor modification and heterojunction construction can be employed to adjust the properties of individual materials, thereby enhancing the transfer and utilization efficiency of photogenerated carriers and facilitating charge separation between semiconductors. Hossain et al. successfully constructed heterojunctions using TiO_2 and g-C_3N_4, which significantly inhibit the recombination of photogenerated carriers, leading to a marked improvement in photocatalytic activity under UV/visible light irradiation [27]. In polluted water, the flow is rich in mechanical energy, which can be harnessed through the piezoelectric effect to convert mechanical force into electric field force, thereby facilitating the migration of photogenerated carriers. This migration, during the photocatalytic reaction, is also influenced by the Coulomb force. The integration of the piezoelectric effect with photocatalysis offers a viable solution that aligns with the green "dual carbon" initiative. Semiconductor material complexes can establish a built-in electric field through the construction of heterojunctions, promoting the transfer of carriers near the interface. Furthermore, the direction of the polarized electric field generated by the piezoelectric effect periodically changes with the expansion and contraction of the device during the simple harmonic vibration of the piezoelectric material. Under the combined influence of the built-in electric field and the polarized electric field, the migration of carriers can be effectively driven, thereby minimizing the ineffective aggregation of carriers on the surface and enhancing the performance of the catalytic reaction.

This review systematically summarizes recent research progress on graphitic carbon nitride and its composites within the domains of photocatalysis and the emerging field of photo-pressure synergistic catalysis. It emphasizes the regulatory mechanisms associated with various material design strategies, including the construction of composite structures, control of morphology, surface modification, and element doping, and their effects on light absorption, charge carrier dynamics, and catalytic activity. Furthermore, it explores the construction principles, performance advantages, and fundamental synergistic mechanisms of g-C_3N_4-based photo-pressure synergistic catalysts. A comparative analysis of the efficiency of different systems in energy conversion and environmental purification applications is also provided. Additionally, this review critically addresses current research challenges, such as the necessity for a deeper understanding of complex coupling mechanisms, the large-scale synthesis of efficient and stable catalysts, and the adaptability to practical application scenarios. Finally, potential future development directions are discussed, including the creation of intelligent responsive materials, exploration of multi-field coupling mechanisms, and enhancement of energy utilization efficiency. It is anticipated that this review will serve as a valuable reference for future explorations in this field and for the design and development of high-performance catalysts, ultimately advancing green catalytic technology towards practical applications.

2. Introduction to Photocatalysis

2.1. Principles of Photocatalytic Technology

The pioneering technology of photocatalysis emerged in 1972 [28], marked by the discovery of hydrogen generation through water splitting by Fujishima and Honda. They proposed a theory suggesting that solar-powered photocatalysts, specifically oxide semiconductors with an appropriate band gap (E_g), can efficiently absorb light and generate photogenerated electron holes [29]. Since then, the application of photocatalytic technology to the degradation of organic pollutants has garnered significant attention. Over time, the applications of this technology have expanded extensively, encompassing not only the degradation of organic pollutants but also the photolysis of water to produce hydrogen, CO_2 reduction, heavy metal remediation, and disinfection and sterilization [30,31]. The application of photocatalytic technology in water treatment enhances water quality safety and addresses the global issue of water scarcity [32].

Photocatalysis refers to a series of photochemical reactions that occur under the influence of visible, ultraviolet, and even infrared light, facilitated by photocatalysts. This process involves the use of semiconductor materials to generate active species with oxidation and reduction capabilities under light exposure, resulting in the transformation of pollutants [33]. To date, research on the fundamental mechanisms of photocatalysis has reached a relatively advanced stage. Essentially, photocatalysis can be categorized into four primary stages: light collection, charge excitation, charge separation and transfer, and surface electrocatalytic reactions [34,35]. As illustrated in Figure 2, the core of photocatalytic degradation of organic matter lies in the interaction between active species and reactants. Thus, the effective utilization of visible light, efficient separation of photogenerated carriers, and the enhancement of the exposed specific surface area at various reaction sites are crucial for increasing the number of active species. These aspects have become focal points for further research by many experts and scholars, as well as challenges faced by photocatalytic technology.

2.2. Problems Faced by Photocatalytic Technology

Despite extensive research on photocatalytic technology, its practical application is still hindered by low quantum efficiency, complex catalyst design, and limited light response ranges [36]. The key challenges include low quantum efficiency resulting from the rapid recombination of photogenerated electron–hole pairs on picosecond to nanosecond timescales, which reduces carrier lifetimes below the requirements for effective reactions [37,38]; insufficient light utilization due to the wide band gaps of most materials, which restricts absorption primarily to ultraviolet light while underutilizing the visible and near-infrared spectra [39]; complex catalyst design characterized by ambiguous active sites and unclear synergistic mechanisms in composite materials; difficulties in energy band regulation, necessitating comprehensive control over composition, structure, and phase state; and practical application barriers such as high preparation costs, poor reproducibility, and limitations in scalability.

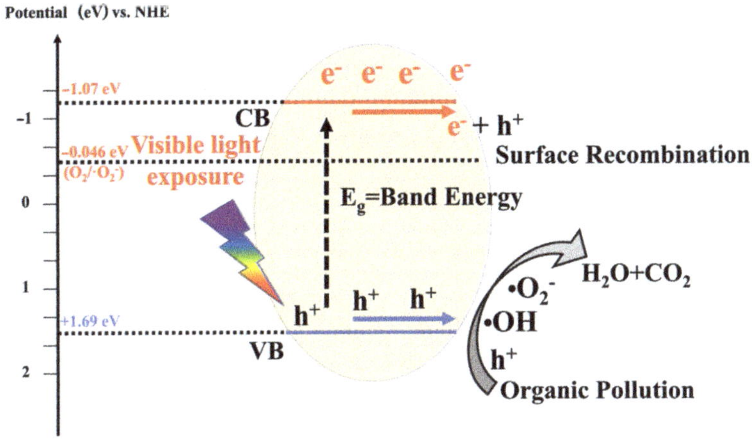

Figure 2. Photocatalytic reaction mechanism.

Addressing these issues necessitates a robust foundation in the research of structure–activity relationships and reaction mechanisms. It also requires the utilization of computational simulations and advanced analytics to facilitate precise catalyst design. Furthermore, fostering interdisciplinary collaboration is essential to integrate cross-domain knowledge for the advancement of material development.

2.3. Classification and Research Status of Common Photocatalysts

The advancement of photocatalytic technology requires a careful balance between enhancing light utilization efficiency and preserving the redox capabilities of materials. Photocatalytic materials have traditionally been classified into four categories: oxides, sulfides, nitrides, and emerging nanomaterials.

Although metal oxides (e.g., TiO_2, ZnO) are widely utilized, their development is hindered by the toxicity of certain metallic elements and low reaction selectivity. Sulfides enhance visible-light absorption due to their narrow band gaps; however, they are plagued by severe photocorrosion phenomena, which necessitate improvements through strategies such as heterojunction construction. Metal sulfides are a class of highly dynamic semiconductor photocatalytic materials, characterized by cations primarily composed of metals with d10 electronic configurations, such as Zn, Cd, Cu, Ag, In, and Sn (Figure 3a). In metal sulfides, the cations hybridize with the atomic orbitals of $S2^{-}$, resulting in numerous closely spaced molecular orbitals, which form conduction bands (CBs) and valence bands (VBs) [40]. The band edge, defined as the bottom of the CB and the top of the VB, is primarily characterized by the outermost orbitals of atoms. The atomic energy of the outermost orbital (np) for elements in the sulfur family typically increases as one moves down the group (as illustrated in Figure 3b, with energy values sourced from the NIST database). Figure 3c presents the electronic structure of cadmium sulfide (CdS). Notably, the conduction band energy of most metal sulfide semiconductors is negative, indicating their strong reducing capability, akin to that of metal oxides. Furthermore, the oxidation capacity of most metal sulfides is generally weaker than that of metal oxides, as the valence band position contributed by the sulfur 3p orbitals is higher than that of the oxygen 2p orbitals (refer to Figure 3d). Consequently, metal sulfides are characterized by a strong reducing ability, relatively weak oxidizing ability, and a narrow band gap. Nitrides (e.g., C_3N_4) can generate highly reactive electron–hole pairs owing to their wide band gap structures, demonstrating significant potential in catalytic efficiency. They have been extensively studied for applications including water treatment, air purification, and bacterial disinfection (Figure 4). Nevertheless, challenges persist, including a limited light-response range, a high charge carrier recombination rate, and elevated preparation costs. The applications of emerging nanomaterials (e.g., MOFs, MXenes) are also expanding.

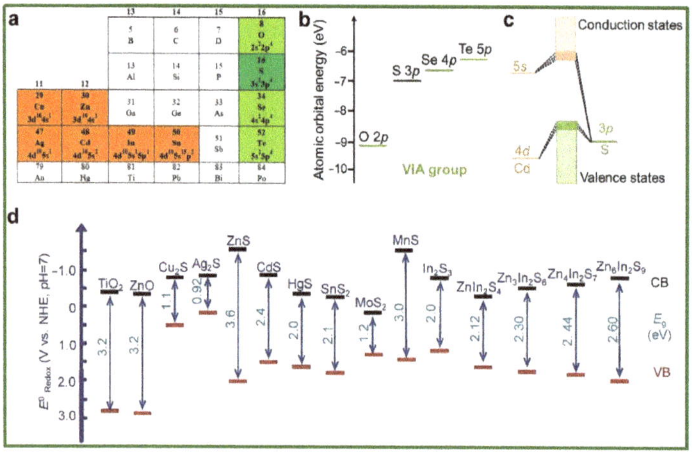

Figure 3. (**a**) The main group of metal sulfides (highlighted); (**b**) atomic orbital energies of O, S, Se, and Te; (**c**) the electronic structure of CdS; and (**d**) the energy band positions of some typical oxides and sulfides [40] (reprint from Wu, X.; Xie, S.; Zhang, H.; et al. Metal Sulfide Photocatalysts for Lignocellulose Valorization. *Adv. Mater.* 2021, 33, 2007129).

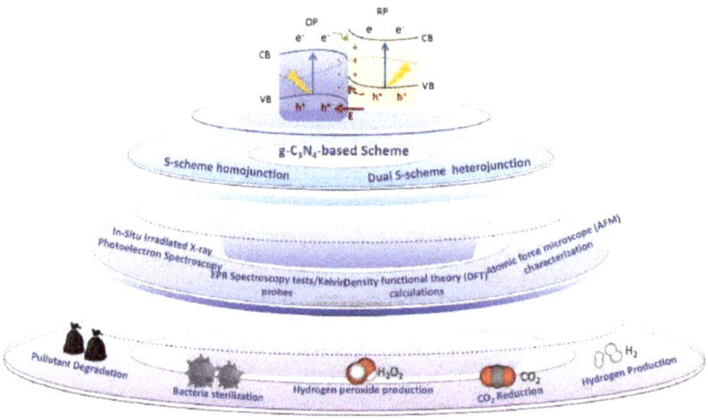

Figure 4. S-type heterostructured g-C_3N_4 photocatalysts for photocatalytic environmental pollutant abatement and sterilization.

Among numerous catalysts, graphitic carbon nitride (g-C_3N_4) has emerged as a focal point of research due to its cost-effective raw materials, namely, urea and melamine, as well as its straightforward single-step thermal polymerization preparation method. Additionally, g-C_3N_4 exhibits a remarkable combination of exceptional redox performance and structural stability. Future research will focus on optimizing its structural design to achieve a synergistic enhancement of light absorption and redox activity.

3. Study Progress of the g-C_3N_4 Photocatalysts

3.1. Introduction to g-C_3N_4 Photocatalysts

Carbon nitride is composed of carbon (C) and nitrogen (N), both of which are abundant on Earth [41]. Active carbon and nitrogen in soil serve as crucial components of the soil's carbon and nitrogen reservoirs, playing a vital role in the carbon and nitrogen cycles within ecosystems [42]. Moreover, C and N are essential elements in living organisms, combining with hydrogen (H) and oxygen (O) to form organic molecular compounds such as proteins, nucleic acids, and amino acids. According to covalent bonding theory, C_3N_4 represents the optimal chemical formula for the formation of nitrides. Graphitic carbon nitride (g-C_3N_4) is a novel class of non-metallic polymer semiconductors that function as a visible light-responsive photocatalyst due to their unique optical properties [43]. In contrast to traditional semiconductor photocatalytic materials, the intentional synthesis of g-C_3N_4 photocatalysts primarily emphasizes the construction of catalytic sites and the modulation of charge transfer dynamics [44]. g-C_3N_4 offers several advantages, including a relatively simple preparation method, low-cost and readily available raw materials, corrosion and high-temperature resistance, stability, good chemical and thermal stability, and non-toxicity, indicating a promising application in energy and environmental fields. The band gap of g-C_3N_4 is approximately 2.7 eV, with the conduction band minimum around −1.1 eV and the valence band maximum at approximately +1.6 eV. g-C_3N_4 demonstrating excellent responsiveness to visible light, making it superior to other semiconductors and increasingly significant in photocatalytic technology.

g-C_3N_4 exhibits a distinctive two-dimensional structure composed of carbon and nitrogen atoms, imparting unique structural characteristics essential for its photocatalytic efficiency. The core structure consists of stacked layers of carbon and nitrogen atoms arranged in a planar hexagonal lattice, akin to that of graphite, thus earning the designation "graphite-like" [45]. The structure consists of layers formed by tri-s-triazine (C_3N_3) units. Each layer is composed of these tri-s-triazine units, which are held together by weak van der Waals forces (Figure 5). This lamellar structure enhances the specific surface area, thereby providing a greater number of active and reactive sites for catalytic reactions [46]. In the g-C_3N_4 lattice, the carbon atoms are sp^2 hybridized, which contributes to the planar triangular geometry of the crystal structure. g-C_3N_4 consists of two fundamental structural units: the triazine ring and the heptazine ring. Notably, calculations based on density functional theory indicate that the heptazine ring (triple-s-triazine) exhibits greater stability than the triazine ring. Furthermore, g-C_3N_4 can adsorb various functional groups, including amino, hydroxyl, and carboxyl groups, on its surface, allowing for modification through adsorption. These functional groups enhance the surface charge, hydrophilicity, and chemical reactivity of the material, rendering it suitable for specific photocatalytic applications [47–49]. The band gap of g-C_3N_4 is measured at 2.7 eV at pH 7.0, which significantly enhances its applicability. This band gap enables g-C_3N_4 to be excited by visible light with an excitation wavelength of approximately 475 nm, a crucial factor for photocatalytic reactions. Recent research has established effective control over the growth mechanism of g-C_3N_4 and the customized design of its structure. By adjusting parameters such as layer spacing, band gap, and surface chemistry, g-C_3N_4 emerges as a formidable candidate for addressing future environmental and energy challenges, thereby contributing significantly to a sustainable and cleaner future.

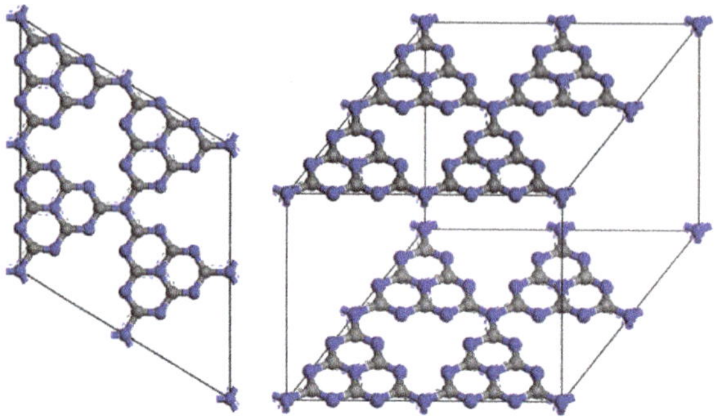

Figure 5. Crystal structure of g-C_3N_4.

3.2. *Preparation Method of g-C_3N_4 Photocatalysts*

The synthesis of graphitic phase carbon nitride (g-C_3N_4) has experienced significant advancements since its initial proposal, enabling researchers to produce g-C_3N_4 materials with various morphologies and structures. These developments allow for the tailoring and design of desired properties for photocatalytic applications through diverse preparation methods. Notably, g-C_3N_4 has not been discovered in nature and must be synthesized using artificial methods. The mainstream synthetic preparation techniques, illustrated in Figure 6, include thermal polycondensation [50–52], chemical vapor deposition (CVD) [53,54], solvothermal versus hydrothermal methods [55,56], and the template-assisted synthesis method [57]. The raw materials utilized in these processes include melamine, urea, cyanamide, dicyandiamide, thiourea, and cyanuric acid.

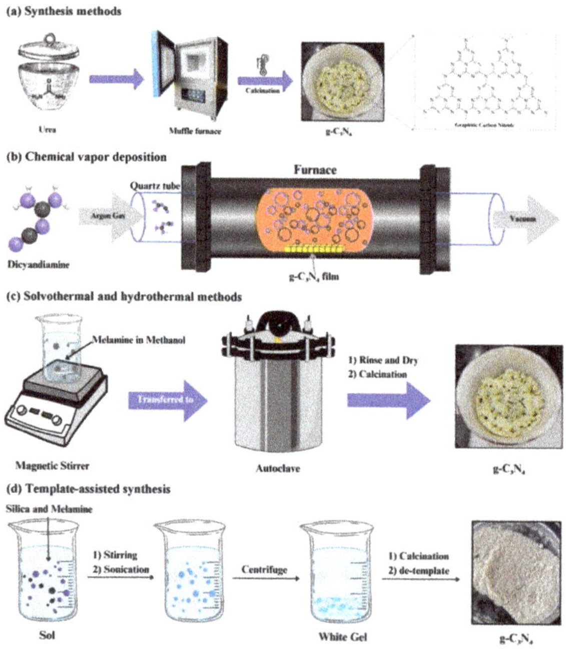

Figure 6. Synthesis of graphite-phase carbon nitride (g-C_3N_4) by (**a**) thermal polycondensation [50–52]; (**b**) chemical vapor deposition [53,54]; (**c**) solvothermal versus hydrothermal [55,56] and (**d**) template-assisted synthesis [57].

Chemical vapor deposition (CVD) involves the vaporization of precursors containing carbon and nitrogen in a vacuum environment, allowing for chemical reactions to occur on the surface of a heated substrate, resulting in the deposition of a g-C_3N_4 film. CVD can produce g-C_3N_4 samples with exceptionally high crystallinity, and these samples exhibit carrier mobility that is several times greater than that of samples prepared using traditional thermal polymerization methods [58]. The high crystallinity of the material significantly reduces both internal and surface defects, effectively inhibiting the recombination of photogenerated electron–hole pairs, thereby greatly enhancing photocatalytic performance. Additionally, g-C_3N_4 synthesized via the CVD method exhibits excellent uniformity and potential for large-area fabrication. Researchers have demonstrated that a uniform film with a thickness fluctuation of less than 5% can be achieved through CVD deposition on a planar substrate. This uniformity is crucial for the development of high-performance optoelectronic devices [59]. However, traditional chemical vapor deposition (CVD) systems necessitate high vacuum conditions, sophisticated gas flow control systems, and high-temperature heating devices, leading to significantly higher equipment investment and maintenance costs compared to other methods. Furthermore, the CVD process typically demands high energy consumption to sustain the elevated temperature and vacuum environment, which further escalates production costs. Another limiting factor is the relatively low growth rate associated with CVD. Due to the reliance on surface reactions of gas-phase precursors, the deposition rate of g-C_3N_4 produced via the CVD method is generally only a few microns per hour. This results in the need for extended processing times to prepare thick film materials, thereby reducing production efficiency. In practical applications, g-C_3N_4 synthesized through CVD is primarily utilized in high value-added fields, such as high-performance optoelectronic devices, precision sensors, and efficient catalytic systems.

The thermal polycondensation method is the most traditional and widely utilized technique for the preparation of g-C_3N_4. This method involves the thermally induced polycondensation of organic precursors containing carbon and nitrogen—such as urea, dicyandiamide, and melamine—at elevated temperatures typically ranging from 500 to 600 °C, resulting in the formation of a g-C_3N_4 polymer network. A significant advantage of thermal polycondensation lies in its simplicity and minimal equipment requirements. In contrast to the conventional CVD method, which necessitates a complex vacuum system, the thermal polycondensation method can be executed using a standard muffle furnace or tube furnace. This considerably reduces both equipment investment and maintenance costs, thereby enabling the preparation of g-C_3N_4 materials even in laboratories with limited research facilities, which in turn fosters extensive research on this material. However, g-C_3N_4 synthesized via the traditional thermal polycondensation method exhibits issues such as low crystallinity and a restricted specific surface area. Various structural defects, including nitrogen vacancies, carbon vacancies, and incomplete heptazine units, are likely to arise during high-temperature polymerization. These defects serve as recombination centers for photogenerated electron–hole pairs, consequently diminishing photocatalytic efficiency [58]. The bulk g-C_3N_4 synthesized through traditional methods typically exhibits a limited specific surface area, generally less than 10 m^2/g, and a reduced number of exposed active sites. These characteristics significantly constrain its catalytic performance [25]. The thermal condensation method is currently the sole approach for achieving large-scale industrial production of g-C_3N_4. This method is particularly significant in the domain of photocatalytic environmental remediation, where g-C_3N_4 synthesized through thermal polymerization has demonstrated substantial practical application value. Results indicate that the g-C_3N_4 photocatalyst, produced by an enhanced thermal condensation method, operates steadily in a custom-designed continuous flow device for a duration of 20 days under real outdoor sunlight, consistently maintaining a pollutant removal efficiency of nearly 100% [60]. Future research on thermal condensation will concentrate on optimizing the process further and minimizing energy consumption. This includes the development of low-temperature polymerization processes, the utilization of microwave-assisted heating to enhance thermal efficiency, and the investigation of more economical precursor sources.

The solvothermal and hydrothermal methods are categorized under liquid-phase synthesis technologies. By heating the precursor solution in a high-pressure reactor, these methods leverage the unique properties of the solvent under high-temperature and high-pressure conditions to facilitate the condensation reactions of carbon and nitrogen precursors, ultimately leading to the formation of g-C_3N_4. A significant advantage of the solvothermal and hydrothermal methods is their capability to synthesize g-C_3N_4 at relatively low temperatures. In contrast to the traditional thermal condensation method, which necessitates temperatures ranging from 500 to 600 °C, the hydrothermal and solvothermal methods typically achieve precursor polycondensation within the temperature range of 200 to 300 °C, thereby considerably reducing energy consumption. However, despite their numerous advantages, the low yield and crystallinity associated with the solvothermal and hydrothermal methods pose challenges for large-scale applications. The reaction occurring in a liquid-phase environment limits the processing capacity due to the volume of the solvent. Typically, laboratory-scale hydrothermal reactors can process only tens to hundreds of milliliters of precursor solution at a time, which restricts the yield of g-C_3N_4 [59]. In terms of application, g-C_3N_4 prepared via solvothermal or hydrothermal methods primarily demonstrates potential in high-performance fields. For instance, studies indicate that g-C_3N_4 nanosheets treated using the hydrothermal method exhibit significant activity in photocatalytic hydrogen production, achieving H_2 yields of 20–60 μmol·h^{-1}·g^{-1}. Regarding CO_2 photoreduction, g-C_3N_4-based catalysts synthesized through the solvothermal method have shown remarkable product selectivity, with CH_4 yields increasing threefold [61]. However, these applications remain in the laboratory research phase, and there is still a significant gap to bridge before achieving large-scale industrial implementation.

Template-assisted synthesis is a preparation method that facilitates the formation of specific structures in g-C_3N_4 by utilizing a template. This method can be categorized into hard template and soft template approaches based on the type of template employed. Hard templates typically consist of inorganic materials with regular pore structures, such as mesoporous SiO_2, while soft templates utilize organic molecules, including surfactants or block copolymers. The template method enables precise control over the pore structure and morphology of g-C_3N_4 at the nanoscale, making it the predominant technique for preparing porous g-C_3N_4. A significant advantage of template-assisted synthesis is its ability to accurately regulate the pore structure and specific surface area of g-C_3N_4. However, this method is also associated with certain limitations, including a complex processing workflow and high costs, as well as the potential for structural damage to g-C_3N_4 during the template removal phase. Consequently, this technique is primarily suited for laboratory research and high value-added applications.

The introduction of the four preparation methods reveals that the thermal polycondensation method has emerged as a more widely utilized approach in practical applications, owing to its straightforward process, favorable preparation conditions, and cost-effectiveness. Future large-scale production should prioritize process optimization, the selection of precursors (such as the development of economical precursor systems), the design of specialized reactors, waste management during production, and the establishment of recycling processes.

3.3. Defects of g-C_3N_4 Photocatalysts

g-C_3N_4 has garnered significant attention due to its remarkable thermal and chemical stability, unique electronic structure, responsiveness to visible light, and various other advantages [6]. This material has a wide range of applications in photocatalysis, bioimaging, optoelectronic sensors, and photovoltaic solar cells. It is undoubtedly a photocatalyst with exceptional capabilities and potential. However, as a standalone photocatalyst, it faces several challenges in transitioning from experimental settings to practical applications: (1) Limited utilization of the effective visible light spectrum: The band gap of 2.7 eV is broader than that of typical semiconductor photocatalysts, rendering it more sensitive to visible light. This wider band gap restricts its responsiveness to visible light. (2) Low quantum efficiency: While many semiconductors exhibit photocatalytic properties, their quantum efficiency is typically low. This limitation arises because the photogenerated conduction band electrons and valence band holes in semiconductors tend to recombine easily, making it difficult for them to react with organic substances on solid surfaces [62]. The photogenerated carriers in g-C_3N_4, following excitation by visible light, are prone to spontaneous complexation, which complicates the assurance of photostability. (3) Photocorrosion and aggregation present significant challenges: g-C_3N_4, typically a yellowish powder, is susceptible to hole oxidation or electron reduction when exposed to visible light. This photocorrosion and aggregation can lead to catalyst deactivation, as the oxidation of holes or reduction of electrons in g-C_3N_4 under visible light irradiation occurs readily, resulting in a decrease in specific surface area due to increased aggregation and stacking of the layered morphology. (4) Achieving high selectivity in complex reaction systems remains challenging; the fine-tuning of the material's surface design and optimization limitations hinder g-C_3N_4 from attaining exceptional selectivity during catalytic reactions. In summary, considering these challenges, researchers continue to strive towards developing innovative approaches aimed at cost-effective fabrication methods and seamless integration with existing infrastructures. Currently, various modifications of g-C_3N_4 have been explored to increase its specific surface area, broaden its photoresponsive range, and enhance the efficiency of photogenerated carrier separation and migration.

4. g-C_3N_4 Photocatalyst Modification Measures

In recent years, semiconductor photocatalysis has emerged as a prominent research focus within the fields of functional ceramic materials, photochemistry, environmental protection, and biotechnology. However, the short lifetime of photogenerated carriers, low quantum efficiency, and limited visible light absorption of monolithic semiconductor photocatalysts continue to impede their further development [63]. Over the past decade, researchers have developed various strategies to enhance the photocatalytic performance of g-C_3N_4, including defect engineering, elemental doping, structural modulation, and the construction of heterojunctions.

4.1. Formal Modification

The morphology control method is one of the most commonly employed techniques; however, g-C_3N_4 synthesized via the traditional thermal polymerization method typically exists in bulk form, resulting in a limited specific surface area and a paucity of photocatalytic active sites. By fabricating g-C_3N_4 into spherical, nanosheet, or porous structures, one can significantly enhance both the specific surface area and the number of active sites, thereby improving its photocatalytic performance. Furthermore, varying the calcination temperatures allows for the regulation of the final exposed morphology and pore size of g-C_3N_4, with a positive correlation generally observed between pore size and specific surface area. As shown in Figure 7, Li et al. [64] fabricated porous g-C_3N_4 nanosheets

(CNHS) with a thickness of 9.2 nm, featuring abundant micropores (1–2 nm), mesopores (2–50 nm), and macropores (50–100 nm) by thermally exfoliating native g-C_3N_4 (CNB) for a continuous duration of 6 h at 520 °C. The resulting specific surface area was 277.98 m^2/g, nearly 26 times greater than that of CNB, which measured 10.89 m^2/g.

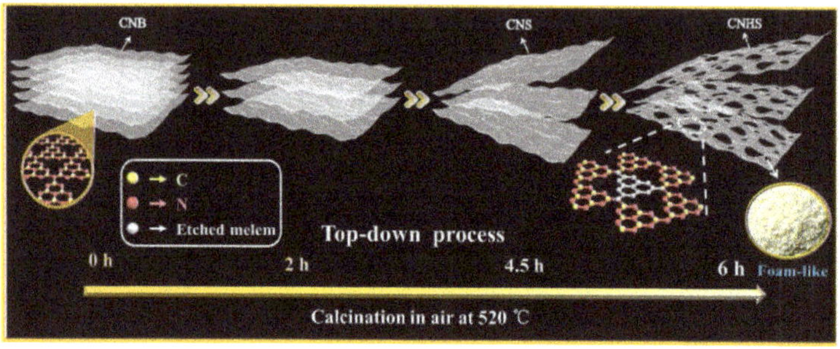

Figure 7. Preparation of foamy porous ultrathin g-C_3N_4 nanosheets by top-down method [64] (reprint from Li, Y.; Jin, R.; Xing, Y.; et al. Macroscopic Foam-Like Holey Ultrathin g-C_3N_4 Nanosheets for Drastic Improvement of Visible-Light Photocatalytic Activity. *Adv. Energy Mater.* 2016, 6, 1601273).

The significant presence of in-plane holes greatly enhances mass transfer and improves the mobility of photogenerated charges. Additionally, the larger pore size facilitates electron transfer orbitals, further improving the mobility of photogenerated charges. As shown in Figure 8, Xiao et al. [65] demonstrated that by controlling the photocatalytic pathway, efficient inactivation of Escherichia coli was achieved using amino acid-rich porous ultrathin g-C_3N_4 nanosheets (AHUCN). This method yielded the highest inactivation efficiency among monolithic g-C_3N_4 structures. The unique characteristics of the porous ultrathin nanosheets, combined with their amino-acid-rich surface, provided synergistic advantages by offering more active reaction sites, facilitating the separation of photogenerated electrons and holes, accelerating photoelectron transfer, and enhancing the yield of photocatalytic H_2O_2.

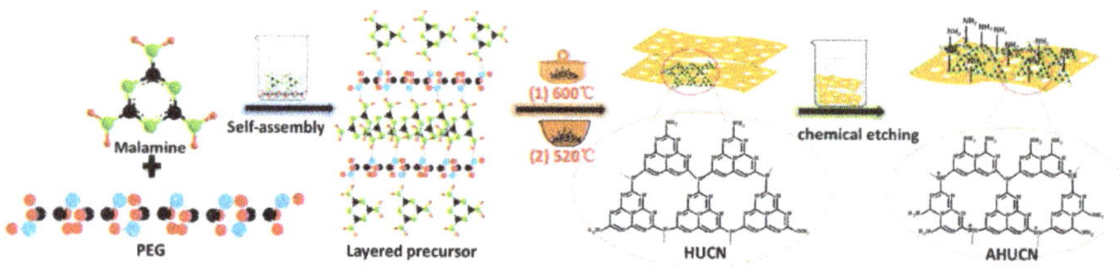

Figure 8. Schematic diagram of the synthesis process of AHUCN [65] (reprint from Xiao, J.; Liu, Q.; Song, M.; et al. Directing Photocatalytic Pathway to Exceedingly High Antibacterial Activity in Water by Functionalizing Holey Ultrathin Nanosheets of Graphitic Carbon Nitride. *Water Res.* 2021, 198, 117125).

4.2. Defect Engineering

Defect engineering in g-C_3N_4 primarily involves the creation of nitrogen vacancies (N-vacancies) and carbon vacancies (C-vacancies). The introduction of nitrogen vacancies generates defect energy levels above the valence band of g-C_3N_4, thereby narrowing the band gap and enhancing its light capture capability. Conversely, the introduction of carbon vacancies can establish an intermediate energy level below the conduction band, facilitating long-wavelength photon absorption. Gao et al. [66] synthesized carbon defects (carbon vacancies) and amino groups in situ on the surface of ultrathin g-C_3N_4 nanosheets via thermal polymerization of urea aqueous solution and gas shock stripping. These carbon defects elevate the conduction band potential, promoting the reduction of protons by photogenerated electrons to generate hydrogen. The amino group enhances hydrophilicity and dispersibility, and its hydrogen production activity is 57 times greater than that of bulk g-C_3N_4. Xue et al. [67] developed g-C_3N_4 (g-C_3N_4-N_{3C}) with three coordinated N vacancies. These N vacancies contribute to narrowing the band gap, improving light capture efficiency, accelerating the separation and transfer of photogenerated carriers, and providing additional active sites. Experimental results confirm that g-C_3N_4-N_{3C} exhibits a commendable photocatalytic N_2 fixation rate of 1915 μmol/h·g and a photocatalytic H_2O_2 production rate of 1098 μmol/h·g. Notably, at λ = 370 nm, the corresponding apparent quantum efficiencies (AQE) are 7.79% and 12.43%, respectively, surpassing those of most known photocatalysts.

4.3. Element Doping

The element doping strategies for g-C_3N_4 primarily encompass non-metal doping, metal doping, and co-doping (multi-element synergy). These strategies aim to optimize the electronic structure, light absorption capacity, and catalytic activity of g-C_3N_4. Metal doping enhances photocatalytic reactions by refining the active sites and energy band structure. For instance, Li et al. [68] employed a transition metal double anchoring configuration to induce asymmetric distortion in the g-C_3N_4 skeleton via sulfur atoms, thereby promoting electron delocalization and forming multiple active sites. Non-metal doping, on the other hand, broadens light absorption and facilitates charge propagation by adjusting the band gap. Additionally, the introduction of non-metal elements can create defects within the semiconductor structure, alter its crystallinity and light absorption characteristics, and provide traps or recombination centers for capturing electrons or holes. According to Mottammal et al. [69], sulfur–potassium co-doping, as well as sulfur doping alone, effectively reduced the band gap to 2.1 eV, resulting in a 50 nm red shift in the visible light absorption edge. Furthermore, the quantum efficiency of the modified g-C_3N_4 under 420 nm illumination reached 15.7%, which is 12 times greater than that of pristine g-C_3N_4. Fahim A. Qaraah et al. innovatively synthesized hexagonal porous g-C_3N_4 supported by Ag-Ni bimetallic sites through hydrothermal and calcination processes. The high porosity of g-C_3N_4, the synergistic effect of Ag-Ni bimetallic sites, and the increased surface area demonstrate excellent photocatalytic performance. The optimal results indicate that the yield of CO is 77.65 μmol/g, while the yield of CH_4 is 17.89 μmol/g [70].

4.4. Heterojunctions Structure Construction

Heterojunctions can be classified into conventional types: type I, type II, type Z, and type S. The construction of semiconductor heterojunctions has garnered significant attention due to its remarkable effectiveness in improving photocatalytic activity. By forming semiconductor heterostructures, the separation of photogenerated electron–hole pairs can be significantly enhanced, and the visible light absorption range of carbon nitride can be broadened, thereby improving its catalytic activity. For g-C_3N_4, the development of heterojunction materials represents an effective approach to enhance its photocatalytic performance. The formation of g-C_3N_4-based heterojunction photocatalysts using g-C_3N_4 and other semiconductor materials can effectively improve the separation efficiency of photogenerated carriers, thereby enhancing the photocatalytic activity of the catalyst [71]. Shreerang Mishra et al. [72] constructed a double S-type heterojunction ternary system of g-C_3N_4-loaded $NiTiO_3$ and $NiSnO_3$ by ultrasonic calcination method, forming a two-way charge transfer path. The composite material achieved a hydrogen production rate of 137 μmol·g^{-1}·h^{-1} under visible light, which was 5.49 times higher than that of a single component.

Due to the differences in band gap among various materials, semiconductors and semiconductor heterojunctions can convert light energy into electricity more efficiently, thereby enhancing photoelectric conversion efficiency. Traditionally, semiconductor heterostructures have garnered significant attention. As illustrated in Figure 9, semiconductor heterojunction systems can generally be categorized based on semiconductor type into two categories: p-n type and non-p-n type. Among these, the p-n junction in semiconductors serves as an effective structure for efficient charge collection and separation. Typically, when p-type and n-type semiconductors come into contact, the diffusion of electrons and holes leads to the formation of a p-n junction with a space charge region at the interface. This generates a built-in electric field that facilitates the movement of electrons and holes in opposite directions.

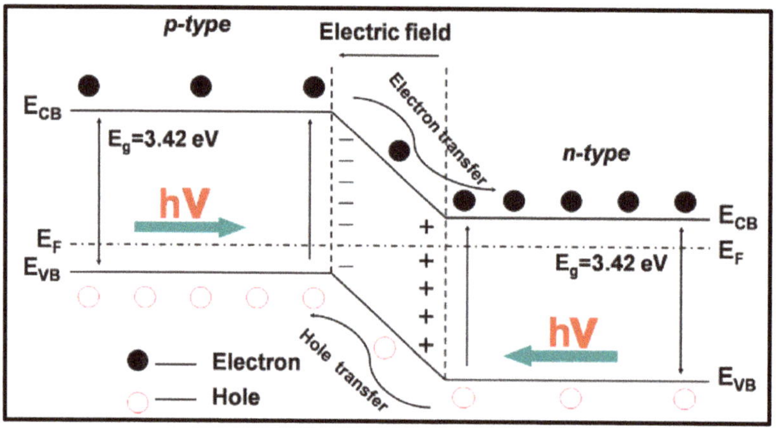

Figure 9. Schematic representation of energy band structure and electron–hole pair separation in p-n heterojunction [73].

Semiconductor heterojunctions can be classified into three structural types: straddling gap (Type I), staggered gap (Type II), and broken gap (Type III) [73]. As illustrated in Figure 10, the Type I semiconductor heterojunction structure, under visible light irradiation, allows for the easy aggregation of excited electron–hole pairs within the valence band and conduction band of the same semiconductor. This results in photogenerated carriers being unable to achieve effective separation. Conversely, the energy band position of the Type III semiconductor heterojunction structure is staggered, preventing the overlap of band gaps between the two semiconductors. Consequently, carriers excited under visible light cannot migrate effectively between the semiconductors, failing to enhance the separation of electron–hole pairs. In contrast, the Type II heterojunction facilitates spatial separation of photogenerated electron–hole pairs due to the distinct positions of the valence band and conduction band. This characteristic makes Type II heterojunctions the most effective conventional heterojunctions for improving photocatalytic activity, as they are well-suited for the spatial separation of electron–hole pairs. Recent studies have increasingly focused on Type II heterostructures based on g-C_3N_4. Wang et al. [74] reported that g-C_3N_4 with quantum thickness was polymerized onto the surface of anatase titanium dioxide (TiO_2) nanosheets to form TiO_2@g-C_3N_4 (TCN) core–shell quantum II heterojunctions. The structural composition of these heterojunctions facilitated the effective transfer of electrons between the two materials, thereby enhancing the photocatalytic degradation activity of tetracycline. This performance was found to be 2-fold higher than that of TiO_2 and 2.3-fold higher than that of pure g-C_3N_4 (see Figures 10 and 11).

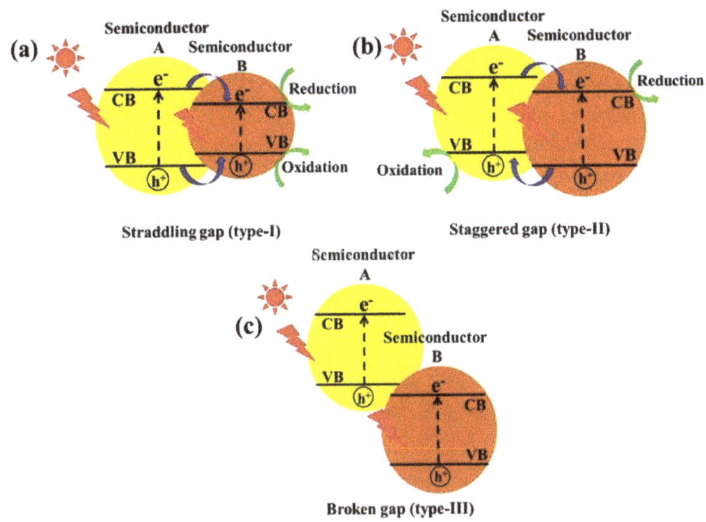

Figure 10. Schematic representation of three different types of separation of electron–hole pairs in the case of heterojunction photocatalysts: (**a**) Type I; (**b**) Type II; and (**c**) Type III heterojunction [73] (Reprint from Low, J.; Yu, J.; Jaroniec, M.; et al. Heterojunction Photocatalysts. *Adv. Mater.* 2017, 29, 1601694).

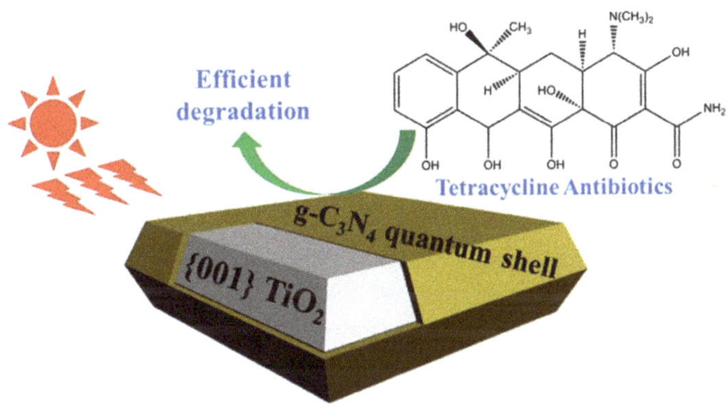

Figure 11. Structures of prepared TiO_2@g-C_3N_4 (TCN) core–shell quantum type II heterojunctions [74] (reprint from Wang, W.; Fang, J.; Shao, S.; et al. Compact and Uniform TiO_2@g-C_3N_4 Core–Shell Quantum Heterojunction for Photocatalytic Degradation of Tetracycline Antibiotics. *Appl. Catal. B* 2017, 217, 57–64).

Although conventional type II and p-n heterojunctions effectively enhance carrier separation, their oxidation and reduction properties are compromised. Consequently, the photogenerated carriers ultimately reside in the valence band of the semiconductor, which has limited oxidation capability, and in the conduction band of the semiconductor, which has limited reduction capability after separation. To address these challenges, Bard et al. (1979) designed the Z-type heterojunction to enhance redox capacity by studying and simulating the photosynthesis process in plants [75]. As research progresses, various types of Z-heterojunctions have emerged, including conventional Z-heterojunctions, all-solid-state Z-heterojunctions, indirect Z-heterojunctions, and direct Z-heterojunctions. This work focuses on the conventional Z-heterojunction system, where the intermediate functions as both an electron acceptor and donor. This configuration effectively traps holes and electrons within the valence and conduction bands, respectively, thereby compensating for the weak redox capacity and facilitating the development of robust redox capabilities and spatially segregated redox reaction sites. Adan et al. [76] directly complexed WO_3 with g-C_3N_4 using a straightforward sonication-assisted method (Figure 12). The resulting complexes exhibited excellent material homogeneity, effective interfaces, and appropriate energy levels. Furthermore, their Z-type heterojunction structure facilitated the formation of superoxide radicals, significantly enhancing the degradation of pollutants and achieving a degradation rate four times greater than that of pure g-C_3N_4.

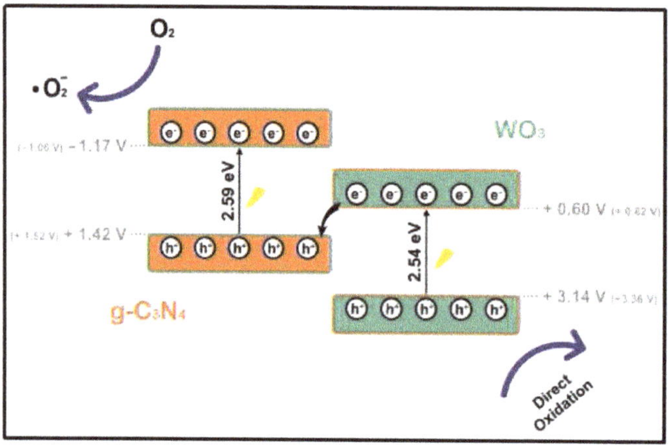

Figure 12. Photocatalytic mechanism of g-C_3N_4/WO_3 heterostructured photocatalysts (positions of theoretical CBs and VBs are shown inside the figure) [76].

Despite their potential advantages, the structural configuration and complex design of Z-type heterojunctions limit their large-scale preparation. Additionally, several shortcomings exist: (1) the presence of intermediate trapped electron–hole pairs makes the newly generated carriers susceptible to thermodynamic reversal; (2) the reaction environment is restricted to the liquid phase, which limits practical application scenarios; (3) during the migration process, photogenerated electron–hole pairs must traverse multiple energy levels, hindering carrier migration and resulting in poor stability of the photocatalytic reaction. Consequently, we shift our focus to a novel S-type heterojunction, characterized by two semiconductors with distinct valence and conduction band positions. When constructing the heterojunction in close contact, a potential energy difference in the energy band structure induces the energy band bending at the contact surface, thereby facilitating the efficient separation of electron–hole pairs. Chen et al. [77] synthesized sulfur-doped g-C_3N_4 and nitrogen-doped MoS_2 using a one-step thermal polycondensation method, resulting in the formation of an S-type heterojunction structure. This structure effectively broadens the absorption of visible light and enhances the separation of photogenerated carriers, achieving a hydrogen generation rate of 658.5 μmol/g/h. This rate is significantly higher than that of pure g-C_3N_4 (28.8 μmol/g/h) and MoS_2 (17.4 μmol/g/h), surpassing them by factors of 23 and 38, respectively (Figure 13). The convergence of Fermi energy levels within the S-type heterojunction generates built-in electric fields at the contact surfaces, leading to the bending of energy bands and Coulombic attraction. This phenomenon acts as a driving force for electrons in the conduction band of the composite and facilitates the recombination of holes in the valence band. Consequently, these mechanisms significantly reduce the ineffective recombination of photogenerated carriers, allowing for the retention of strong photogenerated carriers to participate in the photocatalytic reaction.

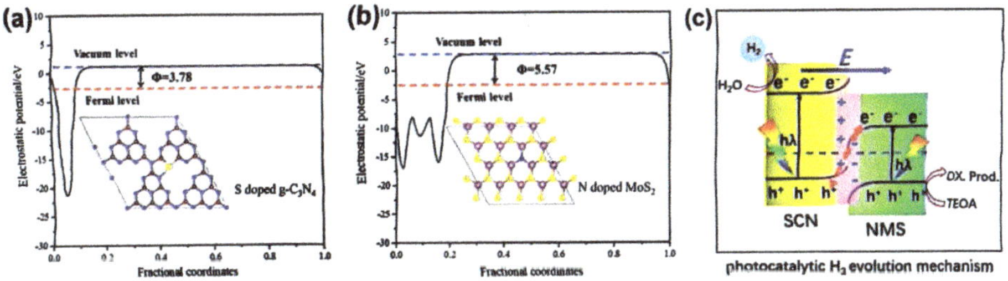

Figure 13. (**a**) Electrostatic potentials of CN (001) surface and (**b**) MoS_2 (002) surface; (**c**) schematic of NMS/SCN photocatalytic mechanism S-type heterojunction [77] (reprint from Chen, Y.; Su, F.; Xie, H.; et al. One-Step Construction of S-Scheme Heterojunctions of N-Doped MoS_2 and S-Doped g-C_3N_4 for Enhanced Photocatalytic Hydrogen Evolution. *Chem. Eng. J.* 2020, 404, 126498).

5. Integration of Multiple Catalytic Technologies: Piezoelectric–Photocatalytic Technology

Photocatalysis is recognized as a green technology that effectively mitigates excessive energy consumption while maintaining the long-term stability of aquatic ecosystems. Laboratory studies demonstrate that photocatalysis can completely oxidize a wide range of organic contaminants in both water and air streams at or near room temperature [78]. However, the low transport efficiency of charge carriers and the high recombination rate of photogenerated electron–hole pairs hinder the enhancement of degradation catalysis performance in conventional semiconductor materials. Currently, photocatalysis is being increasingly integrated with other catalytic technologies to achieve more efficient and environmentally friendly solutions. For instance, the combination of photocatalysis with electrocatalysis and thermal catalysis allows for more effective energy utilization. This synergistic approach, particularly in sewage treatment and organic waste degradation, can significantly enhance degradation rates and overall efficiency. Moreover, piezoelectric–photocatalytic technology utilizes the synergistic effect of piezoelectric and optical processes to effectively incorporate the piezoelectric effect into the photocatalytic process [79]. Piezoelectric–photocatalytic technology is a method for the catalytic degradation of organic pollutants that utilizes the synergistic effects of mechanical force and light in conjunction with a semiconductor catalyst. This technology offers significant advantages, including high efficiency and environmental sustainability [80].

In addition, various forms of energy are readily available in the surrounding environment, including mechanical vibrations, thermal energy, fluid flow, light, and electromagnetic radiation, such as radio waves. These energy sources can be harnessed to generate electrical energy for electronic devices through diverse energy conversion methods. Among these, mechanical vibrations are particularly prevalent in daily life, occurring in contexts such as human movements, vehicular transport, and the operation of industrial machinery [81,82]. Mechanical energy can be stored and converted into electrical energy, which can then be utilized to power electronic devices. This process provides a clean source of electrical energy. Currently, researchers are focusing on various fields, including electromagnetism [83], electrostatics [84], piezoelectrics [85], friction electricity [86], and thermoelectrics [87]. Various energy harvesting strategies at the meso-, micro-, and nanoscales have been proposed using piezoelectric materials, each exhibiting unique polarization properties. Notably, the self-polarization characteristic of piezoelectric materials enables them to convert external mechanical energy into electrical energy without requiring an external magnetic field or the addition of other substances. Furthermore, these materials demonstrate high sensitivity to external forces, making them particularly effective for energy harvesting applications [88]. Furthermore, the energy harvested by piezoelectric materials is not only stable but also exceptionally dense, yielding a higher energy output per unit area compared to other energy collection methods [89–91]. Therefore, the application of piezoelectric technology for harvesting external mechanical energy to enhance the degradation of pollutants has emerged as one of the environmentally friendly solutions for efficient energy utilization.

5.1. Materials Research for Piezoelectric–Photocatalytic Technology

Among the various catalytic technologies, the integration of piezoelectric technology is particularly significant. Piezoelectricity refers to a material property that enables the conversion of mechanical energy into electrical energy, finding extensive applications in sonar, sensors, and energy conversion. The synergy between photocatalysis and piezoelectric technology not only facilitates chemical reactions through photocatalysis but also harnesses piezoelectric technology to transform mechanical energy generated during these reactions into electrical energy. This approach enables the multi-level utilization of energy and contributes to the establishment of a theoretical framework for photoelectrochemical three-phase coupling catalysis. The choice of materials for piezoelectric energy harvesting plays a crucial role in determining performance. Consequently, a diverse range of materials, including inorganic, organic, and composite substances, has been explored in the research of piezoelectric energy harvesting. In the field of piezoelectric catalysis, commonly used piezoelectric materials include ceramic materials such as zirconate–titanate ($PbTiO_3$-$PbZrO_3$, PZT) [92], composite materials like zinc oxide (ZnO) [93], and polymers such as polyvinylidene fluoride (PVDF) [94]. Piezoelectric ceramic materials represent a novel category of intelligent materials that exploit their piezoelectric and inverse piezoelectric effects [95]. These ceramic materials can be classified into two main categories: lead-containing piezoelectric ceramics and lead-free piezoelectric ceramics. They have found extensive applications in various fields, including aerospace, ship sonar, high-speed trains, automobiles, precision instrument control, mobile communications, and both office and home electronic products [96]. Lead-containing ceramics exhibit toxicity due to the presence of lead, which contradicts the principles of green environmental protection. In contrast, lead-free ceramics demonstrate resistance to high pressure and high temperature; however, they are characterized by extreme brittleness, which poses challenges for safe daily use.

5.2. Research on Flexible Films of Polyvinylidene Fluoride

Flexible and non-polluting polymer polyvinylidene fluoride (PVDF) and its copolymers represent the most promising candidates for the preparation of piezoelectric materials. This is attributed to their unique electroactive properties, high flexibility, excellent processing characteristics, and long-term stability [97–99]. Bayan et al. [100] developed a self-polarized PVDF-based nanogenerator by incorporating pristine g-C_3N_4 nanosheets as filler materials within a PVDF matrix. The researchers demonstrated that this nanogenerator could efficiently generate a maximum voltage of approximately 2.3 V and a maximum power output of around 110 $\mu W/cm^2$ when activated by a finger tap. Furthermore, the integration of an additional layer of plasmonic silver nanoparticle-loaded g-C_3N_4 nanosheets resulted in a significant enhancement in optical response. The hybrid plasma nanogenerator, exhibiting a strain of approximately 0.021%, achieved self-powered photodetection with a light-to-dark current ratio of approximately 60 (Figure 14), in contrast to the unstrained device, which exhibited a ratio of about 2.0. Dai et al. [101] prepared a sample of PVDF-$Na_{0.5}Bi_{0.5}TiO_3$-$BiOCl_{0.5}Br_{0.5}$ (PV-NB) based on a flexible composite film of PVDF-$Na_{0.5}Bi_{0.5}TiO_3$. The conversion of mechanical energy from flowing water into piezoelectric potential energy was utilized to enhance charge transfer in the photocatalytic composite film, resulting in a 2.33-fold increase in photocatalytic ability. The durability of the film material was verified through repeated degradation performance tests (Figure 15).

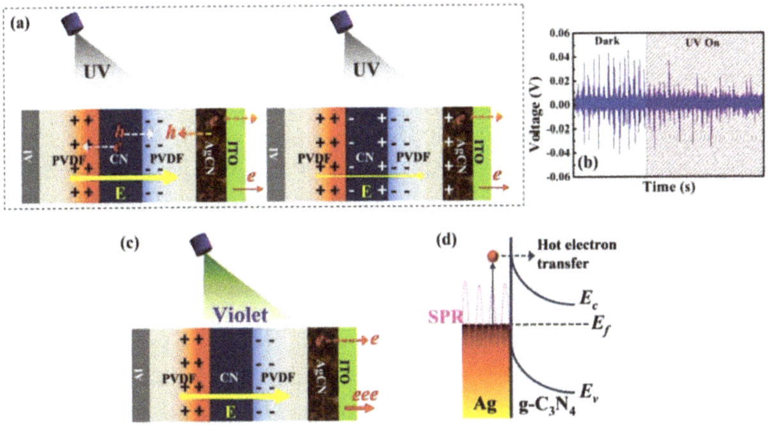

Figure 14. Schematic diagram of the photoresponse mechanism by PENG: (**a**) charge redistribution under UV light under strain conditions; (**b**) output voltage generated by impacts under dark and UV irradiation conditions; (**c**) charge redistribution under UV light irradiation under strain conditions; (**d**) thermal electron transfer mechanism of silver nanoparticle [96].

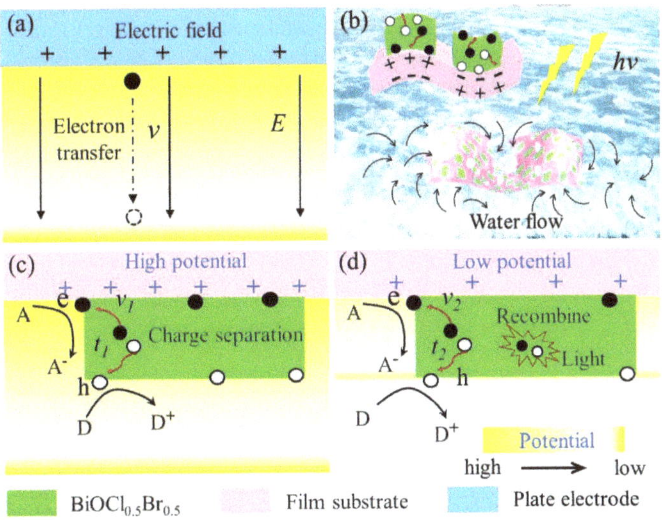

Figure 15. Mechanism of piezoelectric potential enhancing photocatalytic activity: (**a**) electron acceleration along electric field E with transfer rate v; (**b**) light-induced deformation of flexible composite membrane and generation of photoexcited carriers; (**c**,**d**) carrier transfer behavior at high and low potentials (v_1, v_2: carrier migration rates; t_1, t_2: response times) [101] (reprint from Dai, B.; Huang, H.; Wang, F.; et al. Flowing Water Enabled Piezoelectric Potential of Flexible Composite Film for Enhanced Photocatalytic Performance. *Chem. Eng. J.* 2018, 347, 263–272).

5.3. Synergistic Effect of g-C_3N_4 in Piezoelectric Photocatalysis

Piezophotocatalysis, as an emerging coupled catalytic strategy, offers an innovative approach to address the critical challenge of rapid recombination of photogenerated carriers in traditional semiconductor photocatalysis. This method combines photoexcitation with the piezoelectric effect induced by mechanical stress. g-C_3N_4 and its composites have emerged as an ideal platform for constructing efficient piezophotocatalytic systems due to their unique structural and physicochemical properties, which demonstrate significant synergistic enhancement effects.

5.3.1. Theoretical Basis of Synergistic Mechanism

The photoexcitation process involves the transition of electrons from the valence band to the conduction band of g-C_3N_4 under visible light irradiation, resulting in the generation of electron (e^-)–hole (h^+) pairs. In the piezoelectric effect process, when the system experiences external mechanical stress—such as ultrasonic vibration, stirring, or fluid shear force—the piezoelectric component (g-C_3N_4 itself or modified/composite strong piezoelectric materials) with a non-centrosymmetric structure undergoes deformation. This deformation leads to the separation of internal positive and negative charge centers, creating a piezoelectric polarization field and potential. The coupling and synergy effects manifest primarily in three aspects: (1) Directional driving of charge carrier separation: The piezoelectric polarization field functions as a robust built-in electric field, with its direction dynamically changing according to the stress state. This field effectively drives photogenerated electrons and holes to migrate in opposite directions, significantly suppressing their spatial recombination. (2) Band bending and surface charge modulation: The piezoelectric potential can induce significant bending of the catalyst's surface energy bands, thereby influencing the driving force of redox reactions. Concurrently, the polarized charges generated on the surface of the piezoelectric material, with positive and negative charges enriched on different crystal planes, facilitate the directional adsorption and activation of reactant molecules (such as H_2O, O_2, and pollutants) on the catalyst surface. (3) Broadening the light response range (in certain systems): Strong electric fields may indirectly enhance the material's light-harvesting capabilities through the Franz–Keldysh effect or by promoting defect states to participate in light absorption. The synergistic conversion of light and mechanical energy into chemical energy significantly enhances overall catalytic efficiency, often surpassing the mere sum of individual photocatalysis and piezoelectric catalysis.

5.3.2. Construction Strategy of g-C_3N_4-Based Piezoelectric–Photocatalytic System

(1) Intrinsic piezoelectric regulation of g-C_3N_4 can be achieved through structural optimization by adjusting high-temperature polycondensation conditions, selecting appropriate precursors such as melamine, urea, and thiourea, and introducing pore-forming agents. These modifications enhance the layered structure of g-C_3N_4, increase interlayer spacing, and introduce structural defects, such as nitrogen vacancies, which may marginally improve its inherent piezoelectric response. The primary sources of its piezoelectric contribution are interlayer shear slip and out-of-plane deformation. Additionally, elemental doping serves as a technique to modify the lattice of g-C_3N_4 [102]. The introduction of metal ions (e.g., K^+, Na^+) [103] or non-metal elements (e.g., P, B, O) into the lattice can disrupt the symmetry of g-C_3N_4 and alter its electronic structure, thereby enhancing both light absorption and piezoelectric properties. Liang et al. [104] synthesized g-C_3N_4 nanoplates by a new self-template process, and then ultrasonically exfoliated to produce porous g-C_3N_4 nanoplates with a thickness of 1.5–2.0 nm. Piezoelectric response force microscopy shows that the material exhibits excellent piezoelectric response. In addition, the material exhibits obvious piezoelectric–photocatalytic degradation of RhB, showing the best first-order rate constant. Under visible light and ultrasonic excitation, it is 2.3 times under visible light irradiation alone and 9.7 times under ultrasonic excitation alone. The

piezoelectric field generated by ultrasonic excitation drives the opposite direction of photogenerated electrons and holes to migrate to the surface of the nanosheets, resulting in more reactive free radicals, thereby accelerating the degradation of RhB.

(2) Composite materials incorporating strong piezoelectric substances, such as perovskite types (e.g., $BaTiO_3$ [105], $BiFeO_3$ [106], $Bi_4Ti_3O_{12}$ [107]), exhibit notable piezoelectric properties, and their band structures can readily align with g-C_3N_4 to form effective heterojunctions. Wurtzite-type materials, including ZnO and ZnS, are straightforward to synthesize and possess commendable piezoelectric characteristics. Layered transition metal sulfides, such as MoS_2 and WS_2 [108], not only display piezoelectric properties but also exhibit outstanding photoelectric and electrochemical behaviors. In designing composite structures, a critical factor is the construction of tight heterointerfaces (e.g., 0D/2D, 2D/2D, 3D/0D) [109], which ensures efficient transfer of photogenerated carriers and effective penetration of the piezoelectric polarization field at the interface. Core–shell structures and Z-type heterojunction [110] designs have been demonstrated to utilize synergistic effects more efficiently. Cai Jing et al. [111] synthesized a Z-type heterojunction/g-C_3N_4/BiOCl material incorporating graphene quantum dots using a one-step hydrothermal technique. The material demonstrated high photocatalytic stability. The researchers assessed the piezoelectric–photocatalytic performance of the material by immobilizing N_2 under simulated sunlight. The results indicated that the photocatalytic activity of the material was significantly superior to that of pure g-C_3N_4 and BiOCl, with a formation rate of 1773.8 μmol/(h·g), which is 7.3 times greater than that of pure g-C_3N_4 and 5.2 times greater than that of BiOCl, respectively.

(3) Organic Piezoelectric Composite: Utilizing the excellent piezoelectric properties and flexibility of ferroelectric polymers, such as PVDF and PVDF-TrFE [112], these materials are combined with g-C_3N_4 to prepare flexible piezoelectric–photocatalytic films or fibers. This combination allows for the efficient collection of low-frequency mechanical energy from the environment, including wind energy, water flow energy, and human motion energy. Huang et al. [113] developed a modified g-C_3N_4/PVDF membrane, referred to as the MCU-g-C_3N_4/PVDF membrane, based on the original PVDF membrane. The experimental results demonstrated that the MCU-g-C_3N_4/PVDF membrane achieved 84.24% and 71.26% removal rates for rhodamine B and tetracycline hydrochloride, respectively, thereby confirming its excellent photocatalytic performance.

5.3.3. Application of g-C_3N_4 in Piezoelectric Photocatalysis

Furthermore, numerous studies have confirmed that the g-C_3N_4-based piezoelectric–photocatalytic system demonstrates exceptional synergistic properties across a variety of reactions.

Pollutant degradation under ultrasonic-assisted visible light irradiation has been shown to significantly enhance the degradation rates of organic dyes (e.g., rhodamine B, methyl orange) [114], antibiotics (e.g., tetracycline, ciprofloxacin), and persistent organic pollutants. Systems such as g-C_3N_4/$BaTiO_3$, g-C_3N_4/$BiFeO_3$ [115], and g-C_3N_4/MoS_2 exhibit degradation rates that are typically 2–5 times greater, or even higher, than those achieved through single photocatalysis or piezoelectric catalysis. The piezoelectric effect not only accelerates the separation of charge carriers but also facilitates the continuous and efficient generation of reactive oxygen species, including ·OH and $\cdot O_2^-$. Kang et al. [116] constructed a Z-type heterojunction with a core–shell structure by growing Bi_2WO_6 nanosheets on g-C_3N_4 coated ZnO nanorods. This design has more active sites and carrier excitation concentration, showing excellent piezoelectric–photocatalytic efficiency, and the degradation efficiency of RhB is increased to 0.217 min^{-1}. Huang et al. [117] designed Ag/$NaNbO_3$/g-C_3N_4 photocatalytic material by doping Ag into $NaNbO_3$ and g-C_3N_4 composites, which significantly reduced the carrier recombination efficiency. The photocatalytic performance was improved by using the plasma resonance polymerization effect of Ag nanoparticles and the piezoelectric effect of $NaNbO_3$. The complete degradation of methyl orange was achieved within 1h after the addition of appropriate amount of Ag nanoparticles.

The combination of piezoelectricity and photocatalysis has demonstrated that graphite-phase carbon nitride is highly effective in air purification, particularly in the removal of nitrogen oxides and formaldehyde. The primary mechanism underlying this process is photocatalytic oxidation, wherein material modifications optimize the electronic structure, surface activity, and reaction pathways. Liu et al. synthesized a g-C_3N_4/$BaTiO_3$ composite with nitrogen vacancies. The introduction of a built-in electric field enhanced the band gap, facilitated carrier separation, and improved the adsorption and activation of reactants, resulting in remarkable NO removal and NO_2 piezoelectric–photocatalytic properties [118]. Di et al. [119] utilized wind energy to enhance the degradation of formaldehyde using piezoelectric materials. They prepared heterojunction composites of $BiFeO_3$ nanocubes and oxygen vacancy g-C_3N_4 nanosheets. The concentration of formaldehyde was reduced from 1.5 ppm to 0.68 ppm within 90 min, demonstrating a degradation efficiency that is double that of pure carbon nitride. This efficient degradation behavior is attributed to the synergistic interaction between the piezoelectric field and the heterojunction containing oxygen vacancies, providing a novel approach for air purification.

In the realm of photocatalytic nitrogen fixation and energy conversion, carbon nitride—particularly graphite-phase carbon nitride—based photocatalysts exhibit significant potential when subjected to the synergistic effects of light and pressure [120]. Hydrogen production via water splitting is enhanced by the piezoelectric polarization field, which effectively mitigates the insufficient driving force for proton reduction (H^+ to H_2) in the conduction band of g-C_3N_4 while simultaneously inhibiting the reverse reaction. Systems such as g-C_3N_4/ZnO [121] and g-C_3N_4/$BaTiO_3$ demonstrate markedly improved hydrogen production activity due to the synergistic effects of light and pressure, which extend carrier lifetimes by several orders of magnitude. Additionally, Zhang et al. [122] developed a novel TiO_2/g-C_3N_4 composite catalyst. The piezoelectric properties of carbon nitride, when combined with ultrasonic and visible light, induce mechanical vibrations that alter molecular distribution and band bending, thereby significantly enhancing photocatalytic hydrogen production performance. This resulted in a hydrogen production rate of 43.57 mmol/h/g, with a quantum yield of 87.13% at 365 nm. Deng et al. [123] employed a combination of piezoelectric and photocatalytic strategies to construct an S-type heterostructure using Bi_2MoO_6 and C_3N_4, which effectively enhanced carrier migration and demonstrated an impressive nitric acid yield under ultrasound and light. Guo et al. [124] synthesized g-C_3N_4 with a surface pit structure through thermal polymerization, resulting in a 2.3-fold enhancement in piezoelectric response. Under piezoelectric photocatalysis conditions, a H_2O_2 yield of 189.8 μM/h was achieved in pure water, representing an increase of approximately 14.4 times compared to g-C_3N_4 under traditional photocatalysis.

5.3.4. Summary of g-C_3N_4 Photocatalytic Application Cases

Currently, numerous efforts have been directed towards the practical application of carbon nitride. To enhance the material's properties, we present detailed parameters in Table 1. This material exhibits significant potential for development in the field of sewage treatment. The subsequent table illustrates parameters such as pollutant degradation, treatment efficiency, and the optimal reaction conditions for g-C_3N_4 in practical applications.

Table 1. Summary of the practical application of g-C_3N_4 in the degradation of pollutants.

Modification Strategy	Target Pollutant	Degradation Product	Degradation Efficiency	Optimum Conditions	Ref.
Mn/O co-doped g-C_3N_4 (MOCN)	Sulfadiazine (SDZ)	Small molecular organic acids (oxalic acid, acetic acid), CO_2, H_2O	Compared with Mn-doped g-C_3N_4 (MCN), the rate is increased by 3.5 times.	Visible light, high iodine hydrochloric acid activation	[125]
Additional O-doped g-C_3N_4	Tetracycline	Demethylation/ hydroxylation products→ring-opening fatty acids	After 90 min, the removal rate was 90%, and the TOC removal rate was 83.4%.	Neutral pH, visible light	[125]
HCl/phytic acid co-modified polyaniline-g-C_3N_4	Cr(VI), rhodamine b	Cr(III), decolorizing small molecules, CO_2	The reduction rate of Cr (VI) was 98%. The decolorization rate of dye is 95%.	Visible light, pH = 3	[126]
3D g-C_3N_4/TiO_2 heterojunction	Methylene blue (MB), phenol	Hydroxylated cyclic intermediate→long chain alkane→CO_2	Compared with pure g-C_3N_4, the degradation rate of MB increased by 4 times.	Static visible light irradiation	[127]
Oxygen vacancy $In(OH)_3$/g-C_3N_4	Acid orange 7 (AO7)	Deazo product→aniline→ ring-opening carboxylic acid→CO_2	180 min 90% removal	Visible light, wide pH adaptability	[128]
g-C_3N_4-P25/photosynthetic bacteria complex	Reactive brilliant red X-3B	Aniline→phenols→ long-chain alkanes→ CO_2+H_2O	The decolorization rate was 94% and the COD removal rate was 84.7%. 84.7%	Visible light-microorganism synergy	[129]

In addition to the modifications of the material itself, the application of piezoelectric photocatalysis can effectively degrade pollutants. Khadim et al. [130] prepared a Z-type heterojunction of AgBr and Br-doped g-C_3N_4 (AgBr/Br-g-C_3N_4). The catalyst, comprising 20 wt% AgBr/Br-g-C_3N_4, was designed in a simple structure utilizing a split-plate airlift contactor. Under visible light irradiation and ultrasonic conditions, its piezoelectric–photocatalytic degradation ability for methylene blue (MB) was evaluated. The results indicated that 99.7% of MB degradation could be achieved within 60 min. To intuitively illustrate the performance of g-C_3N_4 in the piezoelectric–photocatalytic degradation of organic matter, Table 2 summarizes the performance and mechanism of g-C_3N_4 in degrading pollutants.

Table 2. Summary of the performance and mechanism of some g-C_3N_4 piezoelectric–photocatalytic degradation of pollutants.

Modification Strategy	Target Pollutant	Degradation Efficiency	Piezoelectric Synergy Conditions	pH Effect	Kinetic Equation Parameter	Ref.
AgBr/Br doped g-C_3N_4 Z-type Heterojunction	Methylene blue (MB)	Degradation 99.7% in 60 min.	Visible light, ultrasound, airlift reactor	Wide range adaptability	k = 0.082 min^{-1}	[130]
Ag-PCN/SnO_2−g-C_3N_4 Z-type Heterojunction	Tetracycline (TC)	After 90 min, the removal rate was 90%, and the TOC removal rate was 83.4%.	40 kHz ultrasound (0.5 W/cm^2)	Neutral is the best	k = 0.025 min^{-1}	[130]

Table 2. *Cont.*

Modification Strategy	Target Pollutant	Degradation Efficiency	Piezoelectric Synergy Conditions	pH Effect	Kinetic Equation Parameter	Ref.
Fe doped g-C_3N_4 (FeCN)	Methylene blue (MB)	97.7% degradation in 60 min.	Visible light, ultrasound, neutral	High efficiency at pH 5–9	k = 0.051 min^{-1}	[131]
Porous tubular g-C_3N_4 (PTCN)	Tetracycline (TC)	Degradation of 89% in 30 min.	Ultrasound, visible light	Neutrality	First-order kinetics ($R^2 > 0.98$)	[132]
$BaTiO_3$/tubular g-C_3N_4 heterojunction	Tetracycline (TC)	The degradation rate was 91.0% in 60 min.	Ultrasound, visible light	Unspecified	The rate increased by 3.2 times	[133]
g-C_3N_4/PDI-g-C_3N_4 homojunction	Ateatrazine (ATZ)	The degradation rate was 94% after 60 min.	Magnetic stirring, visible light	pH = 2.97	1.6 times higher than that of pure photocatalysis	[134]
N-vacancy g-C_3N_4/ $BaTiO_3$ (CNVB)	Nitric oxide (NO)	The removal rate of piezoelectric photocatalysis is 77.9%.	Ultrasound, visible light	Unspecified	The rate constant is increased by 2.21 times	[119]

According to Table 2, g-C_3N_4-based piezoelectric photocatalysts have both opportunities and challenges to move towards large-scale practical applications. For the optimization of the material itself, the activity and stability of the g-C_3N_4-based catalyst have been significantly improved by strategies such as element doping, heterojunction construction, and special morphology design. Many studies have also shown that it has good cycle stability (usually the efficiency remains above 90% after five cycles) [131]. This is the premise of large-scale application. At the same time, efficient catalysts require matching reactors to achieve large-scale efficiency. The split-plate airlift reactor used in the study is a positive direction [130]. It uses the principle of gas lifting to realize liquid circulation and mixing, which consumes less energy than traditional mechanical stirring, and is more conducive to the contact between catalyst and pollutants and the transfer of ultrasonic energy. How to efficiently input mechanical energy to induce piezoelectric effect is the key. In addition to ultrasound, low-frequency vibration can also effectively drive g-C_3N_4/PVDF composite membrane to degrade pollutants, which provides an idea for the utilization of low-grade mechanical energy such as flowing water energy and wave energy widely existing in the environment and is expected to reduce operating costs. For powder catalysts, the preparation of magnetic composite materials or thin film catalysts also avoids the difficulty of powder recovery. Despite the remarkable achievements in the laboratory, there are still many challenges to realize the large-scale application of g-C_3N_4-based piezoelectric–photocatalytic technology, which are as follows:

1. At present, the understanding of basic scientific problems such as how to coordinate piezoelectric effect and photocatalytic effect, how to separate and transfer interface charge, and accurate identification of active species still needs to be deepened, especially in complex real water environment, the mechanism of catalytic process may be more complicated.
2. Most of the research is still carried out in a single pollutant solution configured in the laboratory. The actual wastewater composition is complex, and there may be anions, natural organic matter, suspended solids, etc., which may compete with pollutants for active sites or quench active species, thereby significantly inhibiting catalytic efficiency. The anti-interference ability and long-term stability of the catalyst need to be more fully verified in the real environment.

3. Although solar energy is free, the input of mechanical energy (ultrasound) still needs to consume electricity. Although airlift reactors and the use of environmental mechanical energy contribute to energy conservation, the overall energy consumption and economic costs of large-scale applications still need to be carefully evaluated. The preparation cost of high-performance catalysts also needs to be considered.
4. Scale-up of laboratory reactors is a key step towards engineering applications. This involves the optimization of a series of engineering problems such as mass transfer, heat transfer, light distribution, and sound field distribution, which requires the cross-cooperation of materials, chemistry, environment, machinery, and other disciplines.
5. At present, many studies focus on the removal rate of pollutants, but the identification of degradation intermediates, the degree of mineralization, and the environmental toxicity analysis of the final products are not sufficient. Ensuring that pollutants are completely mineralized or converted into non-toxic and harmless substances is the bottom-line requirement for environmental technology applications.

Future research should focus on the treatment effects of actual water bodies, the amplification effects of reactors, the energy efficiency of the system, and the environmental economic evaluation throughout the entire life cycle. Close collaboration across multiple disciplines and fields is essential for advancing this promising technology from the laboratory to practical water environment management.

The piezoelectric photocatalysis of g-C_3N_4 demonstrates notable efficacy in hydrogen production. To facilitate a comparison with conventional photocatalytic hydrogen production and to analyze the distinctions between the two methodologies, we summarize and contrast several representative studies on g-C_3N_4 piezoelectric photocatalysis and photocatalytic hydrogen production in Table 3.

The peak yield of piezoelectric photocatalysis, recorded at 5464 $\mu mol \cdot h^{-1} \cdot g^{-1}$, is slightly lower than that of conventional photocatalysis, which achieves 6095.1 $\mu mol \cdot h^{-1} \cdot g^{-1}$. However, piezoelectric photocatalysis significantly alleviates the limitations associated with conventional photocatalysis, which typically requires high light intensity conditions. By harnessing mechanical energy conversion, piezoelectric photocatalysis diminishes dependence on light and can sustain catalytic activity even in dark environments, facilitated by an ultrasonic independent drive. This capability positions piezoelectric photocatalysis as a superior alternative to conventional methods.

Table 3. Summary of some relevant data of g-C_3N_4 piezoelectric photocatalysis and photocatalytic hydrogen production.

Material Systems	Hydrogen Generation Rate	Optimum Conditions	Reaction Mechanism	Ref.
g-C_3N_4/ $LiNbO_3$/PVDF membrane	136.02 $\mu mol \cdot h^{-1} \cdot g^{-1}$ (19.6875 cm^2 membrane)	White light irradiation, fluid disturbance (Flow rate 0.5 m/s)	The fluid-induced piezoelectric potential enhances carrier separation, and the membrane structure improves mass transfer efficiency.	[135]
The tubular g-C_3N_4/$BaTiO_3$	Unspecified	40 kHz ultrasound, visible light	The $BaTiO_3$ piezoelectric layer is superimposed with the built-in electric field to suppress the e^--h^+ recombination.	[135,136]
1D/2D isomorphic g-C_3N_4	Unspecified	20 kHz ultrasound, $\lambda > 420$ nm visible light	Ultrasonic excitation strain optimizes carrier migration path.	[135]

Table 3. *Cont.*

Material Systems	Hydrogen Generation Rate	Optimum Conditions	Reaction Mechanism	Ref.
Co_3O_4/g-C_3N_4 nanocomposite	5464 μmol·h^{-1}·g^{-1}	40 kHz ultrasound, 3 wt% Co_3O_4 load	Ultrasound forms a close interface and reduces the charge transfer resistance.	[137]
2D/2D Co_3O_4/g-C_3N_4 Z-junction	370 μmol·h^{-1}·g^{-1}	>400 nm light, 0.1 M Na_2S/Na_2SO_3 sacrificial agent	Van der Waals heterojunction accelerates interfacial charge transfer	[138]
$ZnIn_2S_4$-S/CN van der Waals heterojunction	6095.1 μmol·h^{-1}·g^{-1}	Visible light, 10% TEOA sacrificial agent	S vacancies enhance light absorption, and the 2D/2D interface promotes charge separation.	[139]
Cyanide modified g-C_3N_4	4.5 times of the original g-C_3N_4	$\lambda \geq 420$ nm, triethanolamine sacrificial agent	The cyano group forms a polarized electric field and inhibits carrier recombination.	[140]

g-C_3N_4 exhibits significant potential for applications not only in piezoelectric–photocatalytic hydrogen production but also in hydrogen peroxide synthesis. To this end, we have compiled comprehensive data regarding the production of hydrogen peroxide by g-C_3N_4 under both pure photocatalytic and piezoelectric–photocatalytic conditions, followed by a comparative analysis. The detailed data is presented in Table 4.

Table 4. Data summary of hydrogen peroxide production by g-C_3N_4 under partial photocatalytic/piezoelectric photocatalysis.

Material Systems	H_2O_2 Yield	Reaction Condition	Reaction Mechanism	Ref.
g-C_3N_4/PCN-224 core–shell heterojunction	4.86 mmol·g^{-1}·h^{-1}	Ultrasound (40 kHz) [1], visible light	Piezoelectric field drives Z-type charge transfer, and $\cdot O_2^-$ production is increased by 30%.	[141]
MOC-AuNP/g-C_3N_4 hydrogen bond assembly	120.21 μmol·g^{-1}·h^{-1}	Ultrasound (no light)	The interfacial electric field of AuNP accelerates e^- transfer, and MOC inhibits H_2O_2 decomposition.	[142]
Phosphorous modified g-C_3N_4 (CN-P)	1.46 times pure photocatalysis	Ultrasound, visible light	Piezoelectric coefficient d_{33} = 12.4 pm/V (PFM verification).	[143]
Boron doped g-C_3N_4/PDI-Ag (PDI/BCN-Ag)	143 μmol·g^{-1}·h^{-1}	Visible light (without sacrificial agent)	Cascade S-type charge transfer, Ag nanoparticles promote the $\cdot O_2^-$ path.	[144]
Ultrathin porous P-doped g-C_3N_4	1083 μmol·g^{-1}·h^{-1}	Visible light	Phosphorus doping optimizes energy band structure and mesoporous accelerates mass transfer.	[145]
N-vacancy g-C_3N_4/$BaTiO_3$ (CNVB)	703.4 μmol·g^{-1}·h^{-1}	Visible light	The cyano group broadens the light absorption, and SnO_2 O_2/H^+ adsorption sites are provided.	[146]

[1] Piezoelectric systems are clearly dependent on ultrasonic mechanical stress input (mainly 40 kHz).

Through the comparison presented in the table above, it is evident that the highest yield of the piezoelectric–photocatalytic system reaches 4.86 mmol·g^{-1}·h^{-1}, significantly surpassing the maximum yield of the pure photocatalytic system, which is

1083 μmol·g^{-1}·h^{-1}. Furthermore, piezoelectric photocatalysis can reduce the requirements for light conditions and intensity, maintaining catalytic activity even in dark environments, thus demonstrating a highly efficient and versatile potential for H_2O_2 production.

5.3.5. Challenges and Prospects

In-depth mechanism analysis necessitates a more profound exploration of the microscopic dynamic processes involved in light-force-chemical energy conversion. Specifically, it is crucial to understand how the piezoelectric polarization field precisely regulates the behavior of charge carriers at heterointerfaces and surface reactions at the nano/atomic scale. This exploration still requires the application of in situ and operando characterization techniques, such as in situ X-ray photoelectron spectroscopy (XPS), in situ Kelvin probe force microscopy (KPFM), and time-resolved spectroscopy, alongside theoretical simulations.

Designing efficient and stable materials presents a significant challenge, particularly in the development of novel g-C_3N_4-based composite materials that incorporate strong piezoelectric properties, a broad spectral response, high carrier mobility, and excellent chemical and mechanical stability. Achieving precise control over heterogeneous interface structures and optimizing stress transfer efficiency are essential for enhancing performance. Currently, efficient mechanical energy harvesting and utilization are impeded by the reliance on high-energy-density ultrasound in laboratory settings, which contrasts with the low-frequency, irregular mechanical energy found in real-world environments, such as natural water flow, wind, and ambient noise. Consequently, it is crucial to design specialized structures that can effectively capture and convert environmental mechanical energy, including flexible devices and integrated vibration energy harvesting systems, for practical applications.

Fabrication at scale and integration of systems are critical challenges that must be addressed to develop cost-effective and scalable processes for material synthesis and device manufacturing. Furthermore, focusing on the design of effective synergistic reactor systems that seamlessly integrate photonic and mechanical components is essential. Overcoming these challenges is vital for realizing practical applications that can advance the field and lead to innovative solutions. By addressing these issues, researchers can pave the way for more efficient production methods and improved technologies, ultimately contributing to advancements across various industries.

Exploration of multi-field-coupling future research has the potential to delve deeper into the interactions between light pressure synergy and other domains of study, such as thermocatalysis, electrocatalysis, and magnetic catalysis. By investigating these multi-field couplings, researchers aim to enhance energy conversion efficiency. This exploration focuses not only on improving the fundamental processes involved but also on expanding the practical applications of these technologies. By pushing the boundaries of these interdisciplinary connections, it is possible to develop solutions that operate effectively under a wider range of complex conditions. The ultimate goal is to create more efficient and versatile catalytic systems capable of addressing various energy challenges.

6. Conclusions

g-C_3N_4 has demonstrated significant potential as a low-cost, stable, and environmentally friendly photocatalyst. Through optimized preparation methods—including precursor selection, thermal modulation, and nanostructuring—along with strategic modifications such as forming heterojunctions with semiconductors (e.g., TiO_2 or MoS_2), doping with nonmetals or metals, and creating porous architectures, its photocatalytic efficiency has been markedly enhanced. These advancements have facilitated a wide range of applications, including solar fuel generation (hydrogen production and CO_2 conversion), environmental remediation (degradation of organic pollutants and bacterial disinfection), and organic synthesis. Recent advancements in piezoelectric photocatalysis have also demonstrated significant potential. The combination of g-C_3N_4 with piezoelectric materials such as $BaTiO_3$, ZnO, and MoS_2, or the enhancement in intrinsic piezoelectricity through structural engineering, leads to the generation of built-in electric fields under mechanical stress (e.g., ultrasound or fluid flow). This phenomenon substantially suppresses the charge recombination rate and enhances redox kinetics. As a result, this innovative approach has achieved pollutant degradation activities that exceed those of traditional photocatalysis by more than threefold.

Despite significant advancements in the field, considerable challenges persist. The dynamics of carriers at various photo–piezo interfaces necessitate a clearer mechanistic understanding. Furthermore, the scalable synthesis of stable, high-surface-area composites continues to present formidable challenges. Mechanical energy harvesting from real-world sources, such as wind and water flow, demands innovative reactor designs. Additionally, the stability of coupled photo/mechanical stress requires thorough assessment. Future research should prioritize in situ characterization of coupling mechanisms, computationally guided material design, and flexible device engineering for ambient energy capture. This should be accompanied by the development of multi-field (photo/electro/piezo/thermal) catalytic systems. g-C_3N_4-based photocatalysts and photo–piezo catalysts are anticipated to demonstrate significant potential for scalable solar-to-chemical energy conversion and the advancement of sustainable environmental technologies through ongoing innovation.

Author Contributions: Conceptualization, M.L., L.Y. and D.L.; methodology, Y.L.; formal analysis, Y.S.; investigation, H.H. and L.X.; resources, D.L.; data curation, Y.F.; writing—original draft preparation, M.L.; writing—review and editing, D.L.; funding acquisition, D.L. and Y.L. All authors have read and agreed to the published version of the manuscript.

Funding: This work was supported by the Natural Science Basic Research Plan in the Shaanxi Province of China (No. 2023-JC-ZD-25) and the Science and Technology Guidance Project Plan of China National Textile and Apparel Council (No. 2022038, 2023018).

Data Availability Statement: No new data were created or analyzed in this study.

Conflicts of Interest: The authors declare no conflict of interest.

References

1. Vasu, D.; Ravi, P.V.; Subramaniyam, V.; Moorthi, P.; You, Y.F.; Chiu, T.W. Green Fluorine-Confined 2D Layered Graphitic Carbon Nitride (g-C_3N_4) Nanosheets for Electrocatalysts and Photocatalysts toward the Detection and Degradation of Metol Drugs in Wastewater. *ACS Appl. Eng. Mater.* **2023**, *1*, 1866–1880. [CrossRef]
2. Liu, J.; Wang, J.; Zhang, W.; Du, G. Effects of Greenhouse Effect on Freshwater Ecosystem. *Beijing Water* **2004**, *1*, 3.
3. Zhao, Y.; Su, Q.; Li, B.; Zhang, Y.; Wang, X.; Zhao, H.; Guo, S. Have Those Countries Declaring "Zero Carbon" or "Carbon Neutral" Climate Goals Achieved Carbon Emissions-Economic Growth Decoupling? *J. Clean. Prod.* **2022**, *363*, 132450. [CrossRef]
4. Li, Y.; Ding, L.; Guo, Y.; Liang, Z.; Cui, H.; Tian, J. Boosting the Photocatalytic Ability of g-C_3N_4 for Hydrogen Production by Ti_3C_2 MXene Quantum Dots. *ACS Appl. Mater. Interfaces* **2019**, *11*, 41440–41447. [CrossRef]
5. Heo, Y.J.; Seong, D.B.; Park, S.J. Synthesis of Polyethylenimine-Impregnated Titanate Nanotubes for CO_2 Capture: Influence of Porosity and Nitrogen Content on Amine-Modified Adsorbents. *J. CO_2 Util.* **2019**, *34*, 472–478. [CrossRef]
6. Lin, B.; Xia, M.; Xu, B.; Chong, B.; Chen, Z.; Yang, G. Bio-inspired nanostructured g-C3N4-based photocatalysts: A comprehensive review. *Chin. J. Catal.* **2022**, *43*, 2141–2172. [CrossRef]
7. Zou, C.; Xiong, B.; Xue, H.; Zheng, D.; Ge, Z.; Wang, Y.; Jiang, L.; Pan, S.; Wu, S. The Role of New Energy in Carbon Neutral. *Pet. Explor. Dev.* **2021**, *48*, 480–491. [CrossRef]
8. Yu, W.; Gao, Y.; Chen, Z.; Zhao, Y.; Wu, Z.; Wang, L. Strategies for Optimizing Electrocatalytic Hydrogen Evolution Performance of Metal Phosphides. *Chin. J. Catal.* **2021**, *42*, 1876–1902. [CrossRef]

9. Liu, Q.; Zhao, Y.; Wang, J.; Zhou, Y.; Liu, X.; Hao, M.; Chen, Z.; Wang, S.; Yang, H.; Wang, X. Application of Single-Atom-Based Photocatalysts in Environmental Pollutant Removal and Renewable Energy Production. *Crit. Rev. Environ. Sci. Technol.* **2023**, *53*, 1064–1109.
10. Zhu, Q.; Zhang, K.; Li, D.; Li, N.; Xu, J.; Bahnemann, D.; Wang, C. Polarization-Enhanced Photocatalytic Activity in Non-Centrosymmetric Materials Based Photocatalysis: A Review. *Chem. Eng. J.* **2021**, *426*, 131681. [CrossRef]
11. Saravanan, A.; Kumar, P.S.; Vo, D.-V.N.; Yaashikaa, P.R.; Karishma, S.; Jeevanantham, S.; Gayathri, B.; Bharathi, V.D. Photocatalysis for Removal of Environmental Pollutants and Fuel Production: A Review. *Environ. Chem. Lett.* **2020**, *19*, 441–463. [CrossRef]
12. Yang, T.; Wang, B.; Zhu, J.; Xia, J.; Li, M. Enhanced Photocatalytic Performance of In_2S_3 Multi-Level Pore Structure Nanomaterials Derived from Sacrificial Metal–Organic Frameworks. *Chin. J. Catal.* **2024**, *59*, 204–213. [CrossRef]
13. Guo, C.; Zhao, G.; Feng, Y.; Chen, D.; Loh, T.P.; Wang, D.; Jia, Z. Recyclable g-C_3N_4-Catalyzed Decarboxylative Alkenylation of N-Aryl Glycines with Vinyl Sulfones under Visible-Light Irradiation. *Org. Chem. Front.* **2025**, *12*, 2409–2414. [CrossRef]
14. Loeb, S.K.; Alvarez, P.J.J.; Brame, J.A.; Cates, E.L.; Choi, W.; Crittenden, J.; Dionysiou, D.D.; Li, Q.; Li-Puma, G.; Quan, X.; et al. The Technology Horizon for Photocatalytic Water Treatment: Sunrise or Sunset? *Environ. Sci. Technol.* **2018**, *53*, 2937–2947. [CrossRef] [PubMed]
15. Ren, H.; Koshy, P.; Chen, W.-F.; Qi, S.; Sorrell, C.C. Photocatalytic Materials and Technologies for Air Purification. *J. Hazard. Mater.* **2016**, *325*, 340–366. [CrossRef]
16. Xu, J.; Antonietti, M.; Shalom, M. Moving Graphitic Carbon Nitride from Electrocatalysis and Photocatalysis to a Potential Electrode Material for Photoelectric Devices. *Chem. Asian J.* **2016**, *11*, 2499–2512. [CrossRef]
17. Wei, H.; Liang, L.; Ma, T.; Han, H.; Zhu, J.; Liu, Y.; Fang, Z.; Yang, Z.; Guo, K. Controllable-Morphology $CoFe_2O_4$/g-C_3N_4 p–n Heterojunction Photocatalysts with Built-In Electric Field Enhance Photocatalytic Performance. *Appl. Catal. B* **2022**, *306*, 121107.
18. Li, Y.; Wang, L.; Gao, X.; Xue, Y.; Li, B.; Zhu, X. Construction of Graphitic Carbon Nitride-Based Photocatalyst with Strong Built-In Electric Field via π–π Stacking Interaction Boosting Photocatalytic CO_2 Reduction. *J. Mater. Chem. A* **2024**, *12*, 7807–7816. [CrossRef]
19. Hama Aziz, K.H.; Mustafa, F.S.; Omer, K.M.; Hama, S.; Hamarawf, R.F.; Rahman, K.O. Heavy metal pollution in the aquatic environment: Efficient and low-cost removal approaches to eliminate their toxicity: A review. *RSC Adv.* **2023**, *13*, 17595–17610. [CrossRef]
20. Zhu, W.; Yue, Y.; Wang, H.; Zhang, B.; Hou, R.; Xiao, J.; Huang, X.; Ishag, A.; Sun, Y. Recent advances on energy and environmental application of graphitic carbon nitride (g-C_3N_4)-based photocatalysts: A review. *J. Environ. Chem. Eng.* **2023**, *11*, 110164. [CrossRef]
21. Wang, X.; Maeda, K.; Thomas, A.; Takanabe, K.; Xin, G.; Carlsson, J.M.; Domen, K.; Antonietti, M. A Metal-Free Polymeric Photocatalyst for Hydrogen Production from Water under Visible Light. *Nat. Mater.* **2009**, *8*, 76–80. [CrossRef]
22. Ong, W.J.; Tan, L.L.; Ng, Y.H.; Yong, S.T.; Chai, S.P. Graphitic Carbon Nitride (g-C_3N_4)-Based Photocatalysts for Artificial Photosynthesis and Environmental Remediation: Are We a Step Closer to Achieving Sustainability? *Chem. Rev.* **2016**, *116*, 7159–7329. [CrossRef]
23. He, B.; Chen, S.; Cui, Y.; Chen, X.; Lei, Y.; Sun, J. Hollow polymeric ionic liquid spheres with hierarchical electron distribution: A novel composite of g-C_3N_4 for visible light photocatalytic water splitting enhancement. *Chem. Eng. J.* **2022**, *440*, 135625. [CrossRef]
24. Rajput, Y.; Kumar, P.; Zhang, T.C.; Kumar, D.; Nemiwal, M. Recent advances in g-C_3N_4-based photocatalysts for hydrogen evolution reactions. *Int. J. Hydrogen Energy* **2022**, *47*, 38533–38555. [CrossRef]
25. Dong, W.; Chang, Q.; Wang, W.; Yang, S. Research Progress in Preparing Methods of g-C_3N_4 Nanosheets by Exfoliation. *J. Chin. Ceram. Soc.* **2023**, *51*, 1868–1882.
26. Sun, X.; Huang, H.; Shi, W.; Sun, L.; Song, Z. Research Progress on Preparation and Application of Modified g-C_3N_4 Photocatalytic Materials. *Prog. Fine Petrochem.* **2025**, *26*, 35–41.
27. Hossain, S.M.; Park, H.; Kang, H.-J.; Mun, J.S.; Tijing, L.; Rhee, I.; Kim, J.-H.; Jun, Y.-S.; Shon, H.K. Synthesis and NO_x Removal Performance of Anatase S–TiO_2/g-C_3N_4 Heterojunction Formed from Dye Wastewater Sludge. *Chemosphere* **2021**, *275*, 130020. [CrossRef] [PubMed]
28. Lau, V.W.H.; Lotsch, B.V. A Tour-Guide through Carbon Nitride-Land: Structure- and Dimensionality-Dependent Properties for Photo(Electro)Chemical Energy Conversion and Storage. *Adv. Energy Mater.* **2022**, *12*, 2101078. [CrossRef]
29. Fujishima, A.; Honda, K. Electrochemical Photolysis of Water at a Semiconductor Electrode. *Nature* **1972**, *238*, 37–38. [CrossRef]
30. Kamat, P.V. A Conversation with Akira Fujishima. *ACS Energy Lett.* **2017**, *2*, 1586–1587. [CrossRef]
31. Suresh, R.; Gnanasekaran, L.; Rajendran, S.; Soto-Moscoso, M.; Chen, W.-H.; Show, P.L.; Khoo, K.S. Application of Nanocomposites in Integrated Photocatalytic Techniques for Water Pollution Remediation. *Environ. Technol. Innov.* **2023**, *31*, 103149. [CrossRef]
32. Yang, Y.; Ling, R.; Zhou, S.; Duan, Y.; Jiang, P.; Wang, K. Research Progress on the Application of Photocatalytic Technology in Water Treatment. *Fine Chem. Ind.* **2024**, *41*, 707–718.

33. Xavier, M.M.; Mohanapriya, S.; Divya, K.S.; Adarsh, N.N.; Nair, P.R.; Mathew, S. Exploring the Effect of Precursors of Polymeric Carbon Nitride Nanosheets on Their Photo- and Electrocatalytic Applications. *ChemistrySelect* **2020**, *5*, 12679–12689. [CrossRef]
34. Cai, T.; Liu, Y.; Wang, L.; Dong, W.; Zeng, G. Recent Advances in Round-the-Clock Photocatalytic System: Mechanisms, Characterization Techniques and Applications. *J. Photochem. Photobiol. C* **2019**, *39*, 58–75. [CrossRef]
35. Xin, X.; Shunheng, T.; Mingli, L.; Zhong, H.; Zheng, C.; Zuo, X.; Nan, J. Discussion on the Reaction Mechanism of the Photocatalytic Degradation of Organic Contaminants from a Viewpoint of Semiconductor Photo-Induced Electrocatalysis. *Appl. Catal. B* **2016**, *198*, 124–132.
36. Zhang, L.; Zhang, J.; Yu, H.; Yu, J. Emerging S-Scheme Photocatalyst. *Adv. Mater.* **2021**, *34*, 2107668. [CrossRef]
37. Agustina, M.; Lucila, M.S.; Orlando, M.A. Photocatalytic Reactors with Suspended and Immobilized TiO_2: Comparative Efficiency Evaluation. *Chem. Eng. J.* **2017**, *326*, 29–36.
38. Jafari Foruzin, L.; Rezvani, Z.; Nejati, K.; Hou, Z.; Dai, H.; Asadpour-Zeynali, K. High Quantum Efficiency of Photocatalytic Water Oxidation over the TiO_2/MMO Nanocomposite under Visible-Light Irradiation. *J. Mol. Liq.* **2019**, *288*, 111035. [CrossRef]
39. Mehtab, A.; Banerjee, S.; Mao, Y.; Ahmad, T. Type-II $CuFe_2O_4$/Graphitic Carbon Nitride Heterojunctions for High-Efficiency Photocatalytic and Electrocatalytic Hydrogen Generation. *ACS Appl. Mater. Interfaces* **2022**, *14*, 44317–44329. [CrossRef]
40. Wu, X.; Xie, S.; Zhang, H.; Zhang, Q.; F.Sels, B.; Wang, Y. Metal Sulfide Photocatalysts for Lignocellulose Valorization. *Adv. Mater.* **2021**, *33*, 2007129. [CrossRef]
41. Samanta, S.; Srivastava, R. Catalytic Conversion of CO_2 to Chemicals and Fuels: The Collective Thermocatalytic/Photocatalytic/Electrocatalytic Approach with Graphitic Carbon Nitride. *Mater. Adv.* **2020**, *1*, 1506–1545. [CrossRef]
42. Cui, Q.; Li, Z.; Zhang, B.; Zhao, Y.; Nan, F. A Meta-Analysis on the Effects of Freeze–Thaw Cycles on Soil Soluble Carbon and Nitrogen and Microbial Biomass Carbon and Nitrogen Content. *J. Ecol. Environ.* **2022**, *31*, e008031.
43. Tian, H.F.; Song, L.M. Research Progress of g-C_3N_4 Photocatalyst. *J. Tianjin Univ. Technol.* **2012**, *31*, 5.
44. Yang, N.; Chen, Z.; Zhao, Z.; Cui, Y. Electrochemical Fabrication of Ultrafine g-C_3N_4 Quantum Dots as a Catalyst for the Hydrogen Evolution Reaction. *New Carbon Mater.* **2022**, *37*, 392–399. [CrossRef]
45. Miao, W.; Liu, Y.; Chen, X.; Zhao, Y.; Mao, S. Tuning Layered Fe-Doped g-C_3N_4 Structure through Pyrolysis for Enhanced Fenton and Photo-Fenton Activities. *Carbon* **2020**, *159*, 461–470. [CrossRef]
46. Zhang, X.; He, S.; Jiang, S.P. WOx/g-C_3N_4 Layered Heterostructures with Controlled Crystallinity towards Superior Photocatalytic Degradation and H_2 Generation. *Carbon* **2020**, *156*, 488–498. [CrossRef]
47. Wang, X.; Sang, L.; Zhang, L.; Yang, G.; Guo, Y.; Yang, Y. Controllable Synthesis for Carbon Self-Doping and Structural Defect Co-Modified g-C_3N_4: Enhanced Photocatalytic Oxidation Performance and Mechanism Insight. *J. Alloys Compd.* **2023**, *941*, 168921. [CrossRef]
48. Liu, C.; Liu, L.; Tian, X.; Wang, Y.; Li, R.; Zhang, Y.; Song, Z.; Xu, B.; Chu, W.; Qi, F.; et al. Coupling Metal–Organic Frameworks and g-C_3N_4 to Derive Graphene-Like Carbon for Peroxymonosulfate Activation: Upgrading Framework Stability and Performance. *Appl. Catal. B* **2019**, *255*, 117763. [CrossRef]
49. Oh, Y.; Hwang, J.; Lee, E.-S.; Yoon, M.; Le, V.-D.; Kim, Y.-H.; Kim, D.; Kim, S. Divalent Fe Atom Coordination in Two-Dimensional Microporous Graphitic Carbon Nitride. *ACS Appl. Mater. Interfaces* **2016**, *8*, 38. [CrossRef]
50. Jung, H.; Pham, T.-T.; Shin, E.W. Effect of g-C_3N_4 Precursors on the Morphological Structures of g-C_3N_4/ZnO Composite Photocatalysts. *J. Alloys Compd.* **2019**, *788*, 1084–1092. [CrossRef]
51. Wang, Q.; Xu, P.; Zhang, G.; Hu, L.; Wang, P. Visible-Light Responsive g-C_3N_4 Coupled with ZnS Nanoparticles via a Rapid Microwave Route: Characterization and Enhanced Photocatalytic Activity. *Appl. Surf. Sci.* **2019**, *488*, 360–369. [CrossRef]
52. Wei, Y.; Liu, Y.; Liu, C.; Li, X.; Song, K.; Wang, R.; Chen, W.; Zhao, G.; Liu, R.; Wang, H.; et al. Oxygen-Doped Porous g-C_3N_4 via Oxalic Acid-Assisted Thermal Polycondensation as a Visible-Light-Driven Photocatalyst for Bisphenol A Degradation. *ACS Appl. Nano Mater.* **2023**, *6*, 16567–16579. [CrossRef]
53. Cui, L.; Song, J.; McGuire, A.F.; Kang, S.; Fang, X.; Wang, J.; Yin, C.; Li, X.; Wang, Y.; Cui, B. Constructing Highly Uniform Onion-Ring-Like Graphitic Carbon Nitride for Efficient Visible-Light-Driven Photocatalytic Hydrogen Evolution. *ACS Nano* **2018**, *12*, 5551–5558. [CrossRef]
54. Ye, L.; Chen, S. Fabrication and High Visible-Light-Driven Photocurrent Response of g-C_3N_4 Film: The Role of Thiourea. *Appl. Surf. Sci.* **2016**, *389*, 1076–1083. [CrossRef]
55. Nguyen, P.A.; Nguyen, T.K.A.; Dao, D.Q.; Shin, E.W. Ethanol Solvothermal Treatment on Graphitic Carbon Nitride Materials for Enhancing Photocatalytic Hydrogen Evolution Performance. *Nanomaterials* **2022**, *12*, 179. [CrossRef]
56. Wang, Z.; Wang, J.; Iqbal, W.; Shi, M.; Yang, L.; Chang, N.; Qin, C. Morphology-Effects of Four Different Dimensional Graphitic Carbon Nitrides on Photocatalytic Performance of Dye Degradation, Water Oxidation and Splitting. *J. Phys. Chem. Solids* **2022**, *173*, 111109. [CrossRef]
57. Sharma, P.; Sarngan, K.; Lakshmanan, A.; Sarkar, D. One-Step Synthesis of Highly Reactive g-C_3N_4. *J. Mater. Sci. Mater. Electron.* **2021**, *33*, 9116–9125. [CrossRef]

58. Zhao, B.; Zhong, W.; Chen, F.; Wang, P.; Bie, C.; Yu, H. High-crystalline g-C_3N_4 photocatalysts: Synthesis, structure modulation, and H2-evolution application. *Chin. J. Catal.* **2023**, *52*, 127–143. [CrossRef]
59. Tang, S.; Xing, Y.; Wang, Y.; Wei, G. Recent advances in graphitic carbon nitride-based nanocomposites for energy storage and conversion applications. *Nanotechnology* **2025**, *36*, 122002. [CrossRef]
60. Liu, H.; Yang, B.; Liao, G.; Huang, B.; Li, J.; Rodriguez, R.D.; Jia, X. Facilitating carrier kinetics in ultrathin porous carbon nitride through shear-repair strategy for peroxymonosulfate-assisted water purification. *Nat. Commun.* **2025**, *16*, 5909. [CrossRef]
61. Ghosh, U.; Pal, A. Graphitic carbon nitride based Z scheme photocatalysts: Design considerations, synthesis, characterization and applications. *J. Ind. Eng. Chem.* **2019**, *79*, 383–408. [CrossRef]
62. Li, D.; Song, H.; Shen, T.; Sun, J.; Han, W.; Wang, X. Effects of Particle Size on the Structure and Photocatalytic Performance by Alkali-Treated TiO_2. *Nanomaterials* **2020**, *10*, 546.
63. Gao, L.; Huang, X. Research Progress on Catalytic Activity of Nano-TiO_2 Semiconductor. *Electron. Compon. Mater.* **2002**, *21*, 4.
64. Li, Y.; Jin, R.; Xing, Y.; Li, J.; Song, S.; Liu, X.; Li, M.; Jin, R. Macroscopic Foam-Like Holey Ultrathin g-C_3N_4 Nanosheets for Drastic Improvement of Visible-Light Photocatalytic Activity. *Adv. Energy Mater.* **2016**, *6*, 1601273. [CrossRef]
65. Xiao, J.; Liu, Q.; Song, M.; Li, X.; Li, Q.; Shang, J. Directing Photocatalytic Pathway to Exceedingly High Antibacterial Activity in Water by Functionalizing Holey Ultrathin Nanosheets of Graphitic Carbon Nitride. *Water Res.* **2021**, *198*, 117125. [CrossRef] [PubMed]
66. Gao, S.; Wang, X.; Song, C.; Zhou, S.; Yang, F.; Kong, Y. Engineering Carbon-Defects on Ultrathin g-C_3N_4 Allows One-Pot Output and Dramatically Boosts Photoredox Catalytic Activity. *Appl. Catal. B* **2021**, *295*, 120272. [CrossRef]
67. Xue, Y.; Ma, C.; Yang, Q.; Wang, X.; An, S.; Zhang, X.; Tian, J. Construction of g-C_3N_4 with Three-Coordinated Nitrogen (N_3C) Vacancies for Excellent Photocatalytic Activities of N_2 Fixation and H_2O_2 Production. *Chem. Eng. J.* **2023**, *457*, 141146. [CrossRef]
68. Li, W.; Zeng, H.; Zhou, Z.; Li, L.; Tang, R.; Ding, C.; Gong, D.; Deng, Y. Sulfur-Synergized Dual-Cobalt Anchoring Configuration in Carbon Nitride: Deciphering Cooperative Mechanisms for Boosted Peroxymonosulfate Activation. *Chem. Eng. J.* **2025**, *520*, 166214. [CrossRef]
69. Mottammal, D.; Cherusseri, J.; Thomas, S.A.; R.S., R.L.; N.Rajendran, D.; Choi, M.Y. Toward Doping in Graphitic Carbon Nitride: Progress and Perspectives on Catalytic Hydrogen Production. *Adv. Mater. Technol.* **2025**, e00667. [CrossRef]
70. Qaraah, F.A.; Mahyoub, S.A.; Shen, H.; Yin, X.; Salah, A.; A.Onaizi, S.; A.Drmosh, Q.; Xin, F. Synergistic Role of Dual-Metal Sites (Ag–Ni) in Hexagonal Porous g-C_3N_4 Nanostructures for Enhanced Photocatalytic CO_2 Reduction. *Carbon* **2025**, *232*, 119735. [CrossRef]
71. Niu, W.; Yang, Y. Graphitic Carbon Nitride for Electrochemical Energy Conversion and Storage. *ACS Energy Lett.* **2018**, *3*, 2796–2815. [CrossRef]
72. Mishra, S.; Kumar, A.; Verma, N. g-C_3N_4-Supported $NiBO_3$ (B = Ti, Sn) Perovskite-Based Dual-S-Scheme Heterostructure for Efficient Hydrogen Production via Water Splitting. *Fuel* **2025**, *399*, 135673. [CrossRef]
73. Low, J.; Yu, J.; Jaroniec, M.; Wageh, S.; Al-Ghamdi, A.A. Heterojunction Photocatalysts. *Adv. Mater.* **2017**, *29*, 1601694. [CrossRef] [PubMed]
74. Wang, W.; Fang, J.; Shao, S.; Lai, M.; Lu, C. Compact and Uniform TiO_2@g-C_3N_4 Core–Shell Quantum Heterojunction for Photocatalytic Degradation of Tetracycline Antibiotics. *Appl. Catal. B* **2017**, *217*, 57–64. [CrossRef]
75. Bard, A.J. Photoelectrochemistry and Heterogeneous Photocatalysis at Semiconductors. *J. Photochem.* **1979**, *10*, 59–75. [CrossRef]
76. Cadan, F.M.; Ribeiro, C.; Azevedo, E.B. Improving g-C_3N_4:WO_3 Z-Scheme Photocatalytic Performance under Visible Light by Multivariate Optimization of g-C_3N_4 Synthesis. *Appl. Surf. Sci.* **2021**, *537*, 147904. [CrossRef]
77. Chen, Y.; Su, F.; Xie, H.; Wang, R.; Ding, C.; Huang, J.; Xu, Y.; Ye, L. One-Step Construction of S-Scheme Heterojunctions of N-Doped MoS_2 and S-Doped g-C_3N_4 for Enhanced Photocatalytic Hydrogen Evolution. *Chem. Eng. J.* **2020**, *404*, 126498. [CrossRef]
78. Zeltner, W.A.; Tompkins, D.T. Shedding Light on Photocatalysis. *ASHRAE Trans.* **2005**, *111 Pt 2*, 523–534.
79. Zheng, Y.; Liu, J.; Liang, J.; Jaroniec, M.; Qiao, S.Z. Graphitic Carbon Nitride Materials: Controllable Synthesis and Applications in Fuel Cells and Photocatalysis. *Energy Environ. Sci.* **2012**, *5*, 6717–6731. [CrossRef]
80. Song, S.; Zhuo, Z.; Huang, M. Research Progress on Piezoelectric-Photocatalytic Technology and Its Application in Degradation of Organic Dye Wastewater. *Green Technol.* **2022**, *24*, 86–94.
81. Wang, Z.; Hu, X.; Liu, Z.; Zou, C.; Wang, G.; Zhang, K. Recent Developments in Polymeric Carbon Nitride-Derived Photocatalysts and Electrocatalysts for Nitrogen Fixation. *ACS Catal.* **2019**, *9*, 10260–10278. [CrossRef]
82. Besharat, F.; Ahmadpoor, F.; Nezafat, Z.; Nasrollahzadeh, M.; Manwar, N.; Fornasiero, P.; Gawande, M. Advances in Carbon Nitride-Based Materials and Their Electrocatalytic Applications. *ACS Catal.* **2022**, *12*, 5605–5660. [CrossRef]
83. Xie, W.; Gao, L.; Wu, L.; Chen, X.; Wang, F.; Tong, D.; Zhang, J.; Lan, J.; He, X.; Mu, X.; et al. A Nonresonant Hybridized Electromagnetic-Triboelectric Nanogenerator for Irregular and Ultralow Frequency Blue Energy Harvesting. *Research* **2021**, *2021*, 5963293. [CrossRef] [PubMed]
84. Ochiai, T. Electrostatics of Optical Rectification in Metallic Particles. *J. Opt. Soc. Am. B* **2022**, *39*, 990–1002. [CrossRef]

85. Toprak, A.; Tigli, O. Piezoelectric Energy Harvesting: State-of-the-Art and Challenges. *Appl. Phys. Rev.* **2014**, *1*, 031104. [CrossRef]
86. Cheng, T.; Gao, Q.; Wang, Z.L. The Current Development and Future Outlook of Triboelectric Nanogenerators: A Survey of Literature. *Adv. Mater. Technol.* **2019**, *4*, 1800588. [CrossRef]
87. Bowen, C.R.; Taylor, J.; LeBoulbar, E.; Zabek, D.; Chauhan, D. Pyroelectric Materials and Devices for Energy Harvesting Applications. *Energy Environ. Sci.* **2014**, *7*, 3836–3856. [CrossRef]
88. Ng, T.; Liao, W. Sensitivity Analysis and Energy Harvesting for a Self-Powered Piezoelectric Sensor. *J. Intell. Mater. Syst. Struct.* **2005**, *16*, 785–795. [CrossRef]
89. Caliò, R.; Rongala, U.B.; Camboni, D.; Milazzo, M.; Stefanini, C.; Petris, G.; Oddo, C. Piezoelectric Energy Harvesting Solutions. *Sensors* **2014**, *14*, 4755–4796. [CrossRef]
90. Ali, F.; Raza, W.; Li, X.; Gul, H.; Kim, K. Piezoelectric Energy Harvesters for Biomedical Applications. *Nano Energy* **2019**, *57*, 879–902. [CrossRef]
91. Wang, J.; Geng, L.; Zhou, S.; Zhang, Z.; Lai, Z.; Yurchenko, D. Design, Modeling and Experiments of Broadband Tristable Galloping Piezoelectric Energy Harvester. *Acta Mech. Sin.* **2020**, *36*, 592–605. [CrossRef]
92. Panda, P.; Sahoo, B. PZT to Lead-Free Piezo Ceramics: A Review. *Ferroelectrics* **2015**, *474*, 128–143. [CrossRef]
93. Le, A.T.; Ahmadipour, M.; Pung, S.-Y. A Review on ZnO-Based Piezoelectric Nanogenerators: Synthesis, Characterization Techniques, Performance Enhancement and Applications. *J. Alloys Compd.* **2020**, *844*, 156172. [CrossRef]
94. Lu, L.J.; Ding, W.Q.; Liu, J.Q.; Yang, B. Flexible PVDF-Based Piezoelectric Nanogenerators. *Nano Energy* **2020**, *78*, 105251. [CrossRef]
95. Piao, H.; Choi, G.; Jin, X.; Hwang, S.; Song, Y.; Cho, S.; Choy, J. Monolayer Graphitic Carbon Nitride as Metal-Free Catalyst with Enhanced Performance in Photo- and Electro-Catalysis. *Nano-Micro Lett.* **2022**, *14*, 55. [CrossRef]
96. Liu, W.; Zhou, C.; Gao, J.; Xue, D.; Zhang, L.; Li, S.; Ren, X. Research Progress of Lead-Free Piezoelectric ceramics. *Prog. Chin. Mater.* **2016**, *35*, 423–428.
97. Kumar, A.; Choudhary, P.; Chhabra, T.; Kaur, H.; Kumar, A.; Qamar, M.; Krishnan, V. Frontier Nanoarchitectonics of Graphitic Carbon Nitride Based Plasmonic Photocatalysts and Photoelectrocatalysts for Energy, Environment and Organic Reactions. *Mater. Chem. Front.* **2023**, *7*, 1197–1247. [CrossRef]
98. Bai, Y.; Zhao, J.; Lv, Z.; Lu, K. Enhanced Piezo-Phototronic Effect of ZnO Nanorod Arrays for Harvesting Low Mechanical Energy. *Ceram. Int.* **2019**, *45*, 15065–15072. [CrossRef]
99. Fu, Y.; Wang, Y.; Zhao, H.; Zhang, Z.; An, B.; Bai, C.; Ren, Z.; Wu, J.; Li, Y.; Liu, W.; et al. Synthesis of Ternary ZnO/ZnS/MoS_2 Piezoelectric Nanoarrays for Enhanced Photocatalytic Performance by Conversion of Dual Heterojunctions. *Appl. Surf. Sci.* **2021**, *556*, 149695. [CrossRef]
100. Bayan, S.; Bhattacharya, D.; Mitra, R.K.; Ray, S. Self-Powered Flexible Photodetectors Based on Ag Nanoparticle Loaded g-C_3N_4 Nanosheets and PVDF Hybrids: Role of Plasmonic and Piezoelectric Effects. *Nanotechnology* **2020**, *31*, 365401. [CrossRef] [PubMed]
101. Dai, B.; Huang, H.; Wang, F.; Lu, C.; Kou, J.; Wang, L.; Xu, Z. Flowing Water Enabled Piezoelectric Potential of Flexible Composite Film for Enhanced Photocatalytic Performance. *Chem. Eng. J.* **2018**, *347*, 263–272. [CrossRef]
102. Nagar, O.P.; Chouhan, N. Non-Metal Doped Graphitic Carbon Nitride (g-C_3N_4): Prospects and Review on Hydrogen Generation via Water Splitting. *Int. J. Hydrogen Energy* **2024**, *96*, 533–565. [CrossRef]
103. Thennarasu, G.; Rajendran, S.; Kalairaj, A.; Rathore, H.; Panda, R.; Senthivelan, T. A Comprehensive Review on the Application of Semiconductor Nanometal Oxides Photocatalyst for the Treatment of Wastewater. *Clean Technol. Environ. Policy* **2025**, *27*, 495–516. [CrossRef]
104. Liang, F.; Chen, Z.; Lu, Z.; Wang, X. Piezoelectric-Enhanced Photocatalytic Performance of Porous Carbon Nitride Nanosheets. *J. Colloid Interface Sci.* **2023**, *630*, 191–203. [CrossRef]
105. Wu, Z.; Zhu, Z.; Ma, J.; Zhou, M.; Wu, Z.; You, H.; Zhang, H.; Li, N.; Wang, F. High piezo-photocatalysis of BaTiO3 nanofibers for organic dye decomposition. *Surf. Interfaces* **2024**, *48*, 104308. [CrossRef]
106. Lan, S.; Yu, C.; Sun, F.; Chen, Y.; Chen, D.; Mai, W.; Zhu, M. Tuning piezoelectric driven photocatalysis by La-doped magnetic BiFeO3-based multiferroics for water purification. *Nano Energy* **2022**, *93*, 106792. [CrossRef]
107. Xie, Z.; Tang, X.; Shi, J.; Wang, Y.; Yuan, G.; Liu, J.-M. Excellent piezo-photocatalytic performance of Bi4Ti3O12 nanoplates synthesized by molten-salt method. *Nano Energy* **2022**, *98*, 107247. [CrossRef]
108. Abdollahpour, R.; Bazyari, A. Synergistic Photocatalysis: Enhanced Degradation of Organic Dyes Using a Heterojunction Nanocomposite of High-Surface-Area g-C_3N_4 and TiO_2–WO_3–Bi_2O_3/SiO_2. *Environ. Sci. Pollut. Res.* **2025**, *32*, 17466–17486. [CrossRef]
109. Tong, H.; Li, F.-F.; Du, M.; Song, H.; Han, B.; Jia, G.; Xu, X.; Zou, X.; Ji, L.; Kai, J.; et al. Interface Engineering, Charge Carrier Dynamics, and Solar-Driven Applications of Halide Perovskite/2D Material Heterostructured Photocatalysts. *ACS Appl. Mater. Interfaces* **2025**, *17*, 23431–23465. [CrossRef]

110. Xu, Y.; Liao, J.; Zhang, L.; Sun, Z.; Ge, C. Dual Sulfur Defect Engineering of Z-Scheme Heterojunction on Ag-CdS_{1-x}@$ZnIn_2S_{4-x}$ Hollow Core–Shell for Ultra-Efficient Selective Photocatalytic H_2O_2 Production. *J. Colloid Interface Sci.* **2023**, *647*, 446–455. [CrossRef] [PubMed]
111. Cai, J.; Maimaitizi, H.; Okitsu, K.; Tursun, Y.; Abulizi, A. Z-Type Heterojunction of Graphene Quantum Dots/g-C_3N_4/BiOCl Has Excellent Nitrogen Fixation Photocatalytic Performance. *Int. J. Energy Res.* **2022**, *46*, 12147–12159. [CrossRef]
112. Abdolmaleki, H.; Haugen, A.B.; Buhl, K.B.; Daasbjerg, K.; Agarwala, S. Interface Engineering of PVDF-TrFE to Achieve Higher Piezoelectric, Ferroelectric and Dielectric Properties for Sensing and Energy Harvesting Applications. *Adv. Sci.* **2023**, *10*, 2205942. [CrossRef] [PubMed]
113. Huang, J.; Hu, J.; Shi, Y.; Zeng, G.; Cheng, W.; Yu, H.; Gu, Y.; Shi, L.; Yi, K. Evaluation of Self-Cleaning and Photocatalytic Properties of Modified g-C_3N_4-Based PVDF Membranes Driven by Visible Light. *J. Colloid Interface Sci.* **2019**, *541*, 356–366. [CrossRef]
114. Wang, G.; Cheng, H. Application of Photocatalysis and Sonocatalysis for Treatment of Organic Dye Wastewater and the Synergistic Effect of Ultrasound and Light. *Molecules* **2023**, *28*, 123456. [CrossRef]
115. Ye, H.; Guo, T.; Zuo, G. Preparation and Photocatalytic Properties of g-C_3N_4/$BiFeO_3$ Composite Visible-Light Catalyst. *China Ceram.* **2016**, *52*, 5.
116. Kang, Z.; Ke, K.; Lin, E.; Qin, N.; Wu, J.; Huang, R.; Bao, D. Piezoelectric Polarization Modulated Novel Bi_2WO_6/g-C_3N_4/ZnO Z-Scheme Heterojunctions with g-C_3N_4 Intermediate Layer for Efficient Piezo-Photocatalytic Decomposition of Harmful Organic Pollutants. *J. Colloid Interface Sci.* **2022**, *607*, 1589–1602. [CrossRef]
117. Huang, S.; Xiong, F.; Yu, M.; Zhou, Y.; Xu, J.; Liu, J. Synthesis of Ag-Loaded $NaNbO_3$/g-C_3N_4 Heterojunction for Enhanced Photocatalytic Degradation of Methyl Orange. *Mater. Sci. Semicond. Process.* **2025**, *192*, 109401. [CrossRef]
118. Liu, G.; Zhao, T.; Wu, J.; Chang, M.; Fei, H.; Li, F.; Yang, S.; Li, Q. Enhanced Removal and Selective Conversion for NO with N-Vacancies g-C_3N_4/$BaTiO_3$ by Piezo-Photocatalysis. *Sep. Purif. Technol.* **2025**, *360*, 130914. [CrossRef]
119. Di, B.; Wang, Z.; Wang, H.; Gong, S.; Zheng, L.; Min, Y.; Li, H. Piezoelectric Photocatalytic Degradation of Formaldehyde Based on BFO@OCN Heterojunctions. *Nano Energy* **2024**, *132*, 110384. [CrossRef]
120. Kishore, A.; Sunil, S.M.; Viswanathan, R.; Kailaperumal, A.; Ali, B. Different Synthetic Routes and Band-Gap Engineering of Photocatalysts. In *Photocatalysts and Electrocatalysts in Water Remediation: From Fundamentals to Full-Scale Applications*; Wiley: New York, NY, USA, 2022; pp. 39–80.
121. Li, H.; Luo, K.Y.; Hu, W.Y.; Li, J.; Wang, Z.; Ma, C.; Pan, Z.; Zhang, Q.; Yuan, H.; Yu, F.; et al. Facile Synthesis of Core–Shell Structure ZnO/g-C_3N_4 Nanocomposites with Enhanced Photocatalytic Activity. *Acta Photonica Sin.* **2021**, *50*, 1116001.
122. Zhang, B.; Jin, X.; Li, X.; Dong, L.; Liu, D.; Zhang, Y.; Zhao, Z.; Ge, Q.; Zhang, F. Piezoelectric Assisted Boosting Photocatalytic Hydrogen Evolution over a Finely Prepared Pt/TiO_2/g-C_3N_4 Composite System Using Lactic Acid as Sacrificial Agent. *Int. J. Hydrogen Energy* **2024**, *84*, 480–490. [CrossRef]
123. Deng, X.; Chen, P.; Cui, R.; Gong, X.; Li, X.; Wang, X.; Deng, C. Synergistic Polarity Interaction and Structural Reconstruction in Bi_2MoO_6/C_3N_4 Heterojunction for Enhancing Piezo-Photocatalytic Nitrogen Oxidation to Nitric Acid. *Appl. Catal. B* **2024**, *351*, 123977. [CrossRef]
124. Guo, Z.; Sun, K.; Zou, S.; Xiong, B.; Wang, L.; Shi, W.; Sun, Y.; Guo, F. Construction of Surface Pit-Structured g-C_3N_4 by Induced SiO_2 Hard Template for Boosted Piezoelectric-Assisted Photocatalytic H_2O_2 Production. *J. Colloid Interface Sci.* **2025**, *699*, 138118. [CrossRef]
125. Sun, P.; Xiong, J.; Sun, P.; Fang, Y.; Liu, H.; Liu, H.; Xiong, B.; Wang, H.; Li, X.A. Additional O doping significantly improved the catalytic performance of Mn/O co-doped g-C_3N_4 for activating periodate and degrading organic pollutants. *Sep. Purif. Technol.* **2024**, *331*, 125593. [CrossRef]
126. Wu, H.-H.; Chang, C.-W.; Lu, D.; Maeda, K.; Hu, C. Synergistic Effect of Hydrochloric Acid and Phytic Acid Doping on Polyaniline-Coupled g-C_3N_4 Nanosheets for Photocatalytic Cr (VI) Reduction and Dye Degradation. *ACS Appl. Mater. Interfaces* **2019**, *11*, 35702–35712. [CrossRef]
127. Sheng, Y.; Wei, Z.; Miao, H.; Yao, W.; Li, H.; Zhu, Y. Enhanced organic pollutant photodegradation via adsorption/photocatalysis synergy using a 3D g-C_3N_4/TiO_2 free-separation photocatalyst. *Chem. Eng. J.* **2019**, *370*, 287–294. [CrossRef]
128. Zhao, Z.; Liu, J.; Fan, J.; Zheng, H. Preparation and performance of $In(OH)_3$/g-C_3N_4 catalyst containing oxygen vacancies for photodegradation of pollutants in water. *Journal of civil and environmental engineering.* **2020**, *42*, 207–208.
129. Zhang, X.; Wu, H.; Lin, L.; Wu, Y.; Wang, M. Degradation mechanism of azo dyes -contaminated wastewater by g-C_3N_4-P25/photosynthetic bacteria composite. *Acta Sci. Circumstantiae* **2018**, *38*, 2632–2640.
130. Khadim, H.J.; Ammar, S.H.; Al-Farraji, A.; Mohammed, M.S.; Jabbar, Z.H.; Alabdly, H.A. Fabrication of AgBr/Br-doped g-C_3N_4 hybrids for efficient piezo-photocatalytic degradation of organic pollutants in split-plate airlift reactor. *J. Water Process Eng.* **2025**, *76*, 108118. [CrossRef]
131. Mohammed, M.S.; Ammar, S.H.; Kareem, Y.S.; Al-Farraji, A. Assembly of Fe-doped g-C_3N_4 as a robust piezophotocatalyst system for degradation of organic dyes. *J. Mol. Struct.* **2025**, *1344*, 142962. [CrossRef]

132. Wu, T.; Liu, Z.; Shao, B.; He, Q.; Pan, Y.; Zhang, X.; Sun, J.; He, M.; Ge, L.; Cheng, C.; et al. Enhanced piezo-photocatalytic degradation of organic pollutants by cambered wall lamellar structure of porous tubular g-C_3N_4. *Nano Energy* **2024**, *120*, 109137. [CrossRef]
133. Gong, S.; Zhang, W.; Liang, Z.; Zhang, Y.; Gan, T.; Hu, H.; Huang, Z. Construction of a $BaTiO_3$/tubular g-C_3N_4 dual piezoelectric photocatalyst with enhanced carrier separation for efficient degradation of tetracycline. *Chem. Eng. J.* **2023**, *461*, 141947. [CrossRef]
134. Tang, R.; Gong, D.; Zhou, Y.; Deng, Y.; Feng, C.; Xiong, S.; Huang, Y.; Peng, G.; Li, L.; Zhou, Z. Unique g-C_3N_4/PDI-g-C_3N_4 homojunction with synergistic piezo-photocatalytic effect for aquatic contaminant control and H_2O_2 generation under visible light. *Appl. Catal. B. Environ.* **2022**, *303*, 120929. [CrossRef]
135. Cui, Y.; Wang, Z.; Li, B.; Yan, Y.; Xu, R.; Meng, M.; Yan, Y. Fluid-induced piezoelectric field enhancing photocatalytic hydrogen evolution reaction on g-C_3N_4/LiNbO3/PVDF membrane. *Nano Energy* **2022**, *99*, 107429. [CrossRef]
136. Wu, J.; Qin, N.; Bao, D. Effective enhancement of piezocatalytic activity of $BaTiO_3$ nanowires under ultrasonic vibration. *Nano Energy* **2018**, *45*, 44–51. [CrossRef]
137. Tong, R.; Sun, Z.; Wang, X.; Yang, L.; Zhai, J.; Wang, S.; Pan, H. Mo incorporated Ni nanosheet as high-efficiency co-catalyst for enhancing the photocatalytic hydrogen production of g-C_3N_4. *Int. J. Hydrogen Energy* **2020**, *45*, 18912–18921. [CrossRef]
138. Xu, H.; She, X.; Fei, T.; Song, Y.; Liu, D.; Li, H.; Yang, X.; Yang, J.; Li, H.; Song, L.; et al. Metal-Oxide-Mediated Subtractive Manufacturing of Two-Dimensional Carbon Nitride for High-Efficiency and High-Yield Photocatalytic H_2 Evolution. *ACS Nano* **2019**, *13*, 11294–11302. [CrossRef]
139. Qin, Y.; Li, H.; Lu, J.; Feng, Y.; Meng, F.; Ma, C.; Yan, Y.; Meng, M. Synergy between van der waals heterojunction and vacancy in $ZnIn_2S_4$/g-C_3N_4 2D/2D photocatalysts for enhanced photocatalytic hydrogen evolution. *Appl. Catal. B Environ.* **2020**, *277*, 119254. [CrossRef]
140. Wei, Y.; Wang, M.; Li, H.; Zhang, J.; Zhang, R. Preparation and Photocatalytic Hydrogen Production Properties of Cyano-modified g-C_3N_4. *J. Synth. Cryst.* **2020**, *49*, 631.
141. Meng, L.; Zhao, C.; Chu, H.; Li, Y.-H.; Fu, H.; Wang, P.; Wang, C.-C.; Huang, H. Synergetic piezo-photocatalysis of g-C_3N_4/PCN-224 core-shell heterojunctions for ultrahigh H_2O_2 generation. *Chin. J. Catal.* **2024**, *59*, 346–359. [CrossRef]
142. Fu, M.; Luo, J.; Shi, B.; Tu, S.; Wang, Z.; Yu, C.; Ma, Z.; Chen, X.; Li, X. Promoting Piezocatalytic H_2O_2 Production in Pure Water by Loading Metal-Organic Cage-Modified Gold Nanoparticles on Graphitic Carbon Nitride. *Angew. Chem. Int. Ed.* **2024**, *63*, e202316346. [CrossRef] [PubMed]
143. Wang, P.; Fan, S.; Li, X.; Duan, J.; Zhang, D. Modulating the Molecular Structure of Graphitic Carbon Nitride for Identifying the Impact of the Piezoelectric Effect on Photocatalytic H_2O_2 Production. *ACS Catal.* **2023**, *13*, 9515–9523. [CrossRef]
144. Liu, P.; Liang, T.; Bian, J.; Li, Z.; Zhang, Z.; Jing, L. Improved photocatalytic H_2O_2 production of boron doped g-C_3N_4 by controllably co-modifying nanosized PDI and Ag. *Mater. Res. Bull.* **2023**, *167*, 112370. [CrossRef]
145. Zhou, L.; Feng, J.; Qiu, B.; Zhou, Y.; Lei, J.; Xing, M.; Wang, L.; Zhou, Y.; Liu, Y.; Zhang, J. Ultrathin g-C_3N_4 nanosheet with hierarchical pores and desirable energy band for highly efficient H_2O_2 production. *Appl. Catal. B Environ.* **2020**, *267*, 118396. [CrossRef]
146. Chu, C.; Miao, W.; Li, Q.; Wang, D.; Liu, Y.; Mao, S. Highly efficient photocatalytic H_2O_2 production with cyano and SnO_2 co-modified g-C_3N_4. *Chem. Eng. J.* **2022**, *428*, 132531. [CrossRef]

inorganics

MDPI

Article

Calcium Phosphate Ceramic Powders Prepared from Mechanochemically Activated Precursors

Kostadinka Sezanova [1], Yordanka Tuparova [1], Pavletta Shestakova [2], Pavel Markov [1], Daniela Kovacheva [1] and Diana Rabadjieva [1,*]

[1] Institute of General and Inorganic Chemistry, Bulgarian Academy of Sciences, Acad. G. Bonchev Str., bl. 11, 1113 Sofia, Bulgaria; ksezanova@svr.igic.bas.bg (K.S.); dani.tuparova@abv.bg (Y.T.); pvlmarkov@svr.igic.bas.bg (P.M.); didka@svr.igic.bas.bg (D.K.)

[2] Institute of Organic Chemistry with Centre of Phytochemistry, Bulgarian Academy of Sciences, Acad. G. Bonchev Str., bl. 9, 1113 Sofia, Bulgaria; pavletta.shestakova@orgchm.bas.bg

* Correspondence: didiarab@svr.igic.bas.bg

Abstract: The chemical and structural similarity of calcium orthophosphates to hard tissues in the human body makes them suitable as biomaterials for bone implants, cements, injection systems, etc., for bone regeneration and reconstruction. Tetracalcium phosphate ($Ca_4(PO_4)_2O$, TTCP) is a promising component for such biomaterials due to its high calcium content and alkaline nature. The former makes it suitable for promoting mineralization, while the latter supports neutralization of the acidic environment, helping to prevent inflammation and improve the biocompatibility of the materials. However, it is the least used calcium orthophosphate due to the difficulties in its synthesis. This study examines the effect of high-energy mechanochemical activation on the phase evolution, particle morphology, and thermal behaviour of equimolar mixtures of $Ca(OH)_2$ and $CaHPO_4$, with the aim of optimizing precursor conditions for the synthesis of (TTCP)-rich ceramic materials. The results demonstrate that mechanochemical activation effectively induces structural disorder, promotes the formation of amorphous and nanocrystalline phases, and facilitates subsequent phase transitions upon calcination. The combined use of solid-state NMR, XRD, TEM, and thermal analysis provides a comprehensive understanding of the transformation pathways. Ultimately, 24 h of activation under the experimental conditions was identified as optimal for producing a precursor with a favorable phase composition for obtaining TTCP-rich ceramic materials after calcination at 1350 °C.

Keywords: calcium phosphates; mechanochemical activation; solid-state NMR; thermal analysis; XRD analysis

1. Introduction

Calcium orthophosphates are essential in obtaining bone-like biomaterials due to their chemical and structural compatibility with hard biological tissues, ability to support bone growth, and to be absorbed by the body [1]. If we assume that the facilitation of cell adhesion, proliferation, and differentiation, which ultimately leads to the formation of new bone, is more related to the morphology of these materials [2], then bioresorbability is an effect primarily of the solubility of the calcium-phosphate compound [3]. For example, hydroxyapatite ($Ca_5(PO_4)_3OH$, HA), closely mimics the mineral component of natural bone and possesses highly biocompatible and osteoconductive properties, but it is the thermodynamically most stable calcium orthophosphate and has the lowest solubility [4]. Thus, monophasic HA is practically insoluble in the body fluids and does not actively participate in the process of bone remodeling [5]. On the other hand, the monophasic α- and β-tricalcium phosphates ($Ca_3(PO_4)_2$, TCP) are more soluble, degrade faster, and in combination with HA balance bone growth and material resorption [6]. Therefore, it is important to produce multiphase calcium phosphate materials to mitigate the disadvantages of individual phases and enhance their advantages.

From a thermodynamic point of view, the most soluble calcium phosphate at physiological conditions (pH 7–8, and 37 °C) is tetracalcium phosphate ($Ca_4(PO_4)_2O$, TTCP) [4]. The high calcium content makes it suitable for promoting mineralization, and the alkaline nature helps neutralize the acidic environment, which can be useful for preventing inflammation and improving biocompatibility [7–9]. However, it is the least used calcium orthophosphate due to difficulties in its synthesis.

Among synthetic biomaterials, calcium phosphate ceramic powders are widely used in orthopedic and maxillofacial surgery for the production of bone grafts and scaffolds and in dentistry as materials for tooth restoration, for remineralization treatment, and coatings for dental implants [10]. As the primary component, they are integrated in injectable calcium phosphate cements, drug delivery systems, and inorganic/organic composites [11,12].

Various methods have been developed for the production of calcium phosphate ceramic materials, such as additive manufacturing techniques (requiring a 3D printer), selective laser sintering, binder jetting technique, laser-assisted bioprinting, etc., which, however, require specific and expensive equipment [13]. The most common, relatively cheap and effective method is the sintering of precursors synthesized from solutions at ambient temperature or under hydrothermal conditions, by using mechanical activation, sonochemically assisted microwave radiation, etc. [14–16]. Although wet soft chemistry methods are the most common methods for obtaining sparingly soluble calcium orthophosphates, they have a number of disadvantages related to the large number of parameters that must be monitored and maintained within fixed limits to obtain a material with specified structural and morphological properties [17].

Mechanochemical high-energy milling is a fast and effective method for obtaining unified nanoscale materials and energetically activated precursors for the synthesis of high-temperature ceramic powders [18]. Reduction of particle size during the process, increasing the defects of the crystal lattices, developing specific surface area, etc., are prerequisites for lowering the temperatures for the synthesis of high-temperature calcium orthophosphates.

The aims of this work are: (i) systematic investigation of the effect of high-energy mechanochemical activation on the phase and thermal characteristics of equimolar mixture of DCPA ($CaHPO_4$) and $Ca(OH)_2$ to obtain precursors for the preparation of ceramic powders with dominant phase of TTCP, (ii) comparison of the effect of sintering mode on the phase composition of the selected powders and (iii) elucidation the phase transformation processes by using solid-state NMR techniques in combination with XRD and DTA-TG-MASS analyses.

The type and ratio between starting materials were chosen to obtain TTCP as the main phase after sintering. As a component of biomaterials for bone remineralization, TTCP is highly effective in forming HA in vivo, closely mimicking the natural bone mineral [19]. The most common method for obtaining TTCP is heating a mixture of calcium carbonate ($CaCO_3$) and DCPA ($CaHPO_4$) with a Ca/P ratio of 2 in the temperature range of 1450–1500 °C for 6–12 h [9]. Several methods have been proposed in the literature to reduce the synthesis temperature: (i) varying the composition of the starting substances. For example, from $NH_4H_2PO_4$ and $CaCO_3$ at 1350 °C [20] or from a coprecipitated mixture of nanoscale HA and $CaCO_3$ by calcination above 1185 °C [21]; (ii) applying a series of grinding and sintering procedures, in which TTCP was obtained from $NH_4H_2PO_4$ and $Ca(CH_3COO)_2{\bullet}H_2O$ at 1230 °C, but with the presence of HA [22]. In addition, Romeo and Fanovich [23] investigated the solid-state reactions between $CaCO_3$ and $(NH_4)_2HPO_4$ through mechanochemical activation and heat treatment, finding that the smaller crystal size of the milled powders increased reactivity and significantly improved the purity of the final product. This identifies mechanochemical activation as a promising method for obtaining precursors for the synthesis of high-temperature calcium phosphate phases such as TTCP.

In only a few articles, the authors use CaO to obtain TTCP in combination with H_3PO_4 [24], followed by sintering at temperatures up to 1600 °C or with P_2O_5 [25] via the combustion method. Sargin et al. [20] also use CaO or $Ca(OH)_2$ in their studies in combination with $Ca_3(PO_4)_2{\cdot}3H_2O$ and α-$Ca_2P_2O_7$ and sintering in the temperature range 1000–1400 °C to obtain TTCP. Only Jeon et al. [26] reported the synthesis of TTCP from $Ca(OH)_2$ and $CaHPO_4$ by a semi-wet stirring method, in which the starting materials were mixed and stirred with ethanol, dried at 80 °C, and then calcined at 1500 °C. The authors obtained TTCP with 99% purity after 24 h of calcination.

However, there are no studies in the literature on the influence of high-energy activation of a $Ca(OH)_2$ and DCPA mixture on the TTCP preparation temperatures, as well as on the use of solid-state NMR analysis for a more in-depth characterization of phase transformations both during high-energy activation and during high-temperature processing. Both reagents are widely available and relatively inexpensive. Furthermore, upon their heating, only H_2O and adsorbed CO_2 are released, which makes the system ecologically friendly. In systems involving $CaCO_3$, the release of CO_2 is dominant, and in systems with $NH_4H_2PO_4$-ammonia is also released.

2. Results

2.1. Effect of High-Energy Activation on the Phase Composition and Particle Morphology of the Activated Mixtures

The evolution of phase composition during high-energy activation of a $Ca(OH)_2$ and $CaHPO_4$ (DCPA) mixture, investigated using X-ray powder diffraction (XRD), is shown in Figure 1, and quantitative phase analysis is summarized in Table 1.

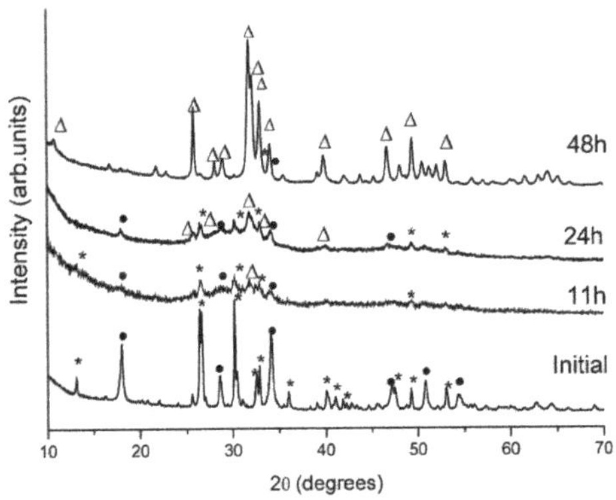

*—$Ca(OH)_2$; •—DCPA; Δ—IIA

Figure 1. X-ray powder patterns of the initial mixture and samples after high-energy activation for 11, 24, and 48 h.

Table 1. Quantitative phase composition calculated from X-ray powder patterns.

Sample	$Ca(OH)_2$ * (%)	DCPA * (%)	HA * (%)	HA Mean Size (nm)	Degree of Crystallinity
Initial	34	66	-		100%
11 h activation	18	45	37	19	8%
24 h activation	11	23	66	20	18%
48 h activation	4	2	94	33	100%

Note: * Phase composition is calculated from the crystalline component.

The results reveal an amorphization of the initial substances and their progressive transformation into hydroxyapatite (HA) with increasing milling time. After 11 h of activation, the degree of sample crystallinity dropped sharply to 8% (Table 1), and the mean crystallite size of the HA was estimated at 19 nm, suggesting the formation of nanocrystalline apatite. We suggest that the amorphous component could be due to both amorphization of the starting crystalline compounds and the formation of amorphous calcium phosphate.

Prolonged activation to 24 h resulted in further conversion, with HA constituting 66% of the crystalline fraction. The crystallinity partially recovered to 18%, and the mean HA crystallite size remained around 20 nm. By 48 h activation, HA accounted for 94% of the crystalline content, and only minor amounts of the precursor phases remained. The degree of crystallinity and crystallite size of HA increased significantly, indicating growth of the apatite phase.

To gain a better understanding of the structural details of the samples, a solid-state NMR spectroscopy analysis was performed.

Figure 2 presents the direct excitation ^{31}P spectra of the initial untreated material and of the samples obtained after high-energy activation times of 11, 24, and 48 h. The spectrum of the untreated sample shows two resonances at −0.3 and −1.5 ppm, which are characteristic of dicalcium phosphate anhydrous (DCPA) [27]. The direct excitation ^{31}P NMR spectra (Figure 2) of all treated samples are dominated by an intense resonance around 2.9 ppm, characteristic of an amorphous calcium phosphate [28]. Additionally, two weak resonances at −0.3 and −1.5 ppm, indicating the presence of a small amount of DCPA, are observed, as well as a weak broad signal around 6.3 ppm. The latter could be attributed to the downfield resonance of β-TCP [29], while the other higher-field resonances of β-TCP are not visible due to overlapping with the broad signal at 2.9 ppm dominating the spectra. However, β-TCP is a high-temperature phase that usually forms upon heating calcium-deficient hydroxyapatite (HA) at temperatures above 800 °C. Osman et al. [30] showed that the chemical shift around 6.3 ppm more likely indicates the presence of surface-deprotonated phosphate groups, whose geometrical bonding distortion is probably similar to that of TCP and may explain the observed chemical shift. With increasing activation time, a narrowing of the signal for the amorphous phase is observed, which indicates ongoing crystallization processes in the sample and the formation of nanocrystalline HA (Figure S1a, Supplementary Material). In parallel to that, a decrease of DCPA resonances is also detected, showing its further transformation to apatite phase upon increasing activation time. In the $^{1}H\rightarrow^{31}P$ cross-polarization magic-angle spinning (CPMAS) spectra (Figure S1b, Supplementary Material), the signals at −0.3 and −1.5 ppm are significantly enhanced compared to their intensity in the direct excitation spectra. This result confirms that the signals are associated with the presence of DCPA, which contains -HPO_4^{2-} structures that allow for effective magnetization transfer. The enhancement of the signal at 2.9 ppm is due to the presence of H_2O molecules in the amorphous phase.

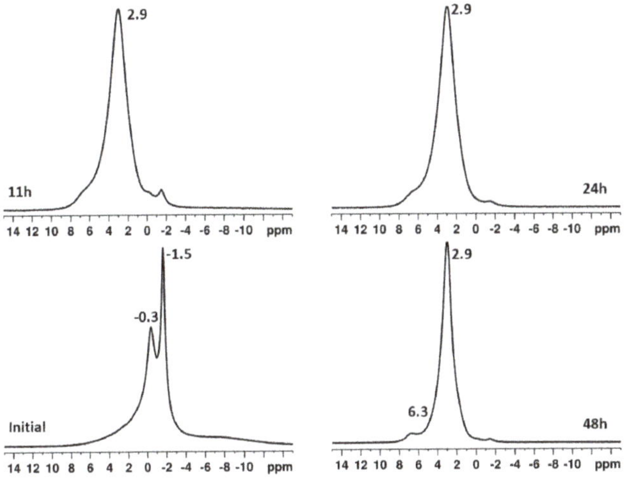

Figure 2. Direct excitation ^{31}P spectra of the initial untreated material and of the samples obtained after high-energy activation times of 11, 24, and 48 h.

TEM analysis reveals that the time of high-energy activation influences the shape and size of the particles. In the starting mixture (Figure 3a), the large >1 μm in length) plate-like particles of DCPA [31] and the aggregated spherulitic particles (<100 nm) of $Ca(OH)_2$ are clearly distinguished [32].

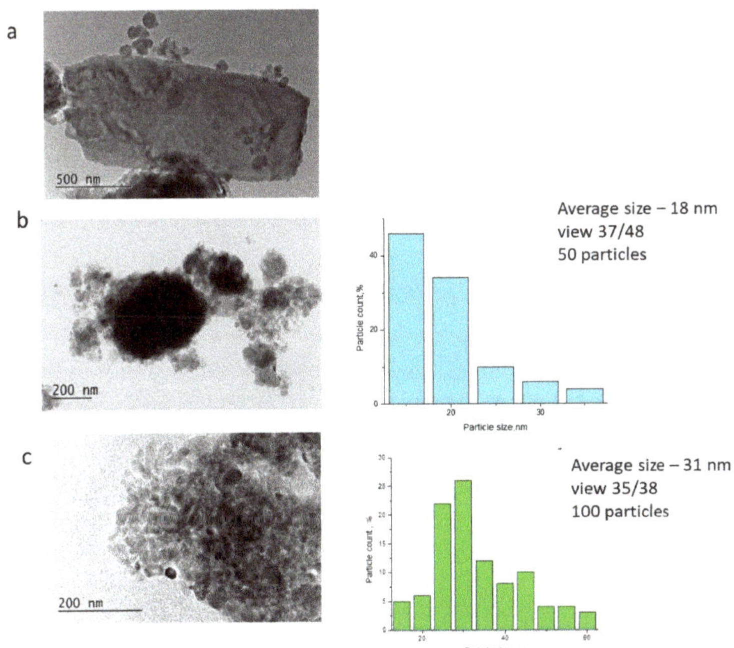

Figure 3. Images from transmission electron microscope and particle size distribution of the: (**a**) initial unactivated mixture; (**b**) 24 h activated mixture; and (**c**) 48 h activated mixture.

After 24 h of activation (Figure 3b), all particles in the mixture ($Ca(OH)_2$, DCPA, and HA, see Figure S2 in Supplementary Material) are below 100 nm and exhibit irregular shapes. Particle size distribution shows that the particle sizes range between 15 and 30 nm, with the highest fraction of particles (46%) with a size of 15 nm, followed by those with a size of 20 nm (34%). After 48 h (Figure 3c), the morphology changes again. Ellipsoidal particles (a shape characteristic of poorly crystalline HA) appear, with clearly defined contours and larger sizes compared to those in the 24 h activated sample. The particle sizes are in the region of 15–60 nm with distribution maximum shifted to 30 nm.

2.2. *Effect of High-Energy Activation on the Thermal Characteristics of the Mixtures*

The thermal characteristics of high-energy activated samples were investigated by DTA-TG-MASS analysis in order to study the influence of activation time on the temperatures of phase transitions and thus to determine the regimes for subsequent sintering of the samples and preparation of ceramic powders.

DTA-TG-MASS curves of an initial not activated mixture, and after activation times of 24 and 48 h, are presented in Figure 4.

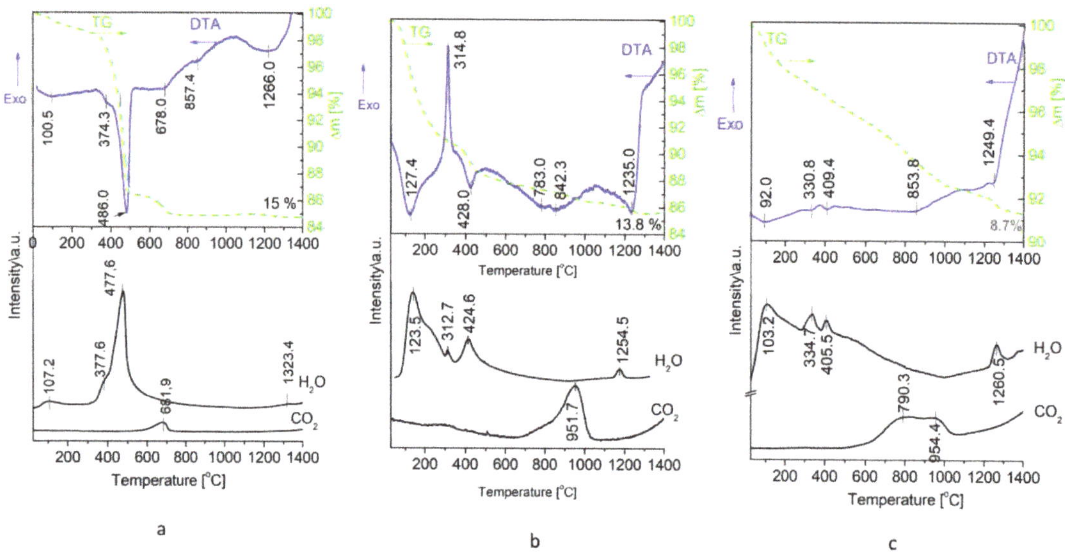

Figure 4. DTA-TG-MASS curves of: (**a**) initialunactivated mixture; (**b**) after 24 h high-energy activation; and (**c**) after 48 h high-energy activation.

The DTA-TG-MASS curves of an unactivated mixture are typical for the decomposition of the two initial substances in the temperature range 25–550 °C (Figure 4a). The first endo-effect in the DTA curve (T_{max} = 100.5 °C), accompanied by a decrease in weight of 1.1% and release of H_2O, can be attributed to the release of adsorbed water. The second strong endo-effect with T_{max} at 486 and a shoulder at 374.3, accompanied by a weight decrease of 12.3%, is due to the decomposition of DCPA and $Ca(OH)_2$, respectively [33,34].

The measured weight loss corresponds very well to the theoretically calculated one of 12.8%. The endo-effect at T_{max} = 678 °C, accompanied by the release of CO_2 and a weight loss of 1.4%, is due to the decomposition of $CaCO_3$ impurities, coming probably from the initial $Ca(OH)_2$. The endo-effect at T_{ma}x = 857.4 °C is not accompanied by gas release and weight loss. It could be attributed either to some polymorphic transition or to the formation of a TCP by the reaction

$$Ca_2P_2O_7 + CaO \rightarrow Ca_3(PO_4)_2$$

The temperature is too high for the polymorphic transition γ-$Ca_2P_2O_7 \rightarrow \beta$-$Ca_2P_2O_7$, which, according to Hazaat et al. [35], occurs at 535 °C, while according to Mulongo-Masamba et al. [36] is at 730 °C.

The last endo-effect with T_{max} = 1266 °C is the result of TTCP formation. The possible reaction is:

$$\beta\text{-}Ca_2P_2O_7 + 2CaO \rightarrow Ca_4(PO_4)_2O$$

The release of a small amount of water, visible in the H_2O release curve with T_{max} at 1323.4 °C, is an indication of the possible occurrence of another type of process.

$$2Ca_5(PO_4)_3OH + 2CaO \rightarrow 3Ca_4(PO_4)_2O + H_2O$$

Evidence for this is the analogous peaks in the water curve in samples activated for 24 and 48 h, in which nanometric HA is dominant.

Significant differences are observed in the DTA-TG-MASS curves of high-energy activated samples (Figure 4b,c). In the DTA curve of the 24 h activated sample (Figure 4b), the first endo-effect at T_{max} = 127.4 °C is much more intense and shifted to higher temperatures, compared to that in the initial unactivated mixture (Figure 4a). The weight loss is also significant (9%), which is an indication of the release of not only adsorbed but also chemically bound water. Such is characteristic of nanocrystalline hydroxyapatite [36]. The next peak is a very strong exothermic effect with T_{max}= 314.8 °C, which can be attributed to the crystallization of amorphous $Ca_2P_2O_7$ [35], obtained from the decomposition of residual DCPA. The following endo-effects are analogous to those described for the previous sample. The maximum temperature characterizing the formation of TTCP (1235.0 °C) is lower by 31 °C than that of an unactivated mixture, and the release of water at 1254.5 °C can be associated with the formation of TTCP by the reaction

$$2Ca_5(PO_4)_3OH + 2CaO \rightarrow 3Ca_4(PO_4)_2O + H_2O$$

In the DTA curve of the 48 h activated sample (Figure 4c), the strong exothermic effect is not detected. Since at 334.7 and 405.5 °C a negligible amount of water is released, we believe that minimal amounts of the starting substances remain, which decompose (DCPA—peak with T_{max} at 330.8 °C and $Ca(OH)_2$—peak with T_{max} at 409.4 °C), but no crystallization process is registered because the amount of amorphous $Ca_2P_2O_7$ is so small and also the decomposition and crystallization temperatures overlap. After that, several weak endothermic effects are observed. These effects can be attributed to the elementary processes in the thermal transformations of the calcium phosphate apatites [37]. The total weight loss is 8.7%, lower than that of the unactivated sample (Figure 4a), which reveals the dehydration and phase transformation processes during energy activation.

Additionally, DTA-TG-MASS analysis was performed on the samples activated for 5, 11, and 120 h (See Figure S3, Supplementary Material). The T_{max} of the peak corresponding to TTCP formation versus activation time is plotted in Figure 5. It can be seen that up to 24 h of mechanochemical activation, the temperature decreases almost linearly, which is due to the dominant processes of fragmentation of the starting materials and formation of amorphous products. After 24 h, crystallization and agglomeration processes of the particles begin, leading to an increase in T_{max} of the peak corresponding to TTCP formation.

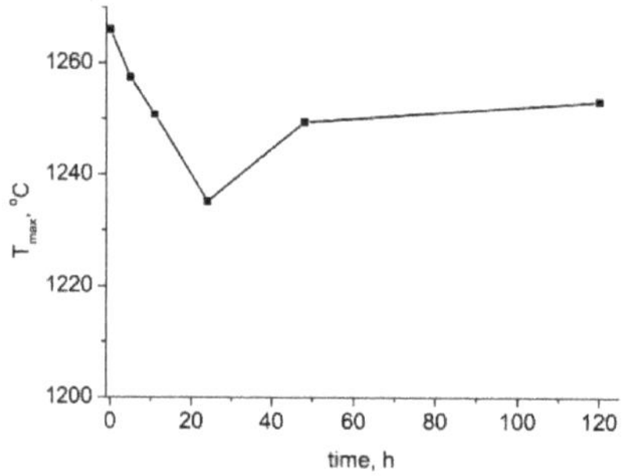

Figure 5. Variation in T_{max} of the peak corresponding to TTCP formation with increasing activation time.

2.3. Effect of High-Energy Activation on the Phase Composition After Calcination

The 24 h high-energy activated sample was selected for the calcination in two modes described in Section 4.3. The final annealing temperature is chosen to be 1350 °C, because in the DTA curves, this is the final temperature of the TTCP obtaining interval. The heating of the unactivated mixture was also done for comparison.

The results from XRD analysis (Figure 6a) show that during slow stepwise heating of the starting unactivated mixture at 550 °C, CaO and β-$Ca_2P_2O_7$ are obtained, which are the result of the decomposition of the starting substances. Minor amounts of $CaCO_3$ and HA were also identified. At 830–1000 °C, HA and β-TCP are obtained, and at 1350 °C—HA (28%), α-TCP (65%), and 5% TTCP (Figures 7a and 8a, 0 h activation time).

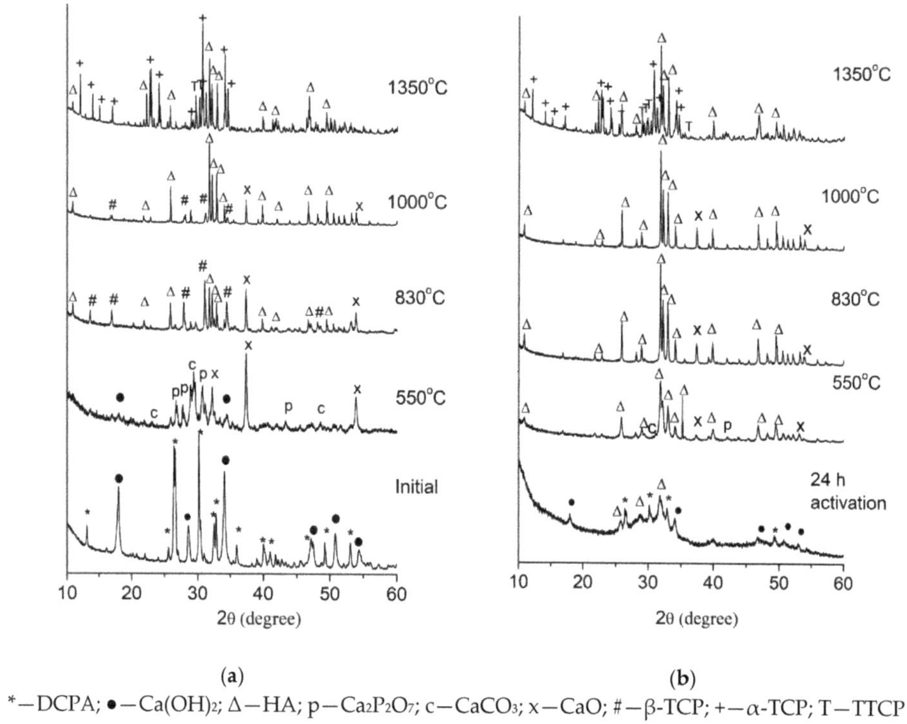

Figure 6. X-ray powder patterns of initial unactivated mixture (**a**) and 24 h activated mixture (**b**) during the stepwise calcined samples at different temperatures.

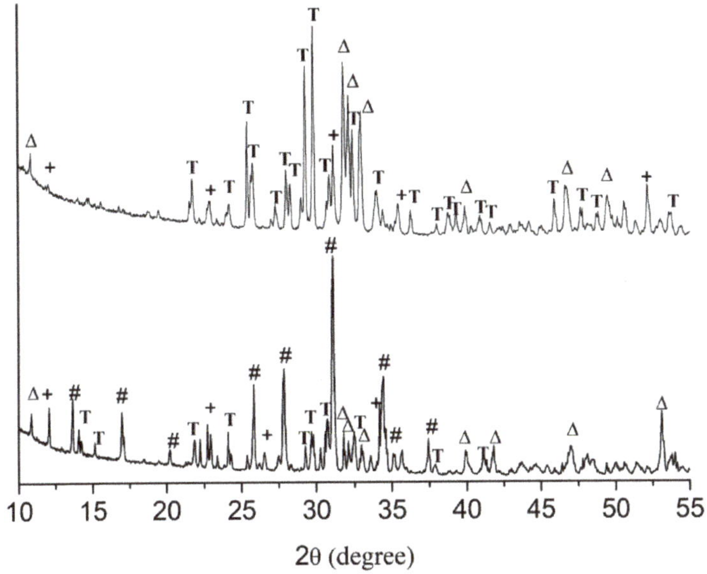

Figure 7. X-ray powder patterns of initial unactivated mixture (**a**) and 24 h activated mixture (**b**) during rapid heating up to 1350 °C and rapid cooling.

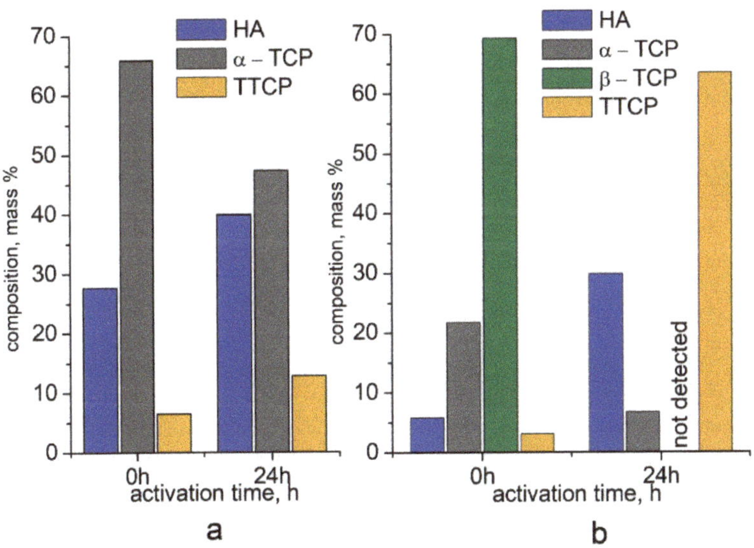

Figure 8. Quantitative composition of the samples calcinated at 1350 °C, calculated from X-ray powder patterns: (**a**) after stepwise calcination; (**b**) after rapid heating and rapid annealing.

In the 24 h activated sample (Figure 6b), HA is dominant up to 1000 °C, and its crystallinity increases with temperature (the intensities of the peaks in the X-ray powder patterns increase). No β-TCP was identified. At 1350 °C, HA, α-TCP, and TTCP were again identified (Figure 7b), but the ratios between the phases differed. The percentages of HA and α-TCP are close, and the amount of TTCP (13%) is higher than in the unactivated mixture (Figure 8a, 24 h activation time).

Rapid heating of an unactivated mixture to 1350 °C at a rate of 10 °C/min followed by rapid annealing resulted in a four-phase material (Figure 7a) consisting of HA, α- and β-TCP, and TTCP, with TCPs dominating over TTCP (Figure 8b, 0 h activation time). Three-phase materials (HA, α-TCP, and TTCP) were obtained from the 24 h mechanochemically activated precursor (Figure 7b), and the amount of TTCP reached 64% (Figure 8b, 24 h activation time).

Solid-state NMR analysis proved the results. Figure 9 shows the direct excitation ^{31}P spectra of the unactivated mixture and the 24 h activated sample, heated to 1350 °C under the two regimes.

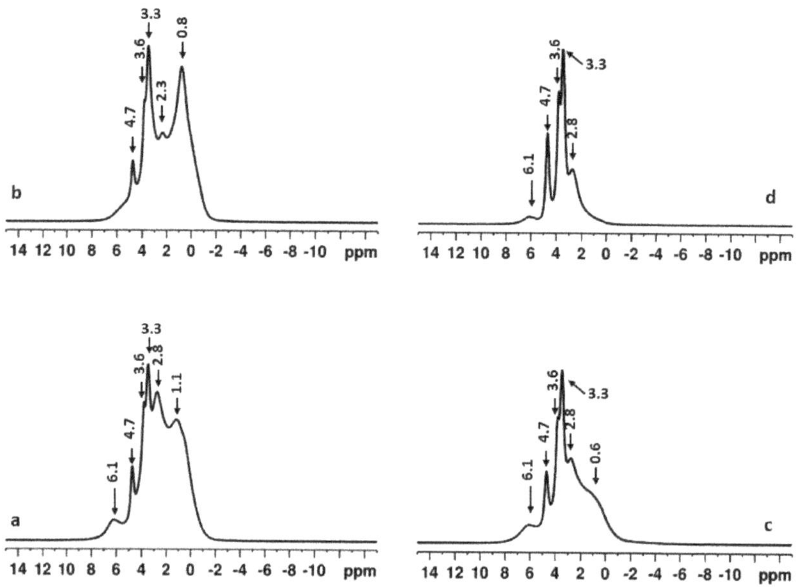

Figure 9. Direct excitation ^{31}P NMR spectra of the unactivated mixture and the activated sample (24 h at 600 rpm), heated to 1350 °C under step-wise and direct regimes: (**a**) step-wise heated unactivated sample; (**b**) rapidly heated unactivated sample; (**c**) step-wise heated 24 h activated sample; (**d**) rapidly heated 24 h activated sample.

In all samples, the presence of a mixture consisting of TTCP, HA, and disordered α- and β-TCP phases was detected. The deconvoluted spectra showing the characteristic signals of the different calcium phosphate phases present in the unactivated and activated samples heated step-wise to 1350 °C, are presented in Figure S4, Supplementary Material. The complex ^{31}P spectral patterns resulting from this complex sample composition do not allow precise quantitative interpretation of the spectra due to the overlap of the resonances of the different calcium phosphate phases. Nevertheless, based on the changes in the relative intensities of the resonances of the characteristic peaks, we can conclude that the amount of HA and TCP components was higher in the unactivated mixture, as indicated by the high intensity of the broad resonances between −0.5 and 3 ppm (Figure 9a,b and Figure S4b, Supplementary Material). In the activated sample (Figure 9c,d and Figure S4a, Supplementary Material), the spectral pattern is dominated by the three sharp resonances between 3 ppm and 5 ppm characteristic of the TTCP, while the relative quantity of the HA and TCP components is much lower, particularly in the sample that was rapidly heated to 1350 °C (Figure 9d).

Since the solid-state NMR method is more sensitive than the XRD analysis for identifying calcium phosphates, NMR spectra were recorded on a 24 h activated sample heated at different temperatures according to the stepwise heating regime (Figure 10).

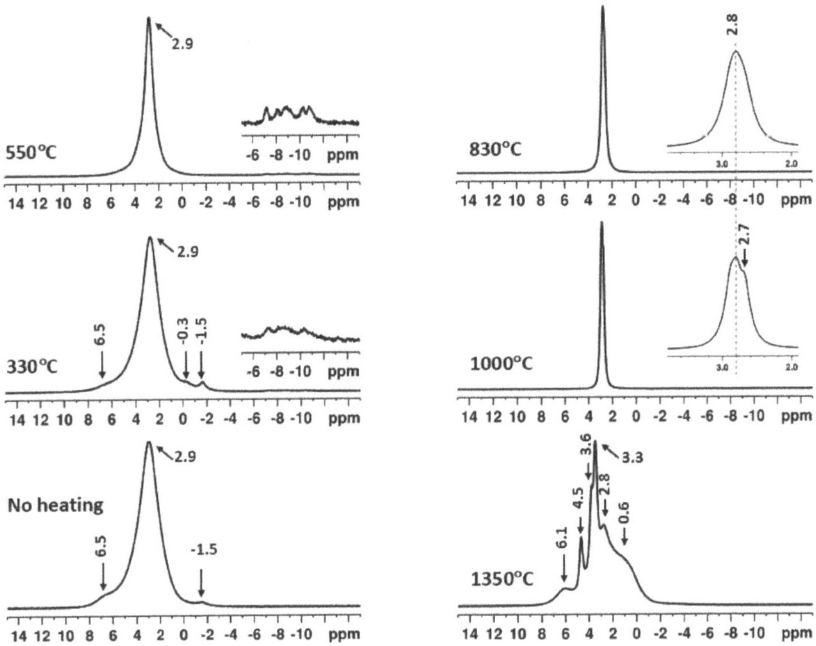

Figure 10. Direct excitation ^{31}P NMR spectra of the samples obtained after 24 h of high energy activation at 600 rpm, followed by step-wise calcination at different temperatures. The spectrum of the initial sample after 24 h of high energy activation at 600 rpm without further thermal treatment is also given for comparison.

The spectrum of the initial sample after 24 h of high energy activation at 600 rpm without thermal treatment is also given for comparison. Heating the 24 h activated sample at 330 °C and 550 °C resulted in narrowing of the main resonance at around 2.9 ppm, indicating the partial transformation of the disordered apatite phase to nano crystalline HA (Figure 9). In parallel, the DCPA resonances and the signal at 6.5 ppm decrease in intensity, and they are no longer observed in the spectra of the samples calcined at temperatures higher than 330 °C. At these two temperatures, there are additional weak resonances in the region from −6 to −13 ppm showing the formation of a small amount of pyrophosphate phase (unidentified in X-ray powder pattern), most probably β-$Ca_2P_2O_7$ [29]. These resonances disappear from the spectra of the samples calcined at higher temperatures. Further calcination up to 1000 °C resulted in a significant decrease of signal width due to the enhanced crystalline nature of the HA phase. At 1000 °C, the main resonance has two partially overlapping components that could be attributed to poorly crystalline HA and calcium-deficient hydroxyapatite (CDHA) (see the insert in the spectrum of the sample calcined at 1000 °C) [38]. Further increase of the temperature up to 1350 °C resulted in the formation of a new phase with the chemical shift signature of tetracalcium phosphate (TTCP) as identified by the relatively sharp resonances at 3.3, 3.6, and 4.5 ppm. The deconvolution of the overall spectral pattern (Figure S4a, Supplementary Material) shows that the sample is a mixture of several phases including TTCP, HA (2.8 ppm) as well as disordered α-TCP (relatively broad resonances at 0.6, 1.8, 3.5 ppm) [39] and β-TCP (relatively broad resonances at 0.1, 1.6, 2.8, 6.1 ppm) [40].

3. Discussion

The effect of high-energy mechanochemical activation on the phase and thermal characteristics of equimolar mixtures of DCPA ($CaHPO_4$) and $Ca(OH)_2$ is investigated in this work to find the optimal conditions for precursor preparation for obtaining ceramic powders with a dominant phase of TTCP. Combining methods of different sensitivity for characterizing solid phases allows us to shed light on the processes taking place.

The results show that high-energy mechanochemical activation under the experimental conditions promotes a decrease in the initial material crystallinity and the formation of nanocrystallized hydroxyapatite, whose crystallinity increases with activation time (Figures 1 and 2 and Table 1). The NMR spectrum of a sample activated for 11 h also shows the formation of an amorphous apatite phase (Figure 2).

The formation of HA from DCPA can be explained by similarities in the structural elements that build up their crystal structures. Francis et al. [41] showed that the calcium ions located on the (010) and (110) planes in the $CaHPO_4{\bullet}2H_2O$ (DCPD) structure have a similar geometric arrangement as the calcium ions on the (010) planes in HA. These networks of calcium ions serve as templates for the growth of HA. The structures of DCPD and DCPA are similar, consisting of parallel to (010) plane distorted versions of the Ca—PO_4 chains [42]. Therefore, we suggest that DCPA, like DCPD, can serve as a natural precursor for the formation of HA. Further, as a result of high-energy activation, simultaneously with the crushing of the starting substances, a process of their dehydration could also occur. The aqueous environment initiates the partial dissolution of DCPA and $Ca(OH)_2$ and formation of the metastable amorphous calcium phosphate that transforms with time into more stable HA. Additionally, as a result of activation, defects are created in the crystal structure, leading to its destruction and energy activation of the particles, which is also a prerequisite for the implementation of phase transformations with time (Figures 1 and 2).

DTA-TG-MASS analysis proved that T_{max} of the effect characterizing the formation of TTCP is lowest in the sample activated 24 h (Figures 4 and 5) under the experimental conditions, which makes it the most promising for further high-temperature treatment.

The calcination results showed that stepwise heating leads to the formation of HA and TCP regardless of whether the sample was mechanochemically activated or not.

We assume that slow heating and slow cooling allow the complete course of low-temperature reactions: (i) decomposition of the starting materials, (ii) interactions between the evolved gases and the existing mineral phases; (iii) interactions between the mineral phases themselves; and (iv) absorption of water from the air during slow cooling, and the occurrence of hydration processes. Holding at intermediate temperatures stabilizes the crystal structure of the newly formed compounds, and especially of HA, thereby increasing its thermal stability. All of this is the reason for the formation of minimal amounts of TTCP in the final product.

As a result of our studies and based on the summarized literature data, the following probable reactions can be assumed (Table 2).

Rapid heating and cooling of the 24 h mechanochemically activated mixture stimulated the formation of TTCP, whereas in the non-activated mixture, the dominant phase was β-TCP (Figures 6–8). During rapid heating, low-temperature reactions occur only partially during the period of reaching the final temperature. A larger part of the starting substances react at 1350°, which stimulates the formation of the high-temperature phases—α-TCP and TTCP.

Table 2. Probable reactions occurring during step heating and slow cooling of the unactivated mixture and the sample activated for 24 h.

Temperature Region	Unactivated Mixture		24 h Activated Sample	
Up to 550 °C	$2CaHPO_4 \rightarrow \beta\text{-}Ca_2P_2O_7 + H_2O\uparrow$	[33]	$Ca_5(PO_4)_3OH$(ncr)—stabilization of the structure, increase of crystallinity	
	$Ca(OH)_2 \rightarrow CaO + H_2O\uparrow$	[34]	$2CaHPO_4 \rightarrow \beta\text{-}Ca_2P_2O_7 + H_2O\uparrow$	[33]
	$2Ca(OH)_2 + 3CaHPO_4 \rightarrow Ca_5(PO_4)_3OH + 2H_2O\uparrow$	[43]	$Ca(OH)_2 \rightarrow CaO + H_2O\uparrow$	[34]
	$CaO + CO_2(atm) \rightarrow CaCO_3$		$2Ca(OH)_2 + 3CaHPO_4 \rightarrow Ca_5(PO_4)_3OH + 2H_2O\uparrow$	[43]
			$CaO + CO_2(atm) \rightarrow CaCO_3$	
550–830 °C	$\beta\text{-}Ca_2P_2O_7 + CaO \rightarrow 2\beta\text{-}Ca_3(PO_4)_2$	[44]	$\beta\text{-}Ca_2P_2O_7 + CaO \rightarrow 2\beta\text{-}Ca_3(PO_4)_2$	[44]
	$Ca_5(PO_4)_3OH$—structure stabilization		$Ca_5(PO_4)_3OH$—stabilization of the structure, increase of crystallinity	
830–1000 °C	$3\beta\text{-}Ca_3(PO_4)_2 + CaO + H_2O = 2Ca_5(PO_4)_3OH$ (during cooling)	[45]	$Ca_5(PO_4)_3OH$—starting the transphormation in which non-stoichiometric HA is formed (NMR, this study)	
	$Ca_5(PO_4)_3OH$—stabilization of the structure, increase of crystallinity			
1000–1350 °C	$2Ca_5(PO_4)_3OH \rightarrow 3\beta\text{-}Ca_3(PO_4)_2 + CaO + H_2O$	[46]	$2Ca_5(PO_4)_3OH \rightarrow 3\beta\text{-}Ca_3(PO_4)_2 + CaO + H_2O$	[46]
	$\beta\text{-}Ca_3(PO_4)_2 \rightarrow \alpha\text{-}Ca_3(PO_4)_2$	[47]	$\beta\text{-}Ca_3(PO_4)_2 \rightarrow \alpha\text{-}Ca_3(PO_4)_2$	[47]
	$2Ca_5(PO_4)_3OH + 2CaO \rightarrow 3Ca_4(PO_4)_2O + H_2O$	[9]	$2Ca_5(PO_4)_3OH + 2CaO \rightarrow 3Ca_4(PO_4)_2O + H_2O$	[9]

Although TTCP is the dominant phase in a sample activated for 24 h and rapidly heated to 1350 °C with a 3 h hold, these conditions are not sufficient to obtain a monophasic product. The hold time should likely be extended, or the calcination temperature should be increased. However, we believe that the material obtained by us is suitable as a component of calcium phosphate cements, since the available HA can act as a nucleating agent, and the presence of α-TCP will regulate the rate of dissolution of the solid phase in the liquid phase of the cements. Similar observations have been reported by Jeon et al. [26].

4. Materials and Methods

4.1. Materials

$Ca(OH)_2$ (Merck, Darmstadt, Germany, p.a.) and $CaHPO_4$ prepared by thermal dehydration of $CaHPO_4{\bullet}2H_2O$ (Sigma Aldrich, St. Louis, MO, USA) at 165 °C for 2 h were used. A high-temperature furnace, type VP 04/17 (LAC Ltd. Company, Stefanikova, Czech Republic) was used. No other phases except DCPA were detected by XRD analysis (Figure S5, Supplementary Materials).

4.2. High Energy Activation

The high-energy activation was carried out in a Fritsch 6 type planetary ball mill (FRITSCH GmbH - Milling and Sizing, Idar-Oberstein, Germany). A mixture of $Ca(OH)_2$ and DCPA was prepared in a molar ratio of $CaHPO_4$:$Ca(OH)_2$ = 1:1 (Ca/P = 2:1). Activation is carried out in an agate vessel with a volume of 80 cm^3. The prepared samples were treated for 5, 11, 24, 48, and 120 h at 600 rpm with agate balls (10 mm diameter) at a constant ratio of sample mass:mass of grinding balls = 1:3. Activation revolutions were selected based on preliminary exploratory experiments with activation at 300 and 450 rpm.

To verify the repeatability of the results, each experiment was performed twice, and each activated mixture was analysed with the appropriate methods described in Section 4.4.

4.3. Calcination of Activated Precursors

Selected samples from the high energy activation were treated in a high-temperature furnace, type VP 04/17 (LAC Ltd. Company, Stefanikova, Czech Republic), in the following operating modes:

- Step heating at a rate of 10 °C/min to a given temperature (determined by DTA-TG-MASS analysis) and hold for 3 h, followed by slow cooling (The samples were left to cool overnight in the furnace itself after it was turned off).
- Rapid heating at a rate of 10 °C/min to 1350 °C and holding for 3 h, followed by rapid annealing (The samples were removed from the furnace at a temperature of 1350°C and placed in a desiccator at room temperature).

Each heating was performed twice to verify the repeatability of the results, and each sample was analysed with the appropriate methods described in Section 4.4.

4.4. Characterisation

4.4.1. X-Ray Diffraction Analysis (XRD)

A Bruker D8 Advance diffractometer with CuKα radiation and a LynxEye detector (Bruker AXS Advanced X-ray Solutions GmbH, Billerica, MA, USA) was used to conduct powder X-ray diffraction. Data were gathered in the range of 10 to 90 2θ with a step of 0.03° 2θ and a counting rate of 57 s/step for the primary phase identification. The phase composition was identified using ICDD-PDF2 (2021) database.

The Bruker Topas v.4.2 program was used to refine the unit cell parameters, mean crystallite size, and quantitative phase composition of the samples using Rietveld quantification with starting models from the ICSD (cif files ## 202233, 14313, 97500, 410782, 20179, 51409, and 56306). For this purpose, powder diffraction patterns were collected at room temperature within the range of 5 to 120° 2θ with a step of 0.02° 2θ and 175 s/step counting time, sample rotation—15 rpm. The standard deviation of phase composition calculations is ±2%.

4.4.2. Differential Thermal Analysis Combined with Detection of Released Gases (DTA-TG-MASS)

The thermal characterization of the samples was performed on a LABSYS^{TM} EVO (Setaram, Caluire, France) apparatus with a Pt/Pt-Rh thermocouple in a corundum crucible at a heating rate of 10 °C/min in the temperature range 25–1400 °C and in an atmosphere of synthetic air with a flow rate of 20 mL/s. The accuracy of temperature measurement was ±1 °C. A corundum crucible was used as a reference one. The weight of the samples was 29 ± 1 mg. The apparatus is equipped with a quadruple mass spectrometer (Pfeiffer vacuum OMNISTAR, GSD 301, Zürich, Switzerland), which serves for the analysis of the escaping gases. In the studied samples, the release of H_2O (m = 18) and CO_2 (m = 44) was determined.

4.4.3. Nuclear Magnetic Resonance (NMR) Studies

NMR spectra were recorded on a Bruker Avance HD III 600 NMR spectrometer (Bruker, Rheinstetten, Germany) operating at 599.90 MHz ^{1}H frequency (242.84 MHz for 31P), using a 4 mm solid state iProbe CPMAS DR-H&F VTN. The samples were loaded in 4 mm zirconia rotors and spun at a magic angle spinning (MAS) rate of 10 kHz for all measurements. The quantitative direct excitation ^{31}P NMR spectra were recorded with a onepulse sequence (Bruker Topspin library), 90° pulse length of 3.3 µs, 7 K time domain data points, spectrum width of 37 kHz, 32 scans, and a relaxation delay of 150 s. The spectra were processed with an exponential window function (line broadening factor 5) and zero filled to 16 K data points. The ^{1}H-^{31}P cross-polarization magic angle spinning (CP MAS) spectra were acquired with the following experimental parameters: ^1H excitation pulse of 2.5 μs, 3.5 ms contact time, 5 s relaxation delay, and 256 scans were accumulated. ^1H SPINAL-64 decoupling scheme was used during the acquisition of CP experiments. All ^{31}P chemical shifts were referenced against the external solid reference $NH_4H_2PO_4$ (δ 0.9 ppm). The DMfit software (release #20220502) was used for the deconvolution, simulation, and fitting of the experimental NMR data [48].

4.4.4. Transmission Electron Microscopy Analysis (TEM)

STEM JEOL JEM 2100 apparatus with an accelerating voltage of 200 kV, a maximum resolution of 0.23 nm between two points, and 0.14 nm in the lattice was used. The device is equipped with a GATAN CCD camera. The powder samples were suspended in ethyl alcohol and sonicated (Ultra Sonic Cleaner, KWAN SWO INSTRUMENTAL Co LTD, Taichung, Taiwan) with a frequency of 45 KHz for 6 min. The suspension is dropped onto a standard carbon-coated copper grid. ImageJ software (version 1.x) was used to calculate particle size distribution

5. Conclusions

High-energy mechanochemical activation significantly influences the phase composition, particle morphology, and thermal behaviour of equimolar mixtures of DCPA ($CaHPO_4$) and $Ca(OH)_2$, making it an effective method for preparing precursors for TTCP-rich ceramic materials.

XRD and solid-state NMR analyses reveal that high-energy milling for 11 to 48 h initiates the amorphization and transformation of the initial crystalline phases into amorphous calcium phosphate phase and nanocrystalline hydroxyapatite, which subsequently increases its crystallinity.

Thermal analysis demonstrates that mechanochemical activation reduces the temperatures at which phase transformations occur.

Calcination studies show that not only mechanochemical activation but also the calcination mode influence the final phase composition. Rapid heating of a 24 h activated precursor produces a material with TTCP as the dominant phase (64%), whereas the unactivated mixture or stepwise heating forms a multi-phase composition with lower TTCP content.

Under the experimental conditions, 24 h activation presents an effective balance between phase transformation, particle refinement, and process efficiency, making it a promising approach for the preparation of TTCP-rich ceramic powders, which are suitable components of calcium phosphate cements. The available HA can act as a nucleating agent, and the presence of α-TCP will regulate the rate of dissolution of the solid in the liquid phase of the cements.

Supplementary Materials: The following supporting information can be downloaded at: https://www.mdpi.com/article/10.3390/inorganics13100313/s1, Figure S1: Experimental HRTEM image of sample activated 24 h: (A) investigated areas; (B) Fourier filtered HRTEM images of CaHPO4, (112)—0.247 nm; (C) Fourier filtered HRTEM images of Apatite-(CaOH) hydroxyapatite, (001)—0.691 nm and (D) Experimental HRTEM image and Fourier filtered HRTEM image of Ca(OH)2, (002)—0.246 nm; Figure S2: (a) Overlaid and vertically scaled direct excitation ^{31}P spectra of the samples activated for 11 h (black line), 24 h (blue line) and 48 h (red line), showing narrowing of the resonance linewidths with increasing activation time; (b) Overlaid ^1H-^{31}P CPMAS of the same samples with identical color code; Figure S3: DTA-TG-MASS curves of the samples obtained after (a) 5 h mechanochemically activation; (b) 11 h mechanochemically activation and (c) 120 h mechanochemically activation; Figure S4: Deconvoluted direct excitation ^{31}P NMR spectra of: (a) activated sample (24 h at 600 rpm) heated step-wise to 1350 °C; (b) unactivated initial mixture heated step-wise to

1350 °C. The experimental spectra are given in black, while the simulated spectra are presented in red lines. The individual contributions of some characteristic lines of the different components obtained after the deconvolution of the spectra are given with colored lines: TTCP (blue), HA (light blue), α-TCP (magenta), and β-TCP (green); Figure S5: X-ray analysis of the initial DCPA.

Author Contributions: Conceptualization, D.R. and K.S.; sample preparation, Y.T. and K.S.; NMR analysis and discussion, P.S.; XRD analysis and calculation, D.K.; TEM analysis, P.M.; DTA-TG-MASS analysis and visualization, D.R. and Y.T.; writing—original draft preparation, D.R., P.S., K.S., D.K. and P.M.; writing—review and editing, D.R., P.S. and K.S. All authors have read and agreed to the published version of the manuscript.

Funding: This research received no external funding.

Institutional Review Board Statement: Not applicable.

Informed Consent Statement: Not applicable.

Data Availability Statement: The original contributions presented in this study are included in the article/Supplementary Material. Further inquiries can be directed to the corresponding author.

Acknowledgments: Research equipment of Distributed Research Infrastructure INFRAMAT, part of Bulgarian National Roadmap for Research Infrastructures, supported by Bulgarian Ministry of Education and Science was used in this investigation.

Conflicts of Interest: The authors declare no conflicts of interest.

Abbreviations

The following abbreviations are used in this manuscript:

HA	Hydroxyapatite
DCPA	Dicalcium phosphate anhydrous
TCP	Thricalcium phosphate
TTCP	Tetracalcium phosphate
NMR	Nuclear Magnetic Resonance
XRD	X-ray diffraction analysis
TEM	Transmission electron microscopy
DTA-TG-MASS	Differential thermal analysis with thermogravimetry and mass spectrometry

References

1. Dorozhkin, S.V. Calcium Orthophosphate ($CaPO_4$)-Based Bioceramics: Preparation, Properties, and Applications. *Coatings* **2022**, *12*, 1380. [CrossRef]
2. Rabel, K.; Kohal, R.J.; Steinberg, T.; Tomakidi, P.; Rolauffs, B.; Adolfsson, E.; Palmero, P.; Fürderer, T.; Altmann, B. Controlling osteoblast morphology and proliferation via surface micro-topographies of implant biomaterials. *Sci. Rep.* **2020**, *10*, 12810. [CrossRef]
3. Wu, V.M.; Uskoković, V. Is there a relationship between solubility and resorbability of different calcium phosphate phases in vitro? *Biochim. Biophys. Acta* **2016**, *1860*, 2157–2168. [CrossRef] [PubMed]
4. Chow, L.C.; Eanes, E.D. (Eds.) *Octacalcium Phosphate*; Karger Medical and Scientific Publishers: Basel, Switzerland, 2001.
5. Jeong, J.; Kim, J.H.; Shim, J.H.; Hwang, N.S.; Heo, C.Y. Bioactive calcium phosphate materials and applications in bone regeneration. *Biomat. Res.* **2019**, *23*, 4. [CrossRef] [PubMed]
6. Bohner, M.; Lemaitre, J. Can bioactivity be tested in vitro with SBF solution? *Biomaterials* **2009**, *30*, 2175–2179. [CrossRef]
7. Ehara, A.; Ogata, K.; Imazato, S.; Ebisu, S.; Nakano, T.; Umakoshi, Y. Effects of α-TCP and TetCP on MC3T3-E1 proliferation, differentiation and mineralization. *Biomaterials* **2003**, *24*, 831–836. [CrossRef]
8. Medvecky, L.; Stulajterova, R.; Giretova, M.; Mincik, J.; Vojtko, M.; Balko, J.; Briancin, J. Effect of tetracalcium phosphate/monetite toothpaste on dentin remineralisation and tubule occlusion in vitro. *Dent. Mater.* **2018**, *34*, 442–451. [CrossRef]
9. Moseke, C.; Gbureck, U. Tetracalcium phosphate: Synthesis, properties and biomedical applications. *Acta Biomater.* **2010**, *6*, 3815–3823. [CrossRef]
10. Samavedi, S.; Whittington, A.R.; Goldstein, A.S. Calcium phosphate ceramics in bone tissue engineering: A review of properties and their influence on cell behavior. *Acta Biomater.* **2013**, *9*, 8037–8045. [CrossRef]

11. Xu, H.; Wang, P.; Wang, L.; Bao, C.; Chen, Q.; Weir, M.D.; Chow, L.C.; Zhao, L.; Zhou, X.; Reynolds, M.A. Calcium phosphate cements for bone engineering and their biological properties. *Bone Res.* **2017**, *5*, 17056. [CrossRef]
12. Lukina, Y.; Safronova, T.; Smolentsev, D.; Toshev, O. Calcium Phosphate Cements as Carriers of Functional Substances for the Treatment of Bone Tissue. *Materials* **2023**, *16*, 4017. [CrossRef] [PubMed]
13. Mahmoud, E.M.; Sayed, M.; Mansour, T.S.; Naga, S.M. Biodegradable ceramic materials for orthopedic and dentistry applications. *Discov. Appl. Sci.* **2025**, *7*, 990. [CrossRef]
14. Balbuena, O.B.F.; Paiva, L.F.S.; Ribeiro, A.A.; Monteiro, M.M.; de Oliveira, M.V.; Pereira, L.C. Sintering parameters study of a biphasic calcium phosphate bioceramic synthesized by alcoholic sol-gel technique. *Ceram. Int.* **2021**, *47*, 32979–32987. [CrossRef]
15. Safronova, T.V.; Putlyaev, V.I. Powder systems for calcium phosphate ceramics. *Inorg. Mater.* **2017**, *53*, 17–26. [CrossRef]
16. Irfa'i, M.A.; Schmahl, W.W.; Pusparizkita, Y.M.; Muryanto, S.; Ismail, A.P.R.; Jamari, J.; Bayuseno, A.P. Hydrothermally synthesized-nanoscale carbonated hydroxyapatite with calcium carbonates derived from green mussel shell wastes. *J. Mol. Struct.* **2024**, *1306*, 137837. [CrossRef]
17. Nikolenko, M.V.; Vasylenko, K.V.; Myrhorodska, V.D.; Kostyniuk, A.; Likozar, B. Synthesis of Calcium Orthophosphates by Chemical Precipitation in Aqueous Solutions: The Effect of the Acidity, Ca/P Molar Ratio, and Temperature on the Phase Composition and Solubility of Precipitates. *Processes* **2020**, *8*, 1009. [CrossRef]
18. Ali, W.A.; Richards, S.E.; Alzard, R.H. Unlocking the potential of ball milling for nanomaterial Synthesis: An overview. *J. Ind. Eng. Chem.* **2025**, *149*, 63–93. [CrossRef]
19. Chen, W.C.; Ju, C.P.; Tien, Y.C.; Lin, J.H. In vivo testing of nanoparticle-treated TTCP/DCPA-based ceramic surfaces. *Acta Biomater.* **2009**, *5*, 1767–1774. [CrossRef]
20. Sargin, Y.; Kizilyalli, M.; Telli, C.; Güler, H. A new method for the solid-state synthesis of tetracalcium phosphate, a dental cement: X-ray powder diffraction and IR studies. *J. Eur. Ceram. Soc.* **1997**, *17*, 963–970. [CrossRef]
21. Brosseau, C.; Liao, J.; Duan, X.; Li, Y.; Zheng, C.; Yang, Z.; Zhou, A.; Zou, D. Synthesis and Mechanism of Tetracalcium Phosphate from Nanocrystalline Precursor. *J. Nanomater.* **2014**, *2014*, 840102. [CrossRef]
22. Jalota, S.; Tas, A.C.; Bhaduri, S.B. Synthesis of HA-seeded TTCP ($CA_4(PO_4)_2O$) powders at 1230 °C from $Ca(CH_3COO)_2.H_2O$ and $NH_4H_2PO_4$. *J. Am. Ceram. Soc.* **2005**, *88*, 3353–3360. [CrossRef]
23. Romeo, H.E.; Fanovich, M.A. Synthesis of tetracalcium phosphate from mechanochemically activated reactants and assessment as a component of bone cements. *J. Mater. Sci. Mater. Med.* **2008**, *19*, 2751–2760. [CrossRef]
24. Nizar, M.S.; Sofiyaningsih, N.; Manullang, R.J.; Wijayanti, R.B.; Rosmayanti, I.; Sumardan, D. Novel fast firing method to synthesize tetracalcium phosphate. *AIP Conf. Proc.* **2021**, *2349*, 020001. [CrossRef]
25. Ayers, R.; Hannigan, N.; Vollmer, N.; Unuvar, C. Combustion synthesis of heterogeneous calcium phosphate bioceramics from calcium oxide and phosphate precursors. *Int. J. Self-Propag. High-Temp. Synth.* **2011**, *20*, 6–14. [CrossRef]
26. Jeon, C.; Chun, S.; Lim, S.; Kim, S. Synthesis and Characterization of TTCP for Calcium Phosphate Bone Cement. *Biomater. Res.* **2011**, *15*, 1–6.
27. Louati, B.; Hlel, F.; Guidara, K.; Gargouri, M. Analysis of the effects of thermal treatments on $CaHPO_4$ by ^{31}P NMR spectroscopy. *J. Alloys Compd.* **2005**, *394*, 13–18. [CrossRef]
28. Edén, M. Structure and formation of amorphous calcium phosphate and its role as surface layer of nanocrystalline apatite: Implications for bone mineralization. *Materialia* **2021**, *17*, 101107. [CrossRef]
29. Yu, Y.; Guo, H.; Pujari-Palmer, M.; Stevensson, B.; Grins, J.; Engqvist, H.; Eden, M. Advanced solid-state $^1H/^{31}P$ NMR characterization of pyrophosphate-doped calcium phosphate cements for biomedical applications: The structural role of pyrophosphate. *Ceram. Int.* **2019**, *45*, 20642–20655. [CrossRef]
30. Ben Osman, M.; Diallo-Garcia, S.; Herledan, V.; Brouri, D.; Yoshioka, T.; Kubo, J.; Millot, Y.; Costentin, G. Discrimination of surface and bulk structure of crystalline hydroxyapatite nanoparticles by NMR. *J. Phys. Chem. C* **2015**, *119*, 23008–23020. [CrossRef]
31. Lee, J.; Bae, J.-S.; Kim, Y.-I.; Yoo, K.-H.; Yoon, S.-Y. Synthesis, Characterization, and Biological Performances of Magnesium-Substituted Dicalcium Phosphate Anhydrous. *Materials* **2024**, *17*, 4605. [CrossRef] [PubMed]
32. Pesce, C.; Luca, G.P.; Molinari, M.; Richardson, A. Effects of organic additives on calcium hydroxide crystallisation during lime slaking. *Cem. Concr. Res.* **2021**, *139*, 106254. [CrossRef]
33. Duff, E.J. Orthophosphates. XIII,* Thermal Decomposition of Secondary Calcium Orthophosphate ($CaHPO_4$) and Secondary Calcium Orthophosphate Dihydrate ($CaHPO_4{\bullet}2H_2O$). *J. Appl. Chem. Biotechnol.* **1971**, *21*, 233–235. [CrossRef]
34. Khachani, M.; El Hamidi, A.; Halim, M.; Arsalane, S. Non-isothermal kinetic and thermodynamic studies of the dehydroxylation process of synthetic calcium hydroxide $Ca(OH)_2$. *J. Mater. Environ. Sci.* **2014**, *5*, 615–624.
35. El Hazzat, M.; El Hamidi, A.; Halim, M.; Arsalane, S. Complex evolution of phase during the thermal investigation of brushite-type calcium phosphate $CAHPO_4{\bullet}2H_2O$. *Materialia* **2021**, *16*, 101055. [CrossRef]
36. Mulongo-Masamba, R.; El Kassri, T.; Khachani, M.; Arsalane, S.; Halim, M.; El Hamidi, A. Synthesis and thermal dehydroxylation kinetic of anhydrous calcium phosphate monetite $CaHPO_4$. *J. Therm. Anal. Calorim.* **2016**, *124*, 171–180. [CrossRef]

37. Mortier, A.; Lemaitre, J.; Rouxhet, P.G. Temmperature-Programmed Characterization of Synthetic Calcium-Deficient Phosphate Apatites. *Thermochim. Acta* **1989**, *143*, 265–282. [CrossRef]
38. Miquel, J.L.; Facchini, L.; Legrand, A.P. Characterisation and Conversion Study into Natural Living Bone of Calcium Phosphate Bioceramics by Solid State NMR Spectroscopy. *Clin. Mater.* **1990**, *5*, 115. [CrossRef]
39. Vogel, J.; Rüssel, C.; Günther, G.; Hartmann, P.; Vizethum, F.; Bergner, N. Characterization of plasma-sprayed hydroxyapatite by ^{31}P-MAS-NMR and the effect of subsequent annealing. *J. Mater. Sci. Mater. Med.* **1996**, *7*, 495. [CrossRef]
40. Guo, H.; Pujari-Palmer, M.; Yu, Y.; Stevensson, B.; Engqvist, H.; Edén, M. Quantitative phase analyses of biomedical pyrophosphate-bearing monetite and brushite cements by solid-state NMR and powder XRD. *Ceram. Int.* **2000**, *46*, 11000. [CrossRef]
41. Francis, M.D.; Webb, N.C. Hydroxyapatite formation from a hydrated calcium monohydrogen phosphate precursor. *Calc. Tis Res.* **1970**, *6*, 335–342. [CrossRef]
42. Dickens, B.; Bowen, J.S.; Brown, W.E. A refinement of the crystal structure of $CaHPO_4$ (synthetic monetite). *Acta Cryst.* **1972**, *B28*, 797–806. [CrossRef]
43. Rodríguez-Lugo, V.; Angeles, C.; De la Isla, A.; Castano, V.M. Effect of bio-calcium oxide on the morphology of hydroxyapatite. *Int. J. Basic Appl. Sci.* **2015**, *4*, 395. [CrossRef]
44. Jinlong, N.; Zhenxi, Z.; Dazong, J. Investigation of Phase Evolution During the Thermochemical Synthesis of Tricalcium Phosphate. *J. Mater. Synth. Process.* **2001**, *9*, 235–240. [CrossRef]
45. Hui, P.; Meena, S.L.; Singh, G.; Agarawal, R.D.; Prakash, S. Synthesis of Hydroxyapatite Bio-Ceramic Powder by Hydrothermal Method. *J. Miner. Mater. Charact. Eng.* **2010**, *9*, 683–692. [CrossRef]
46. Kivrak, N.; Taş, A.C. Synthesis of Calcium Hydroxyapatite-Tricalcium Phosphate (HA-TCP) Composite Bioceramic Powders and Their Sintering Behavior. *J. Am. Ceram. Soc.* **1998**, *81*, 2245–2252. [CrossRef]
47. Böhme, N.; Hauke, K.; Dohrn, M.; Neuroth, M.; Geisler, T. High-temperature phase transformations of hydroxylapatite and the formation of silicocarnotite in the hydroxylapatite–quartz–lime system studied in situ and in operando by Raman spectroscopy. *J. Mater. Sci.* **2022**, *57*, 15239–15266. [CrossRef]
48. Massiot, D.; Fayon, F.; Capron, M.; King, I.; Le Celve, S.; Alonson, B.; Durand, J.O.; Bujoli, B.; Gan, Z.H.; Hoatson, G. Modelling one- and two-dimensional solid-state NMR spectra. *Magn. Reson. Chem.* **2002**, *40*, 70–76. [CrossRef]

inorganics

MDPI

Review

Smart Inorganic Nanomaterials for Tumor Microenvironment Modulation

Zhenqi Jiang *, Hui Xiang and Xiaoying Tang *

School of Medical Technology, Beijing Institute of Technology, Beijing 100811, China
* Correspondence: jiangzhenqi@bit.edu.cn (Z.J.); xiaoying@bit.edu.cn (X.T.)

Abstract: The tumor microenvironment (TME) is characterized by hypoxia; acidic pH; oxidative stress; and immune suppression; all of which severely impair the efficacy of conventional cancer therapies. Recent advances in inorganic nanotechnology have led to the development of smart nanomaterials capable of modulating these abnormal features; thereby reprogramming the TME toward a more therapy-responsive state. Inorganic nanomaterials such as manganese dioxide; iron oxide; and cerium oxide can selectively alleviate hypoxia; buffer acidity; regulate redox balance; and even stimulate anti-tumor immunity through catalytic or structural mechanisms. These materials can further serve as carriers for stimuli-responsive drug delivery; enabling synergistic therapies that include chemodynamic; photothermal; and immunomodulatory treatments. This review summarizes recent developments in smart inorganic nanomaterials for TME modulation; discusses design considerations including biosafety and biodegradability; and evaluates the current translational status and future directions. Such strategies represent a promising leap toward precise and personalized cancer nanomedicine

Keywords: tumor microenvironment; inorganic nanomaterials; hypoxia modulation; redox-responsive nanoparticles; smart nanotechnology

1. Introduction

Cancer is not just a disease of rapidly dividing cells. It occurs within a complex and dynamic setting called the tumor microenvironment (TME) [1,2]. The TME includes not only tumor cells but also various non-malignant components. These include immune cells, fibroblasts, blood vessels, and extracellular matrix proteins [3,4]. The TME is not passive. It plays an active role in promoting tumor growth, invasion, and metastasis. It also protects tumors from therapeutic agents. Among its many characteristics, four stand out as key contributors to treatment resistance: hypoxia, acidity, oxidative stress, and immune suppression [5,6].

Hypoxia, or low oxygen levels, is common in solid tumors. Tumor cells often grow faster than their blood supply can support. This creates regions within the tumor that lack sufficient oxygen [7,8]. Hypoxia leads to the activation of specific genes that help cancer cells survive. It also reduces the effectiveness of therapies such as radiotherapy and some forms of chemotherapy. Acidic pH is another hallmark of the TME. Tumor cells often switch to a type of metabolism called aerobic glycolysis [4,9]. This produces large amounts of lactic acid. As a result, the tumor becomes acidic, especially in its extracellular environment. This acidic environment can degrade therapeutic drugs or prevent them from entering tumor cells effectively. High levels of reactive oxygen species (ROS) are also common in tumors [10,11]. While ROS can damage cells, cancer cells often tolerate

 https://doi.org/10.3390/inorganics13100337

and exploit them. High ROS can promote genetic mutations, drive cancer progression, and trigger resistance to treatment. Finally, the TME is often immunosuppressive. Tumor cells recruit immune cells like regulatory T cells (Tregs) or tumor-associated macrophages (TAMs) that suppress immune responses [12,13]. This makes it difficult for the immune system to recognize and destroy cancer cells. Immunosuppression in the TME is one of the major barriers to the success of immunotherapies.

Traditional cancer therapies include surgery, chemotherapy, radiotherapy, and immunotherapy. Each has made important contributions to cancer care. However, their effectiveness is often reduced by the hostile and abnormal conditions of the TME. Chemotherapy drugs rely on blood flow to reach the tumor [4,14]. But in many tumors, the blood vessels are irregular and leaky. This makes drug delivery inefficient. Even when the drugs reach the tumor, acidic pH and high glutathione (GSH) levels can deactivate them. Radiotherapy is also less effective in hypoxic environments [15,16]. Oxygen is required to generate free radicals that damage cancer cell DNA. Without oxygen, radiation becomes significantly less lethal to tumor cells. Immunotherapies have revolutionized treatment for many cancers [17,18]. But they often fail in "cold" tumors, which lack active immune cell infiltration. The immunosuppressive TME further limits the activation of T cells, even after checkpoint blockade therapies. These challenges reveal a major problem. Conventional therapies are designed for normal tissues, not for the unique biology of tumors. The abnormal features of the TME create a protective shield around cancer cells. Overcoming this shield requires new therapeutic strategies that can adapt to or modify the TME itself.

To address the challenges posed by the TME, researchers are turning to smart inorganic nanomaterials. These are tiny particles, often less than 100 nanometers in size, made from metals or metal oxides. They can be engineered to respond to the specific features of the TME. These materials are "smart" because they are stimuli-responsive. For example, some release their drug payload only in acidic conditions [19,20]. Others activate in the presence of high ROS or reducing agents like glutathione. This responsiveness allows them to deliver therapies directly to tumor sites while sparing healthy tissue. One key advantage of inorganic materials is their catalytic activity. Materials like manganese dioxide (MnO_2) can react with tumor-derived hydrogen peroxide to produce oxygen [21,22]. This can relieve hypoxia and improve the effectiveness of radiotherapy or photodynamic therapy [23]. Iron oxide nanoparticles can trigger Fenton reactions in the tumor. These reactions convert hydrogen peroxide into hydroxyl radicals, which are toxic to cells. This forms the basis of chemodynamic therapy, a new approach that relies on the TME to generate therapeutic effects. Some inorganic nanoparticles also serve as imaging agents. Gold, cerium, and gadolinium-based nanoparticles enhance imaging techniques such as MRI, CT, and photoacoustic imaging. This allows for image-guided therapy, where clinicians can track treatment in real-time. In addition, inorganic nanomaterials can modulate the immune environment. For example, certain iron-based materials can reprogram tumor-associated macrophages from an immunosuppressive M2 type to an inflammatory M1 type. This can help "heat up" cold tumors and make them more responsive to immunotherapy. As research advances, these smart inorganic systems are becoming more multi-functional. They can carry drugs, generate oxygen, regulate redox balance, and support imaging—all in a single platform. This multi-modal approach is especially valuable in tackling the complexity of the TME.

This review focuses on the design, function, and application of smart inorganic nanomaterials for modulating the tumor microenvironment. We explore how these materials are used to target specific features of the TME—such as hypoxia, acidity, redox imbalance, and immune suppression. We provide a detailed overview of the components and characteristics of the TME. This sets the stage for understanding why it is such a critical barrier to effective treatment. We categorize the different types of inorganic nanomaterials used in cancer therapy. This includes metal oxides, noble metals, carbon-based materials, and more. We cover hypoxia relief, pH buffering, redox regulation, and immune reprogramming. We review how these strategies enable multi-modal therapies, including chemo, photo, radio, and immunotherapies. Finally, outlines current challenges and future perspectives. We discuss how nanotechnology, combined with machine learning and personalized medicine, might transform cancer therapy. Through this structured approach, we aim to provide a comprehensive and accessible overview of how smart inorganic nanomaterials are reshaping the future of cancer treatment by overcoming the long-standing barriers of the tumor microenvironment.

2. Overview of the Tumor Microenvironment (TME)

Cancer is not just a disease of rogue cells. It is deeply influenced by the biological environment in which it develops. This environment, known as the tumor microenvironment, plays a central role in cancer progression, treatment resistance, and immune evasion.

2.1. Key Components of the TME

The tumor microenvironment is a complex and dynamic system (Figure 1). It contains both cancerous and non-cancerous elements. Together, they form an ecosystem that supports tumor survival. Cancer cells are the core of the tumor. They undergo uncontrolled growth due to genetic mutations and epigenetic changes [24,25]. However, they do not act alone. They communicate with surrounding cells and adapt to environmental stress. This ability to co-evolve with their surroundings is a major reason why cancer is hard to treat. Cancer cells release signals. These signals recruit and reprogram nearby cells. As a result, the tumor gains access to nutrients, avoids immune attack, and promotes blood vessel formation. These are classic examples of how cancer cells shape their environment. Stromal cells are non-cancerous cells found in the TME. These include fibroblasts, endothelial cells, and mesenchymal stem cells. One of the most important stromal cells in tumors is the cancer-associated fibroblast (CAF) [26,27]. CAFs differ from normal fibroblasts. They are activated by tumor signals and support cancer progression. CAFs secrete growth factors, remodel the extracellular matrix, and influence immune cell behavior [28]. Endothelial cells form the lining of blood vessels. In tumors, they help create abnormal vasculature. These blood vessels are often irregular, leaky, and poorly organized. This limits the delivery of oxygen and drugs into the tumor core.

Immune cells are a major component of the TME. However, their role is complex. In some cases, immune cells fight cancer. Cytotoxic T cells and natural killer (NK) cells can recognize and destroy tumor cells [29,30]. But in many tumors, immune cells are reprogrammed to help the tumor. For example, TAMs often support tumor growth [31,32]. Instead of attacking cancer, they suppress other immune cells, promote angiogenesis, and enhance metastasis. Tregs also help the tumor by preventing effective immune responses. Myeloid-derived suppressor cells (MDSCs) further block immune activity. Together, these cells create an immunosuppressive environment. The ECM is a network of proteins and molecules surrounding the cells [33,34]. It provides structure and biochemical signals. In tumors, the ECM is often dense and disorganized. This stiff matrix can act as a barrier to drug penetration. It can also send signals that promote cancer cell survival and migration. Collagen, fibronectin, and hyaluronic acid are commonly overexpressed in the tumor ECM [35,36]. These molecules interact with integrins and other cell receptors to affect cell behavior. In short, the ECM is not just a scaffold. It is an active regulator of cell fate in the tumor setting.

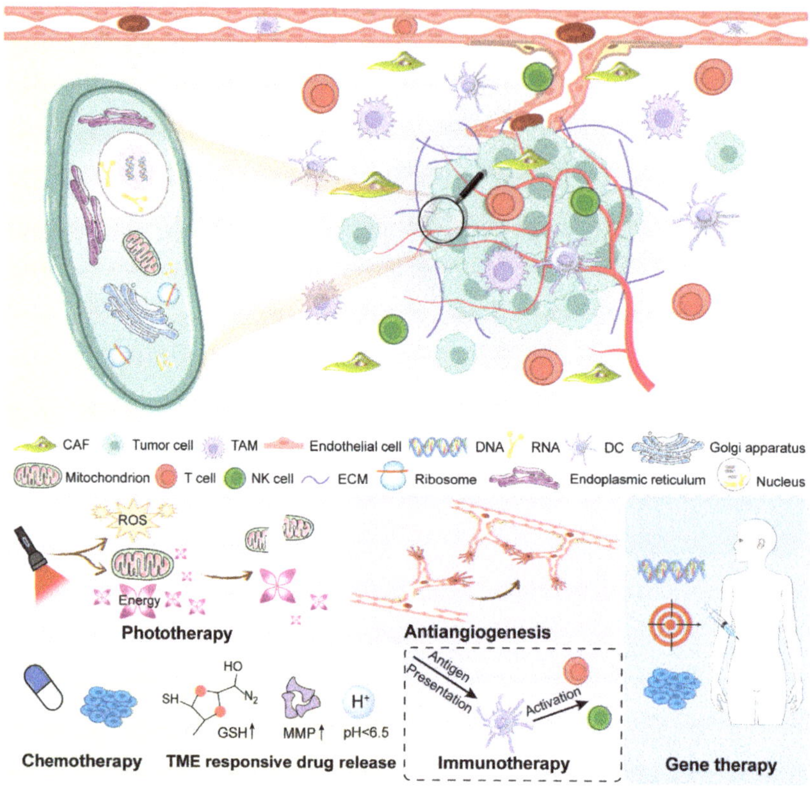

Figure 1. Overview of the tumor microenvironment.

2.2. Major Abnormalities in the Tumor Microenvironment

The TME is not like normal tissue. It is hostile, stressful, and highly altered. These changes affect how the tumor behaves and how it responds to therapy. Here, we discuss the four most prominent abnormalities: hypoxia, acidity, oxidative stress, and glutathione (GSH) overexpression. Hypoxia refers to low oxygen levels. It is a defining feature of many solid tumors [37,38]. As tumors grow, they often outpace their blood supply. This results in regions that receive little to no oxygen. Cells in these regions adapt by activating hypoxia-inducible factors (HIFs). These are transcription factors that help cells survive under low oxygen. HIFs trigger the expression of genes that promote angiogenesis, anaerobic metabolism, and drug resistance [39,40]. They also increase the expression of immune checkpoint molecules, which helps the tumor evade immune detection. Hypoxia makes tumors more aggressive. It also reduces the effectiveness of radiotherapy, which relies on oxygen to generate free radicals that kill cancer cells.

Tumors are acidic. This is mainly due to the Warburg effect, where cancer cells rely on glycolysis even in the presence of oxygen [41,42]. This produces lactic acid, which is then secreted into the extracellular space. The result is a low extracellular pH, often around 6.5 or lower. Normal tissues typically have a pH of 7.4. Acidic pH has multiple effects. It impairs drug uptake, especially for weakly basic drugs. It also alters enzyme activity and immune cell function. Moreover, acidic conditions can promote the activation of certain proteases that help cancer cells invade surrounding tissues. Importantly, acidity is not uniform. Some regions of the tumor are more acidic than others. This creates a patchy landscape that complicates treatment.

Cancer cells often have high levels of ROS. These are chemically reactive molecules that include superoxide, hydrogen peroxide, and hydroxyl radicals [43,44]. ROS are produced during metabolism and signaling. In tumors, their levels are elevated due to mitochondrial dysfunction, rapid growth, and inflammation. High ROS levels can damage DNA, proteins, and lipids. This drives mutations and genomic instability. But cancer cells often adapt. They develop antioxidant defenses to tolerate the oxidative stress [45,46]. ROS also influence signaling pathways. They activate NF-κB, HIF-1, and MAPK pathways, all of which promote survival, proliferation, and invasion. ROS can also affect the immune system [47,48]. Moderate ROS levels can suppress T cell activation and contribute to immune escape.

To counter high ROS, cancer cells increase their levels of antioxidants. The most important of these is glutathione (GSH). GSH is a tripeptide that directly neutralizes ROS. It is present at high concentrations in many tumors, often several times higher than in normal tissue [49,50]. High GSH helps cancer cells resist oxidative damage. It also contributes to drug resistance. Many chemotherapy agents work by generating ROS. When GSH levels are high, these drugs become less effective. GSH also plays a role in detoxification. It conjugates with drugs and toxins to make them easier to eliminate. This further reduces treatment effectiveness. In summary, elevated GSH levels give cancer cells a survival advantage in the harsh tumor environment.

2.3. The Tumor Microenvironment as a Therapeutic Target

The unique features of the TME present both challenges and opportunities. While they help the tumor resist treatment, they also offer targets for therapy (Figure 2). One approach is to deliver oxygen or generate it in situ. Materials like manganese dioxide (MnO_2) can react with tumor hydrogen peroxide to produce oxygen [51,52]. This relieves hypoxia and improves the efficacy of radiotherapy or photodynamic therapy. Another approach is to exploit hypoxia. Certain prodrugs are activated only in low-oxygen conditions. These selectively kill hypoxic tumor cells while sparing normal tissue. Acidic pH can be used as a trigger for drug release. pH-sensitive materials can release their payloads only in acidic conditions. This improves selectivity and reduces side effects. Buffering the tumor pH is another strategy. Compounds like bicarbonate can neutralize acidity, although clinical translation remains difficult.

Chemodynamic therapy (CDT) uses Fenton or Fenton-like reactions to convert tumor hydrogen peroxide into toxic hydroxyl radicals [53,54]. Iron, copper, or manganese-based materials can catalyze these reactions. This increases ROS and kills tumor cells from within. To enhance CDT, researchers are designing nanoparticles that also deplete GSH. Reducing GSH makes tumor cells more vulnerable to ROS. This dual approach maximizes oxidative damage. The TME's immunosuppressive nature can also be reversed. Certain nanoparticles can repolarize tumor-associated macrophages from an M2 (suppressive) state to an M1 (pro-inflammatory) state. Others can deliver antigens or adjuvants to dendritic cells to improve immune activation. Nanomaterials are also being used to deliver checkpoint inhibitors more precisely.

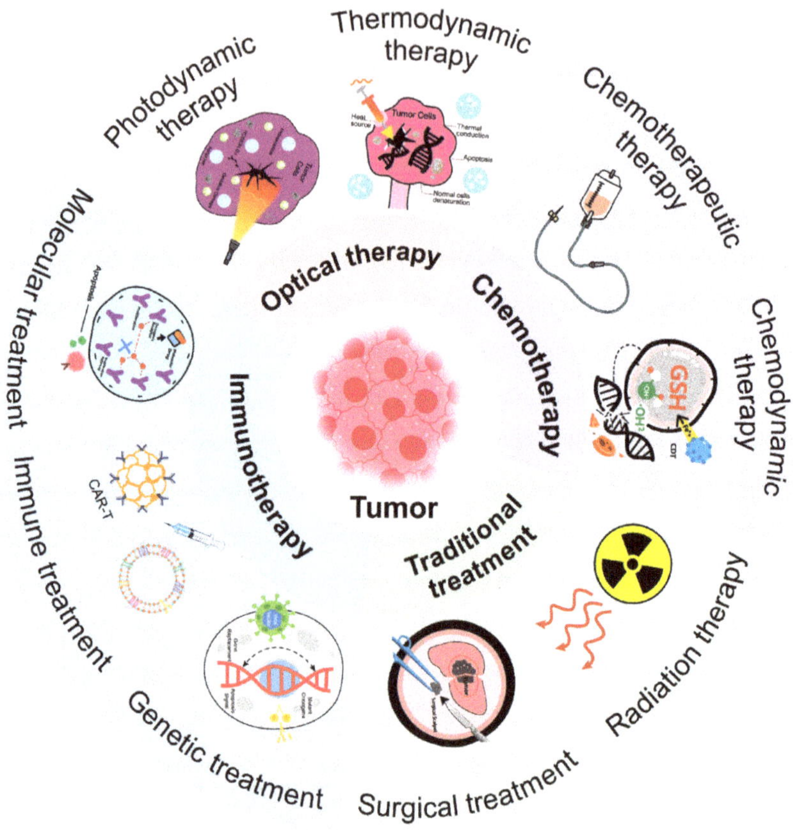

Figure 2. Summary of tumor therapy.

3. Classes of Inorganic Nanomaterials Used in TME Modulation

Inorganic nanomaterials have emerged as powerful tools in the fight against cancer. They offer unique physical and chemical properties that can be harnessed to interact with the TME. These materials can respond to specific features such as acidity, hypoxia, and oxidative stress.

3.1. Metal Oxides

Metal oxides are among the most widely studied materials for TME modulation. They are chemically versatile and often exhibit catalytic or redox activity. Their ability to react with molecules like hydrogen peroxide or glutathione makes them especially useful in hostile tumor environments. Manganese dioxide is a standout in TME-focused research. It reacts efficiently with hydrogen peroxide (H_2O_2), which is abundant in tumors. This reaction produces oxygen, which helps relieve hypoxia. In doing so, MnO_2 can enhance the performance of radiotherapy and photodynamic therapy [55,56] (Figure 3). MnO_2 also degrades in acidic conditions. This is useful because many tumors have a low extracellular pH. When MnO_2 breaks down, it releases Mn^{2+} ions. These ions can serve as contrast agents in magnetic resonance imaging (MRI), enabling real-time monitoring [57,58]. Moreover, MnO_2 can reduce levels of glutathione (GSH). This weakens the tumor's antioxidant defenses and enhances oxidative damage. That is why MnO_2 is often used in chemodynamic therapy (CDT) strategies. In summary, MnO_2 is multifunctional. It relieves hypoxia, lowers

GSH, and enhances both imaging and therapy. These features make it a central material in smart cancer nanomedicine.

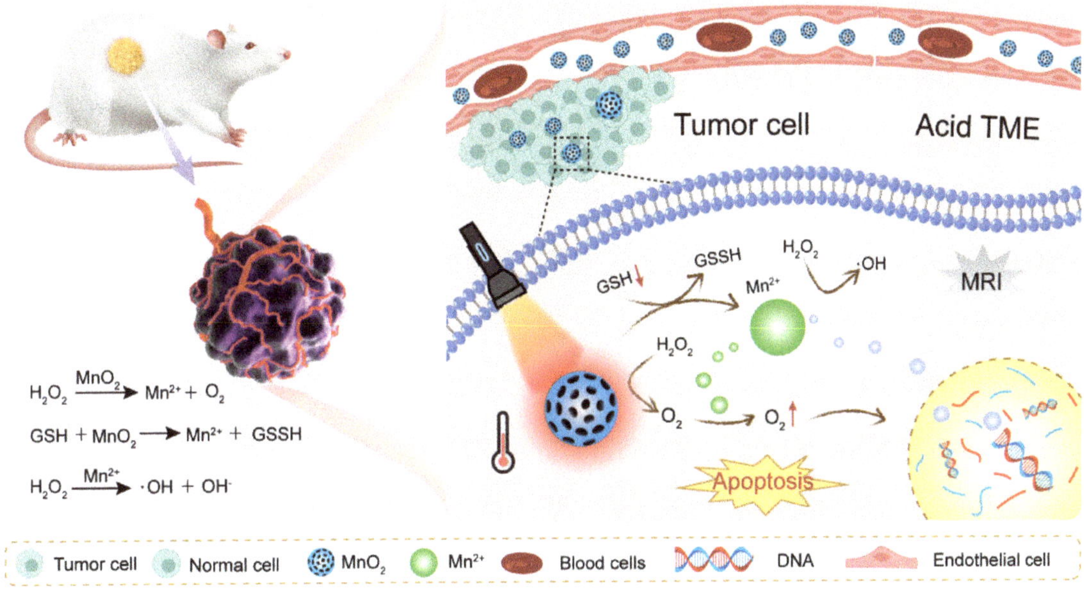

Figure 3. MnO_2 NPs remodel tumor microenvironment by reacting with H_2O_2 to release oxygen and Mn^{2+}.

Iron oxide nanoparticles are already approved for clinical use in imaging. They also show promise in cancer therapy. Their value lies in their ability to catalyze the Fenton reaction. In this reaction, ferrous ions (Fe^{2+}) convert hydrogen peroxide into hydroxyl radicals. These radicals are extremely reactive and toxic to cells [59,60]. By generating reactive oxygen species (ROS) within tumors, iron oxide promotes chemodynamic therapy. This is especially effective in tumors with high H_2O_2 levels [61]. Iron oxide nanoparticles can also modulate immune responses. Some studies suggest they help repolarize tumor-associated macrophages. This can turn the tumor from an immunosuppressive to an immunostimulatory environment. Additionally, Fe_3O_4 can serve as a delivery vehicle for drugs [62,63]. It can be guided to tumors using external magnetic fields. This improves targeting and reduces off-target toxicity. In short, iron oxide is a flexible and clinically relevant material. It works both as a therapeutic agent and an imaging enhancer.

Cerium oxide is known for its redox-switching behavior. It can shift between Ce^{3+} and Ce^{4+} oxidation states. This makes it useful for regulating oxidative stress in tumors [64,65]. Interestingly, CeO_2 shows dual activity. In normal cells, it scavenges excess ROS and protects tissues. In tumor cells, where the redox balance is already disrupted, it may behave differently. Some studies suggest it can actually increase ROS under certain conditions. CeO_2 nanoparticles also influence inflammation [66,67]. They can inhibit pro inflammatory cytokines and reduce tumor-promoting inflammation. This immune-regulatory effect is still being explored. Its antioxidant and pro-oxidant duality give cerium oxide a unique role. It is promising in cases where fine-tuned redox control is needed.

3.2. Noble Metals

Noble metal nanoparticles have distinct surface chemistry and photothermal properties. They are chemically stable and easy to modify. Among these, gold and platinum are the most commonly used in TME-targeted therapies. Gold nanoparticles (AuNPs) are highly popular in cancer nanomedicine. They are biocompatible and easily functionalized. Their surfaces can be modified with drugs, peptides, or targeting ligands [67,68]. One of their key strengths is their optical property. They absorb near-infrared light and convert it into heat. This makes them excellent agents for photothermal therapy (PTT). In tumors, this localized heating can disrupt cancer cells without damaging surrounding tissues. In addition to photothermal effects, AuNPs can act as redox modulators. Their surfaces can be engineered to interact with glutathione or deliver redox-active compounds. This helps disturb the tumor's antioxidant defense. Gold nanoparticles are also valuable in imaging. They enhance contrast in computed tomography (CT) and other imaging modalities. This supports image-guided treatment strategies [69,70]. AuNPs are safe, versatile, and multifunctional. Their ease of use makes them ideal platforms for combined diagnostic and therapeutic applications.

Platinum-based drugs like cisplatin have long been used in chemotherapy. However, platinum nanoparticles offer a new level of control. PtNPs can catalyze ROS production in tumors. Like iron-based materials, they can participate in redox reactions that generate toxic radicals [71,72]. These nanoparticles can amplify oxidative stress and kill tumor cells from within. PtNPs can also degrade in acidic environments. This makes them suitable for pH-sensitive drug delivery. Some formulations combine PtNPs with chemotherapeutic agents to create synergistic systems [73,74]. Moreover, platinum has high X-ray absorption. This enhances its role in radiotherapy. PtNPs can increase the radiation dose absorbed by the tumor while sparing healthy tissues. While toxicity remains a concern, new surface coatings and targeting methods are improving their safety profiles.

3.3. Carbon-Based Nanomaterials

Carbon-based materials have emerged as a versatile category in cancer therapy. They are lightweight, biocompatible, and rich in surface functionalities (Figure 4). Their ability to modulate ROS and serve as photothermal agents is especially valuable in the TME context. Graphene oxide (GO) is a single-atomic-layer material with oxygen-containing groups. It offers a large surface area and good dispersibility in water [75,76]. GO can absorb near-infrared light and generate heat. This makes it a powerful tool for photothermal therapy. It can also carry drugs and release them in response to TME stimuli such as pH or GSH. Additionally, GO has redox-modulating properties. It can interact with ROS and contribute to oxidative stress in tumor cells. Functionalization improves its performance. Modified GO with polyethylene glycol (PEG) or targeting ligands shows improved circulation and tumor accumulation [77,78]. Its 2D structure also allows for easy loading of imaging agents. This opens opportunities for multimodal theranostics.

Carbon dots (CDs) are small, fluorescent carbon-based nanoparticles. They are highly tunable and photostable. Their small size allows for deep tumor penetration. CDs can scavenge ROS or generate ROS, depending on their surface chemistry. This makes them adaptable to different tumor environments. They are also excellent imaging agents. Many CDs exhibit intrinsic fluorescence, which is useful for optical tracking [79,80]. Some carbon dots have also been developed to carry therapeutic cargo. Their tunable surface allows for stimuli-sensitive drug release. Overall, carbon-based materials offer a flexible platform for TME modulation. They combine diagnostic and therapeutic capabilities in compact designs.

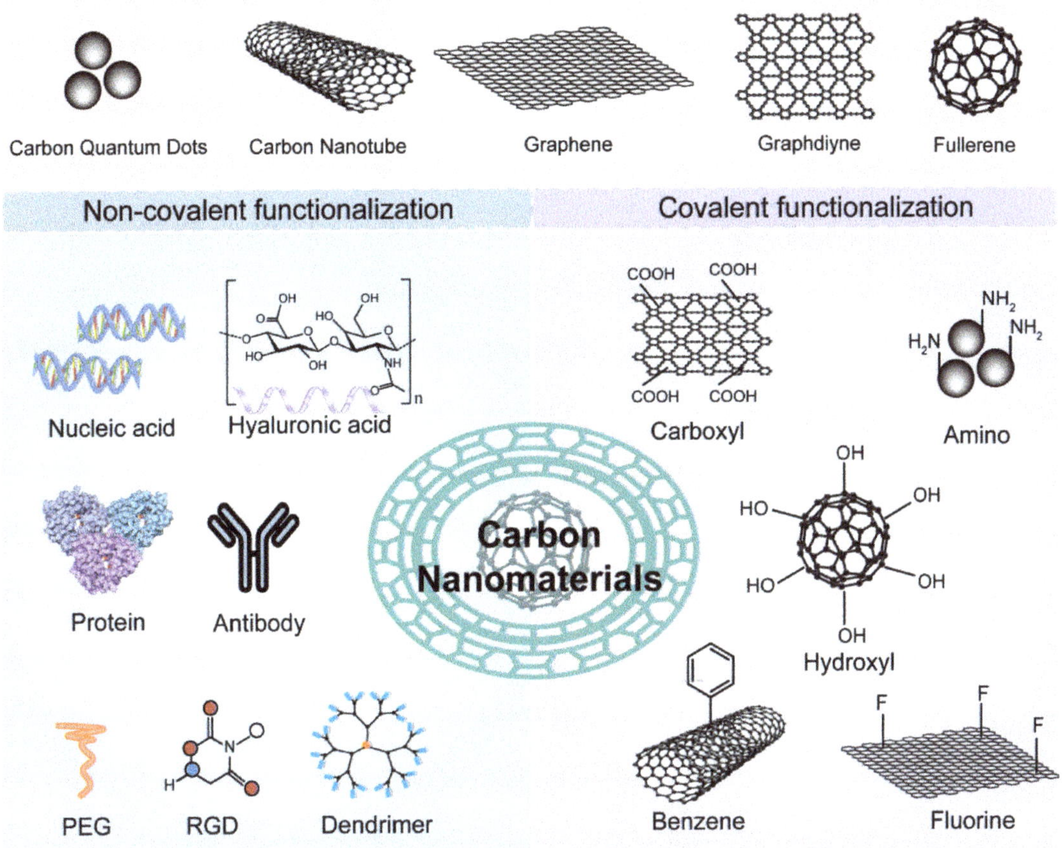

Figure 4. Schematic of Carbon Nanomaterial Functionalization. Carbon nanomaterials can be functionalized via covalent or non-covalent approaches using diverse chemical groups and biomolecules. This modification enhances their anticancer therapeutic efficacy.

3.4. Other Inorganic Materials

Beyond metal oxides, noble metals, and carbon forms, several other inorganic materials are gaining attention. These include calcium carbonate, silica, black phosphorus, and metal sulfides. Calcium carbonate is a naturally occurring, biocompatible material. It dissolves in acidic environments, making it ideal for pH-responsive drug delivery [81,82]. In the acidic TME, $CaCO_3$ particles degrade and release calcium ions. These ions can disrupt calcium signaling in cancer cells and trigger cell death. $CaCO_3$ also helps buffer the acidic TME. This makes it a useful adjunct to therapies that are pH-sensitive. The simplicity and safety of $CaCO_3$ make it appealing for clinical translation. Mesoporous silica nanoparticles (MSNs) are well-studied in nanomedicine. They have a tunable pore structure, large surface area, and good biocompatibility [83]. MSNs can load various drugs and release them in response to TME triggers. Acid-sensitive gatekeepers or enzyme-cleavable coatings allow precise control over release kinetics [84]. Silica is also used to coat other nanomaterials, improving their stability and safety. While inert on their own, MSNs gain functionality through surface modifications. This adaptability allows for customization based on tumor needs.

Black phosphorus is an emerging 2D material with excellent biodegradability. It breaks down into phosphate, a harmless by-product. BP absorbs near-infrared light and efficiently converts it to heat [85,86]. This makes it a potent photothermal agent. It also contributes to ROS production under light exposure, supporting photodynamic therapy. Its soft, layered structure allows it to carry drugs or nucleic acids. However, its instability in aqueous environments remains a technical challenge. Metal sulfides such as copper sulfide (CuS) and bismuth sulfide (Bi_2S_3) are gaining interest. These materials are photoactive and generate heat or ROS under light or ultrasound stimulation. CuS, for instance, is a strong absorber of NIR light. It has been used in photothermal and photoacoustic imaging. Bismuth-based sulfides offer high X-ray attenuation. This makes them excellent for CT imaging and radiosensitization. Metal sulfides can also modulate redox balance. Some can deplete glutathione or generate ROS in situ. These features support TME-targeted therapies.

4. Strategies for Tumor Microenvironment Modulation Using Inorganic Nanomaterials

The tumor microenvironment is an abnormal and hostile landscape. It is shaped by low oxygen levels, acidic pH, oxidative stress, and immune evasion. These features help cancer cells survive and resist therapy. However, these same features can be used as targets. Smart inorganic nanomaterials are designed to exploit these characteristics. They can respond to local cues and activate therapeutic actions only within the tumor.

4.1. Alleviating Hypoxia

Hypoxia means low oxygen. It is one of the most common features in solid tumors. Poor blood supply and rapid tumor growth both contribute to it. Hypoxia makes tumors more aggressive and resistant to treatments like radiotherapy and some chemotherapies. Radiation therapy depends on oxygen to generate free radicals. When oxygen is low, the effect is weakened. Hypoxia also supports immune evasion and promotes metastasis. It is a serious barrier to effective cancer treatment.

To overcome hypoxia, researchers have turned to oxygen-generating or oxygen-carrying nanoparticles. These materials can increase oxygen levels directly at the tumor site [87,88]. One of the most studied materials isMnO_2. MnO_2 reacts with hydrogen peroxide (H_2O_2), which is naturally overproduced in tumors (Figure 5). The reaction produces oxygen: $2\,H_2O_2 + MnO_2 \rightarrow O_2 + 2\,H_2O + Mn^{2+}$. This simple reaction helps supply oxygen locally. The produced oxygen enhances the effects of radiotherapy or photodynamic therapy. At the same time, MnO_2 breaks down in the acidic tumor environment. This improves its biodegradability and biocompatibility.

Other strategies involve perfluorocarbon (PFC)-loaded nanoparticles. PFCs can physically carry and release oxygen [89,90]. These particles are often combined with inorganic carriers like iron oxide or silica for imaging and targeting. After injection, they accumulate in tumors and gradually release oxygen. Some nanoparticles mimic the enzyme catalase, which naturally decomposes hydrogen peroxide into water and oxygen. Materials like platinum, cerium oxide, and iron oxide can perform this reaction. They act as nanozymes—catalytic nanomaterials that behave like enzymes. These materials do not just deliver oxygen. They also relieve oxidative stress and improve cell function. This is helpful for activating immune responses that are otherwise shut down by hypoxia. In short, alleviating hypoxia is a promising way to enhance the success of many therapies. Inorganic nanomaterials offer smart and effective ways to generate oxygen inside tumors.

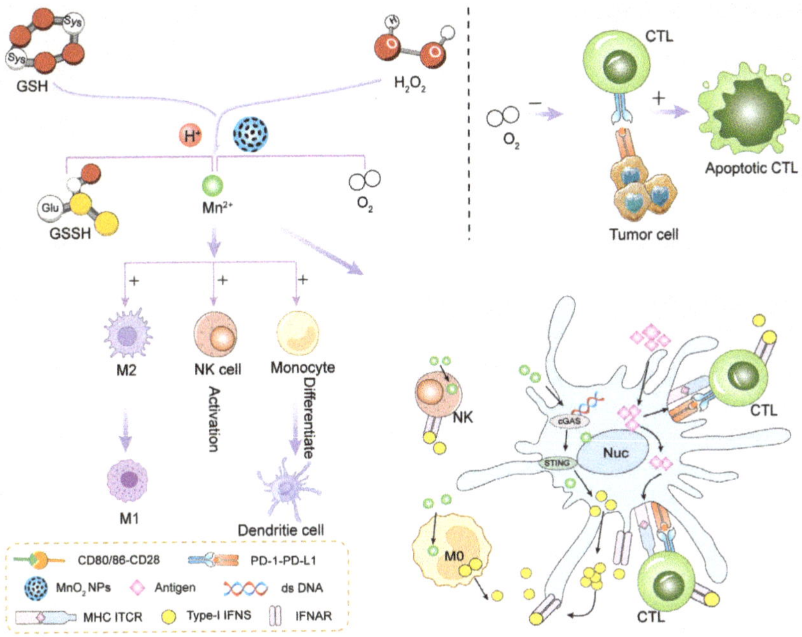

Figure 5. MnO_2 NPs regulate innate and adaptive immune cells against tumor and infections.

4.2. pH Modulation

Tumors are acidic. This acidity results from altered metabolism. Cancer cells rely heavily on glycolysis, even in the presence of oxygen. This leads to lactic acid accumulation. Proton pumps and poor blood flow make the situation worse.

The extracellular pH in tumors can fall below 6.5, compared to 7.4 in normal tissues. This low pH affects drug stability, uptake, and cell signaling. It also helps tumors invade nearby tissues. Inorganic nanomaterials can respond to pH. Some materials degrade or change structure in acidic conditions [91,92]. This feature is used for controlled drug release. For example, calcium carbonate dissolves in low pH. In the acidic tumor, $CaCO_3$ breaks down into calcium ions and carbon dioxide. This neutralizes local acidity and also promotes cell stress.

Another widely used material is ZIF-8, a zinc-based metal–organic framework. ZIF-8 is stable in blood but dissolves in acidic conditions. The main products are Zn^{2+} and 2-methylimidazole, both of which are involved in antitumor effects. This allows it to release drugs specifically inside the tumor. These pH-responsive behaviors improve drug precision. They help minimize damage to normal tissues. Some materials do more than just degrade. They can actively buffer the acidity. Magnesium oxide, $CaCO_3$, and other proton-scavenging materials absorb protons and raise the pH. This can normalize the extracellular environment. A more neutral pH improves the function of T cells and natural killer (NK) cells. It also enhances drug uptake and reduces tumor invasiveness. By altering pH, nanomaterials not only make drugs more effective but also improve immune activation. This dual effect makes pH modulation a powerful therapeutic strategy.

4.3. Redox Balance Regulation

Tumors have an abnormal redox environment. They contain high levels of ROS like hydrogen peroxide. At the same time, they also overexpress antioxidants, mainly GSH. ROS promote mutations and tumor growth. But cancer cells must control ROS to avoid death [93,94]. They do this by boosting GSH levels. High GSH allows cancer cells to survive under oxidative stress. Unfortunately, it also helps them resist chemotherapy and radiation. Smart nanomaterials are being developed to deplete GSH inside tumors [95,96]. This disrupts the redox balance and sensitizes cells to oxidative damage. Materials like copper, manganese, and iron-based nanoparticles can react with GSH. In doing so, they consume GSH and reduce the tumor's ability to handle ROS. Some nanoplatforms are coated with disulfide bonds. These bonds are cleaved by GSH, triggering drug release while depleting GSH at the same time. Lowering GSH increases the effect of chemodynamic therapy and photodynamic therapy. It also helps convert an immunosuppressive tumor into an immunogenic one.

Another strategy is to generate ROS inside the tumor. Many metal-based nanoparticles can catalyze Fenton or Fenton-like reactions. For example, iron-based materials convert H_2O_2 into hydroxyl radicals, which are highly toxic. This is the core of CDT [97,98]. It kills cancer cells without requiring external triggers like light or heat. Copper and manganese-based materials also show Fenton-like activity. They broaden the types of reactions possible in the tumor environment. When ROS levels are pushed too high, cancer cells undergo oxidative stress-induced death. This offers a non-traditional way to kill tumors by exploiting their own metabolism.

4.4. *TME-Triggered Drug Release*

One of the major limitations of chemotherapy is poor selectivity. Many anticancer drugs harm both healthy and cancerous cells. This causes severe side effects. Smart drug delivery systems aim to change that. They use TME-responsive triggers to release drugs only where needed. These triggers include acidic pH, high GSH, and elevated H_2O_2 levels. By linking drug release to these local conditions, therapies become more targeted and safer. As discussed earlier, acidic conditions are ideal for triggering drug release. Materials like ZIF-8, $CaCO_3$, and black phosphorus degrade in low pH. These systems can carry chemotherapeutics or photosensitizers. Once they reach the acidic tumor, they break apart and release their payload.

This reduces systemic toxicity and improves drug accumulation at the tumor site. Many tumors have high levels of GSH. Nanoparticles can be designed with disulfide bonds or redox-cleavable linkers. These break down in reductive environments, releasing drugs specifically inside cancer cells. This method is especially useful for delivering genetic material, peptides, or immunotherapy agents. Some systems respond to high ROS levels. These include boronate- or thioketal-linked carriers. ROS cleave these chemical bonds, which open the structure and release the drug. ROS-triggered systems can be combined with CDT to create a feedback loop. As CDT increases ROS levels, it accelerates drug release, boosting therapeutic efficiency. In short, TME-triggered drug release enhances precision. It ensures that drugs are released where they are needed most.

4.5. *Immune Modulation*

The immune system should be able to destroy cancer cells. However, many tumors avoid immune detection. They build a microenvironment that suppresses immune responses. They recruit regulatory T cells, secrete inhibitory cytokines, and express immune checkpoint molecules. The result is a "cold" tumor that is invisible to the immune system. Immune checkpoint inhibitors have helped. But their success is limited in cold tumors. New approaches are needed to stimulate anti-tumor immunity.

Some therapies can cause immunogenic cell death (ICD) [99,100]. This is a type of cell death that releases signals to alert the immune system. Inorganic nanoparticles like iron oxide, black phosphorus, or manganese dioxide can induce ICD. They do this by generating ROS or delivering drugs that cause endoplasmic reticulum stress. ICD helps turn cold tumors into hot tumors, which are more responsive to immunotherapy. Nanoparticles that trigger ICD may be combined with checkpoint inhibitors for better outcomes.

TAMs are often of the M2 type, which supports tumor growth. Reprogramming them to the M1 type can enhance inflammation and immune activation. Some nanomaterials, like iron oxide or manganese-based particles, help polarize TAMs to the M1 phenotype. They act through ROS generation or delivery of immunomodulatory agents. This shift in macrophage behavior improves T cell recruitment and tumor clearance. Ferroptosis is a non-apoptotic form of cell death driven by iron and lipid peroxidation. It leads to massive oxidative stress and cell rupture. Nanoparticles that deliver iron or inhibit antioxidant enzymes can trigger ferroptosis in tumors. This not only kills cancer cells but also releases danger signals that activate immune cells. Ferroptosis is being studied as a method to both kill tumors and stimulate anti-tumor immunity.

5. Multimodal Therapies Enabled by Tumor Microenvironment (TME) Modulation

Modifying the tumor microenvironment opens new paths for treating cancer. In its natural state, the TME works against therapy. It blocks drug delivery, weakens immune cells, and protects cancer cells from stress. By changing the TME, treatment becomes more effective. Inorganic nanomaterials play a big role in this transformation. They enable therapies to work better by responding to local tumor conditions. They also make it possible to combine different therapies into a single treatment strategy.

5.1. Chemo/Chemodynamic Therapy (CDT)

Chemotherapy remains a standard cancer treatment. It uses drugs that kill fast-growing cells. But in many cases, the TME reduces drug effectiveness. Poor blood flow limits drug entry. High glutathione levels neutralize drug-induced oxidative stress. Acidic pH can degrade drugs before they reach their target. TME modulation helps solve these problems. By delivering drugs in a smart way—such as through pH-sensitive nanocarriers—chemotherapy becomes more precise. Some nanomaterials also deplete glutathione, which makes cancer cells more sensitive to chemotherapy. For example, calcium carbonate or ZIF-8-based systems can release drugs only in acidic environments. This reduces side effects and boosts drug concentration at the tumor site [100,101].

CDT is a newer approach. It uses the TME's own chemistry to kill tumor cells. CDT is based on the Fenton reaction. In this process, transition metals like iron or copper react with hydrogen peroxide to produce hydroxyl radicals. These radicals cause oxidative damage and cell death. Tumors often contain high levels of hydrogen peroxide and low pH. These are perfect conditions for the Fenton reaction. Nanoparticles made from Fe_3O_4, CuO, or MnO_2 can catalyze this reaction within the tumor. CDT is unique because it does not need external activation like light or heat. It uses the tumor's own biochemistry [102]. This makes it useful for deep-seated tumors or patients who cannot tolerate other therapies. CDT is also highly selective. Normal tissues have lower H_2O_2 and higher pH, so the reaction does not happen there [103]. This improves safety. Combining CDT with chemotherapy creates a dual attack. One weakens the tumor's antioxidant system, while the other delivers direct cytotoxic agents.

5.2. Photothermal and Photodynamic Therapy

Photothermal therapy uses light to kill cancer cells. Nanoparticles absorb near-infrared (NIR) light and convert it into heat. This heat kills tumor cells by damaging their membranes and proteins. Gold nanoparticles, carbon nanotubes, and black phosphorus are common photothermal agents. These materials accumulate in the tumor and respond to external light. The TME affects PTT. Hypoxia, acidity, and poor blood supply can limit the effect. But nanomaterials that modulate the TME can overcome this [104,105]. For example, MnO_2 can generate oxygen. This improves local blood flow and oxygenation. The result is better light absorption and stronger thermal effects. Some nanocarriers are also designed to change shape or surface charge after TME exposure. This improves retention and penetration.

PTT is non-invasive and can be precisely controlled. It works well with imaging, making image-guided therapy possible. Photodynamic therapy is similar to PTT, but instead of heat, it uses light to produce ROS. A photosensitizer molecule absorbs light and transfers energy to nearby oxygen. This produces singlet oxygen, a highly reactive form that kills cells. PDT depends on oxygen. This is a problem in hypoxic tumors [106,107] (Figure 6). TME modulation can help. Oxygen-generating nanoparticles like MnO_2 or catalase-mimicking materials supply O_2 where needed. This allows the photosensitizer to work even in low-oxygen conditions. Additionally, pH-sensitive systems can release the photosensitizer only in the acidic tumor. This reduces off-target effects. When combined, PTT and PDT offer synergistic effects. Heat from PTT can increase blood flow, helping with oxygen delivery for PDT. Meanwhile, ROS from PDT can stress tumor cells, making them more sensitive to heat. Many nanoplatforms now carry both photothermal and photodynamic agents. They respond to a single light source, delivering a powerful one-two punch to tumors.

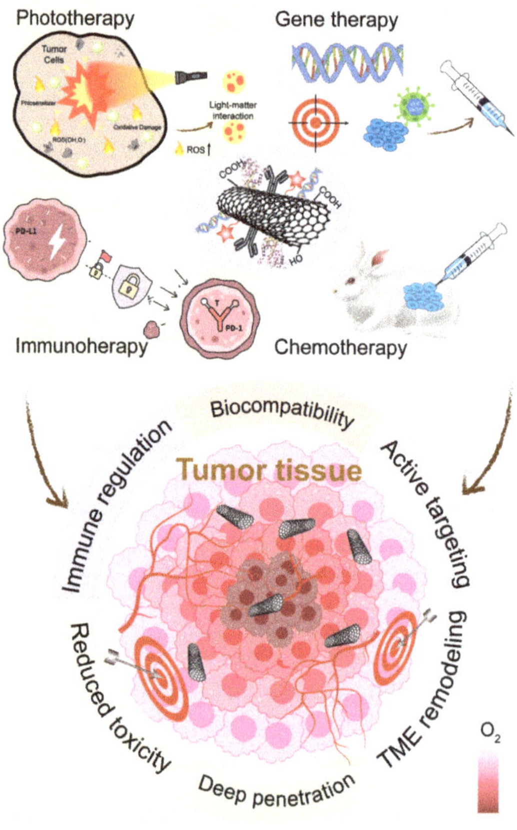

Figure 6. Advantages and Applications of Functionalized CNTs in Cancer Therapy. Functionalized carbon nanotubes (CNTs) are widely applied in cancer treatments, including chemotherapy, phototherapy, immunotherapy, and gene therapy. Surface modification enhances their biocompatibility, tumor penetration, and active targeting, while also reducing inherent toxicity.

5.3. Radiotherapy Enhancement

Radiotherapy is one of the oldest cancer treatments. It works by using ionizing radiation to damage DNA. The goal is to cause lethal mutations in cancer cells. However, the success of radiotherapy depends on oxygen. Radiation creates free radicals from water molecules. These radicals damage DNA. But in hypoxic tumors, this process is less efficient. The DNA damage is not enough to kill the cells. Also, tumors often repair radiation damage quickly. High antioxidant levels like GSH contribute to this resistance. Nanomaterials can enhance radiotherapy in several ways. One method is oxygen generation. As discussed earlier, MnO_2, catalase mimics, and PFC-based carriers improve oxygen levels in tumors. This restores radiosensitivity.

Another approach is using high-Z (atomic number) materials like gold, bismuth, or hafnium. These materials absorb radiation more effectively than soft tissue. When exposed to X-rays, they produce a burst of secondary electrons. This increases DNA damage in nearby cells. Some nanocarriers are also designed to deliver radiosensitizers—agents that make cancer cells more vulnerable to radiation. These may include platinum compounds or ROS-generating agents. Other systems release drugs after radiation triggers them. This turns radiation into both a treatment and a drug-release signal. By integrating nanotechnology, radiotherapy becomes more targeted and effective. Lower doses may be used, reducing damage to healthy tissue.

5.4. Synergistic Immunotherapy

Immunotherapy activates the body's own immune system to fight cancer. Immune checkpoint inhibitors, like anti-PD-1 and anti-CTLA-4, have shown success in melanoma and lung cancer. But many tumors remain unresponsive. These "cold tumors" lack active immune cells. They are surrounded by suppressive signals and immune-silent stromal cells. The TME blocks T cell entry, reduces antigen presentation, and increases regulatory cell activity. TME modulation can make immunotherapy more effective. Nanomaterials can reverse immunosuppression and trigger immune recognition. One key strategy is inducing immunogenic cell death (ICD) [108,109]. This form of death alerts the immune system. Nanoparticles that produce ROS, trigger ferroptosis, or disrupt cell membranes can cause ICD. This turns dying tumor cells into a signal that activates dendritic cells and T cells. Another method is macrophage repolarization. Tumors often have M2-type macrophages, which promote growth and suppress inflammation. Certain nanomaterials—like iron oxide or manganese-based systems—can shift these to the M1 type. M1 macrophages release pro-inflammatory cytokines and help recruit killer T cells.

Smart nanoparticles can deliver checkpoint inhibitors directly into tumors. This avoids systemic exposure and reduces side effects. Combining photothermal or photodynamic therapy with immunotherapy is also promising. Heat or ROS from these therapies not only kill tumor cells but also expose tumor antigens. This boosts the immune response when combined with anti-PD-1 or anti-CTLA-4 drugs. Chemodynamic therapy also supports immune activation. The ROS generated from CDT causes cell damage that is recognized by the immune system. When GSH is depleted, the immune-suppressive environment weakens. These multimodal combinations are now in clinical trials. They offer a more complete way to treat tumors that resist single-mode treatments.

6. Design Considerations for Smart TME-Modulating Nanoplatforms

Designing nanomaterials to modulate the TME is a complex task. The tumor is a harsh and unpredictable space. It is acidic, oxygen-deprived, full of reactive molecules, and often shielded from the immune system. For a nanoparticle to work well, it must be engineered with care.

6.1. Surface Functionalization and Targeting

The surface of a nanoparticle is the first thing it presents to its biological surroundings. It determines how the material interacts with cells, proteins, and the immune system. A poorly designed surface can lead to quick clearance or toxicity. A well-designed one can lead to precise targeting and enhanced therapeutic effects. One common method of surface functionalization is PEGylation. This means coating the nanoparticle with polyethylene glycol (PEG). PEG helps the nanoparticle avoid detection by the immune system [110,111]. It reduces opsonization, a process where blood proteins tag foreign objects for clearance. PEG also improves the stability and circulation time of nanoparticles in the bloodstream. However, PEG is not perfect. Some patients develop anti-PEG antibodies. This can lead to rapid clearance after repeated doses. Newer materials such as zwitterionic polymers or biomimetic coatings are being explored as alternatives.

Another important strategy is active targeting. This involves attaching ligands—such as peptides, antibodies, aptamers, or sugars—to the surface. These ligands recognize specific receptors overexpressed on tumor cells or tumor-associated cells. For example, the folate receptor is overexpressed in several cancers [112,113]. Nanoparticles with folic acid ligands can bind selectively to those tumor cells. Similarly, RGD peptides target integrins, which are common on angiogenic blood vessels in tumors. Targeting does not guarantee entry into every tumor cell, but it helps improve the local accumulation of therapeutic agents. Combining passive targeting (via the enhanced permeability and retention effect) with active targeting improves therapeutic outcomes. More recently, cell membrane-coated nanoparticles have gained attention. These use membranes from red blood cells, platelets, or cancer cells themselves. This gives the nanoparticles a "self" identity, helping them evade the immune system. Cancer cell membranes, for instance, can provide homotypic targeting. This means the particles are more likely to bind to other cells from the same tumor type. These biomimetic systems blend the natural and synthetic worlds for better performance.

6.2. Biodegradability and Biosafety

Biodegradability is essential. After a nanoparticle completes its job, it should break down into safe, easily excretable components. If it does not, it can accumulate in organs and cause long-term toxicity. This is a major concern in clinical translation. Many early nanomaterials showed promise in the lab but failed in trials due to poor clearance and toxicity. Some inorganic materials naturally degrade under TME conditions. Calcium carbonate, for example, dissolves in acidic environments. It breaks down into calcium and carbon dioxide—both safe and easily processed by the body. Black phosphorus is another biodegradable material. It degrades into phosphate, a common and safe ion found in cells. Its degradation is accelerated in water or in the acidic tumor environment.

Iron oxide nanoparticles are metabolized through natural iron pathways in the liver and spleen. This makes them safer than some synthetic metal-based systems. Designers must also consider the speed of degradation. Too fast, and the material may lose its effect. Too slow, and it may linger in tissues. An ideal nanoparticle should remain stable in blood but degrade in the tumor. To achieve this, some materials are coated with pH-sensitive shells. These protect the core during circulation but dissolve when exposed to acidic tumor pH. Researchers must strike a balance. The material needs to be robust enough to reach the tumor, but soft enough to break down afterward.

6.3. Stimuli-Responsiveness (pH, Redox, Enzyme, Light)

Stimuli-responsive systems are at the heart of smart nanomedicine. They allow the nanoparticle to stay dormant during circulation and only activate at the tumor site. This improves specificity and reduces side effects. The TME offers several triggers. These include low pH, high GSH, elevated hydrogen peroxide (H_2O_2), overexpressed enzymes, and even external stimuli like light or ultrasound.

Tumor environments are more acidic than normal tissues. This difference is exploited by pH-sensitive systems. Materials like ZIF-8 or $CaCO_3$ degrade in acidic pH. They release their contents only when they reach the tumor. This makes pH-responsiveness one of the most widely used strategies. Glutathione levels are much higher inside tumor cells than in the bloodstream. Nanoparticles designed with disulfide bonds or redox-cleavable linkers respond to this difference. Once inside the cell, high GSH levels break the disulfide bonds. This leads to the release of drugs or other agents. Redox-responsive systems are especially useful for intracellular delivery.

ROS are elevated in tumors. Some systems use boronate esters or thioketal linkers, which are sensitive to ROS. When triggered, these systems degrade and release their contents. This is useful for combined therapies. For instance, a photodynamic agent may raise ROS levels, which then triggers drug release from nearby nanoparticles. Tumors often overexpress enzymes like matrix metalloproteinases (MMPs) [114,115]. These enzymes break down the extracellular matrix. Enzyme-sensitive coatings can be cleaved by MMPs, triggering release. This adds another layer of selectivity. Light, ultrasound, and magnetic fields can also trigger nanoparticle activation. For example, gold nanoshells respond to near-infrared light for photothermal therapy. External triggers give researchers precise control. They allow activation only in the tumor area, minimizing systemic toxicity. The use of combined stimuli—such as pH and light, or ROS and enzyme—adds further sophistication. These multi-responsive systems are gaining popularity in advanced cancer therapy.

6.4. Size, Charge, and Biodistribution

Even the best-designed nanomaterial can fail if it does not reach the tumor. Size, shape, and charge determine how a nanoparticle moves through the body, how it accumulates in tumors, and how it is cleared. Size is a critical factor. Nanoparticles must be large enough to avoid kidney filtration but small enough to pass through leaky tumor blood vessels. The ideal size for passive targeting is often between 30 and 150 nm. This range allows particles to accumulate in tumors via the enhanced permeability and retention (EPR) effect. Smaller particles (below 10 nm) may be rapidly cleared by the kidneys. Larger particles (above 200 nm) may get trapped in the liver or spleen. Researchers must fine-tune the size for each application. Drug loading, circulation time, and tumor penetration all depend on it.

Charge affects how particles interact with cell membranes and proteins. Positively charged nanoparticles are taken up more easily by cells. However, they are also more likely to bind serum proteins and get cleared quickly. They may also trigger unwanted immune responses. Neutral or slightly negative particles tend to circulate longer. They are less likely to activate immune clearance. PEGylation can help mask the charge and improve behavior in the bloodstream. Charge must also be compatible with targeting strategies. For example, cationic surfaces are good for nucleic acid delivery but not always ideal for systemic circulation. The goal is to maximize tumor accumulation while minimizing exposure to healthy organs. This depends on a combination of size, charge, surface chemistry, and shape. Rod-shaped or disk-like particles may behave differently from spherical ones. Some shapes may penetrate tumors better or avoid clearance longer. Advanced imaging tools like PET, CT, or fluorescence imaging are used to track biodistribution. These tools help refine nanoparticle design over time.

7. Challenges and Future Perspectives

Smart inorganic nanomaterials have shown strong potential in modulating the tumor microenvironment. They offer new tools for cancer diagnosis and therapy. They respond to local stimuli. They improve the precision of treatments. They even help overcome drug resistance. But despite these advances, challenges remain. Several barriers still limit the full clinical translation of TME-modulating nanoplatforms. The path forward requires new ideas, better designs, and more personalized strategies.

7.1. Tumor Heterogeneity in TME Responsiveness

One of the biggest challenges is tumor heterogeneity. This means that tumors differ not just between patients but also within the same tumor mass. Some regions may be hypoxic. Others may have normal oxygen levels. Some areas are acidic. Others are not. This makes it difficult to design one-size-fits-all nanomaterials. A system that works well in one tumor may fail in another. Even in the same patient, different metastatic sites may respond differently to treatment. Nanomaterials often rely on the presence of specific triggers—like low pH, high glutathione, or excess hydrogen peroxide. But these triggers are not consistent. Their levels vary widely across tumor types and stages. This variation affects the efficiency of TME-responsive systems. A nanoparticle designed to release drugs in acidic conditions may not activate in a tumor with near-neutral pH. Similarly, chemodynamic therapy may fail if the tumor lacks sufficient hydrogen peroxide. To address this, future systems must include real-time feedback. Nanomaterials that can both diagnose and treat are needed. These "theranostic" platforms will help determine when and where the therapy should be activated. Advanced imaging or biosensing components could provide this feedback. This will allow doctors to adjust treatment based on the actual TME profile of each patient or tumor site.

7.2. Immune Evasion and Nanoparticle–Immune System Interaction

Nanoparticles are foreign bodies. The immune system is trained to detect and remove anything that does not belong. This includes smart nanocarriers. When injected, nanoparticles may be quickly coated by plasma proteins. This process is called opsonization. It marks them for removal by macrophages. As a result, fewer particles reach the tumor. Some nanoparticles may also activate immune cells unintentionally. This leads to inflammation or cytokine release. In some cases, it may cause serious side effects. Even when particles are designed for immune modulation, the response can be unpredictable. For example, a nanoparticle meant to polarize macrophages to the M1 state may instead get trapped by other immune cells, reducing its therapeutic effect.

To overcome this, surface modification is key. Coating nanoparticles with PEG can reduce recognition. However, repeated use may lead to anti-PEG antibodies. Alternative coatings—such as zwitterionic polymers or cell membrane cloaking—are being explored. Another approach is to use immune-informed design. Nanomaterials should not only avoid immune clearance but also cooperate with immune cells. For instance, biomimetic nanoparticles that use cancer cell membranes can enhance immune compatibility and improve homing to tumors. Future research must aim to balance stealth and immune engagement. The goal is to evade unwanted immune responses while enabling therapeutic immune activation.

7.3. Integration of AI for Nanoparticle Design

Designing effective nanomaterials involves many variables. Material type, size, charge, shape, drug load, and release profile all matter. Changing one feature affects many others. Testing every combination in the lab is time-consuming and expensive. This is where AI can help. AI tools, especially machine learning algorithms, can analyze large datasets. They can find patterns in how nanoparticles behave in different conditions. They can predict outcomes based on design features. For example, by predicting protein corona formation through machine learning, researchers can precisely control the surface properties of nanoparticles, significantly improve their tumor targeting efficiency, and provide a new data-driven approach to personalized nanodrug design. AI can also optimize formulation development. It can speed up screening for new materials. It can suggest new combinations that researchers might not consider. To support this, large and consistent datasets are needed. Many labs collect data differently. Without standardization, AI models may be inaccurate. Creating open-access nanoparticle databases will be essential. These should include detailed information on particle design, in vivo behavior, and therapeutic outcomes. With more data, AI will become an even more powerful tool in nanomedicine.

7.4. Potential Nanoparticle Toxicity

Inorganic nanomaterials are widely used in cancer treatment due to their unique optical, magnetic, and catalytic properties, but their potential toxicity seriously restricts clinical translation. First, soluble nanoparticles release metal ions, leading to oxidative stress and apoptosis directly induce membrane damage, mitochondrial dysfunction and DNA breakage. Second, long-term risks, such as the fibrous structure of carbon nanotubes may cause cancer and metal oxide nanoparticles induce genotoxicity. At present, clinical translation is facing challenges such as lack of standardized toxicity tests, insufficient long-term data, and species differences, and it is necessary to establish an international toxicology framework, develop degradable materials, and integrate organoids and AI to simulate human metabolic responses, and finally promote safe and efficient inorganic nanomaterials from the laboratory to clinical practice through precise modification, intelligent design and AI-assisted prediction.

8. Conclusions

Over the past decade, smart inorganic nanomaterials have transformed the field of cancer research. They have opened new doors in how we think about treating tumors. They do more than carry drugs. They interact with the tumor microenvironment. They respond to it. They change it. This review has explored the various ways these nanomaterials can be used. It has also discussed the strategies behind their design and the barriers that still remain. As the field continues to grow, it is important to reflect on what we have achieved—and where we are heading.

One of the most important contributions of inorganic nanomaterials is their ability to modulate the tumor microenvironment. Traditional cancer treatments often fail because the TME is hostile. It is acidic, oxygen-deprived, and full of antioxidants like glutathione. These conditions protect cancer cells and reduce therapy effectiveness. Smart inorganic nanomaterials are designed to work with these conditions, not against them. They use TME features as triggers for activation. This allows therapies to become more targeted and safer. For example, MnO_2 and similar materials generate oxygen from hydrogen peroxide [116,117]. This helps relieve hypoxia and boosts radiotherapy and photodynamic therapy. Iron and copper-based materials catalyze Fenton-like reactions, generating toxic hydroxyl radicals. This forms the basis of CDT, a treatment that uses the tumor's own biochemistry against it.

Nanoparticles also help manage pH and redox balance. Materials like calcium carbonate or ZIF-8 break down in acidic environments. Others release drugs in response to high levels of glutathione or ROS. These features allow for stimuli-responsive drug delivery, improving precision and reducing harm to healthy tissues. Another key advance is the development of immune-modulating nanoplatforms. Certain inorganic materials reprogram immune cells, turning "cold" tumors into "hot" ones. Some induce ICD. Others promote macrophage polarization or trigger ferroptosis, a unique form of cell death that stimulates immune activity [107,118]. Together, these features make smart nanomaterials versatile tools. They support multimodal therapies that combine chemotherapy, phototherapy, radiotherapy, and immunotherapy. Instead of using one weapon, doctors can now use several—all guided by the same platform. The design principles behind these platforms are also advancing. Nanoparticles are now tailored for specific size, shape, and surface chemistry. They are made biodegradable and safe. They include surface ligands for active targeting. Many are even multi-responsive, reacting to more than one TME feature at once. We also see progress in integrating diagnosis and therapy into the same system. These theranostic platforms allow real-time tracking of treatment. They help doctors adjust strategies as needed. This is a big step toward personalized, adaptive medicine.

Looking ahead, the role of smart inorganic nanomaterials in precision oncology will become even more central. They align perfectly with the goals of personalized medicine. They can adapt to each patient's tumor environment. They can deliver treatment based on real-time biological cues. However, future progress will require addressing key challenges. Tumor heterogeneity remains a major barrier. Not all tumors respond the same way. Some may not have enough hydrogen peroxide for CDT. Others may not be acidic enough for pH-triggered release. Future nanoplatforms must include feedback systems to assess the local TME before acting.

The immune system is another complex factor. Nanoparticles must avoid immune clearance but still interact with immune cells when needed. This balance is difficult but essential. Better biomimetic designs and immune-compatible coatings will be important moving forward.

AI will also play a role. AI can help predict which nanoparticle designs will work best for specific tumors. It can accelerate screening and optimization. It can even help design personalized formulations on demand. The future will also bring modular platforms. These will allow doctors to mix and match drug types, targeting ligands, and surface coatings. One core nanoparticle could be customized for different patients or cancer types. This modularity will reduce cost, improve speed, and enhance treatment outcomes. To truly bring smart nanomaterials into everyday clinical practice, we also need regulatory progress. New evaluation frameworks must be developed for these complex systems. Safety testing must include both short- and long-term effects. Manufacturing must be standardized and scalable. Despite these hurdles, the potential benefits are too great to ignore. Smart inorganic nanomaterials can improve how we deliver drugs. They can enhance how tumors respond to treatment. They can guide therapy in real-time. And most importantly, they can help doctors tailor treatments to individual patients.

Cancer is a complex and evolving disease. It adapts quickly and resists simple solutions. To beat it, we need tools that are just as smart and adaptable. Inorganic nanomaterials are one of those tools. They bring intelligence to therapy. They bring precision to treatment. And they bring hope for better outcomes. As we move forward, collaboration between scientists, clinicians, and engineers will be key. Together, we can refine these materials and bring them from the lab to the clinic. Smart inorganic nanomaterials are not just a research trend. They are part of the future of cancer care.

Funding: This research was funded by Young Elite Scientists Sponsorship Program by BATSA (BYESS2023244), and Beijing Institute of Technology Research Fund Program for Young Scholars.

Acknowledgments: The authors also want to thank the Analysis & Testing Center at the Beijing Institute of Technology.

Conflicts of Interest: The authors declare no conflicts of interest.

References

1. Xiao, Y.; Yu, D. Tumor microenvironment as a therapeutic target in cancer. *Pharmacol. Ther.* **2021**, *221*, 107753. [CrossRef]
2. Elhanani, O.; Ben-Uri, R.; Keren, L. Spatial profiling technologies illuminate the tumor microenvironment. *Cancer Cell* **2023**, *41*, 404–420. [CrossRef] [PubMed]
3. Vitale, I.; Manic, G.; Coussens, L.M.; Kroemer, G.; Galluzzi, L. Macrophages and Metabolism in the Tumor Microenvironment. *Cell Metab.* **2019**, *30*, 36–50. [CrossRef] [PubMed]
4. Jin, M.-Z.; Jin, W.-L. The updated landscape of tumor microenvironment and drug repurposing. *Signal Transduct. Target. Ther.* **2020**, *5*, 166. [CrossRef] [PubMed]
5. Jing, X.; Yang, C.; Shao, F.; Wei, K.; Xie, M.; Shen, H.; Shu, Y. Role of hypoxia in cancer therapy by regulating the tumor microenvironment. *Mol. Cancer* **2019**, *18*, 157. [CrossRef]
6. Chen, S.; Liao, C.; Hu, H.; Liao, J.; Chen, Z.; Li, S.; Zeng, X.; Peng, B.; Shen, S.; Li, D.; et al. Hypoxia-driven tumor stromal remodeling and immunosuppressive microenvironment in scirrhous HCC. *Hepatology* **2024**, *79*, 780–797. [CrossRef]
7. Wu, Q.; You, L.; Nepovimova, E.; Heger, Z.; Wu, W.; Kuca, K.; Adam, V. Hypoxia-inducible factors: Master regulators of hypoxic tumor immune escape. *J. Hematol. Oncol.* **2022**, *15*, 77. [CrossRef]
8. Wang, C.; Xu, S.; Yang, X. Hypoxia-Driven Changes in Tumor Microenvironment: Insights into Exosome-Mediated Cell Interactions. *Int. J. Nanomed.* **2024**, *19*, 8211–8236. [CrossRef]
9. Boedtkjer, E.; Pedersen, S.F. The Acidic Tumor Microenvironment as a Driver of Cancer. *Annu. Rev. Physiol.* **2020**, *82*, 103–126. [CrossRef]
10. Malla, R.; Kumari, S.; Ganji, S.P.; Srilatha, M.; Nellipudi, H.R.; Nagaraju, G.P. Reactive oxygen species of tumor microenvironment: Harnessing for immunogenic cell death. *Biochim. Biophys. Acta Rev. Cancer* **2024**, *1879*, 189154. [CrossRef]
11. Malla, R.; Surepalli, N.; Farran, B.; Malhotra, S.V.; Nagaraju, G.P. Reactive oxygen species (ROS): Critical roles in breast tumor microenvironment. *Crit. Rev. Oncol.* **2021**, *160*, 103285. [CrossRef]
12. Chen, C.; Wang, Z.; Ding, Y.; Qin, Y. Tumor microenvironment-mediated immune evasion in hepatocellular carcinoma. *Front. Immunol.* **2023**, *14*, 1133308. [CrossRef] [PubMed]
13. DePeaux, K.; Rivadeneira, D.B.; Lontos, K.; Dean, V.G.; Gunn, W.G.; Watson, M.J.; Yao, T.; Wilfahrt, D.; Hinck, C.; Wieteska, L.; et al. An oncolytic virus–delivered TGFβ inhibitor overcomes the immunosuppressive tumor microenvironment. *J. Exp. Med.* **2023**, *220*, e20230053. [CrossRef] [PubMed]
14. O'COnnell, B.C.; Hubbard, C.; Zizlsperger, N.; Fitzgerald, D.; Kutok, J.L.; Varner, J.; Ilaria, R.; Cobleigh, M.A.; Juric, D.; Tkaczuk, K.H.R.; et al. Eganelisib combined with immune checkpoint inhibitor therapy and chemotherapy in frontline metastatic triple-negative breast cancer triggers macrophage reprogramming, immune activation and extracellular matrix reorganization in the tumor microenvironment. *J. ImmunoTher. Cancer* **2024**, *12*, e009160. [CrossRef] [PubMed]
15. Jarosz-Biej, M.; Smolarczyk, R.; Cichoń, T.; Kułach, N. Tumor Microenvironment as A Game Changer in Cancer Radiotherapy. *Int. J. Mol. Sci.* **2019**, *20*, 3212. [CrossRef]
16. Ozpiskin, O.M.; Zhang, L.; Li, J.J. Immune targets in the tumor microenvironment treated by radiotherapy. *Theranostics* **2019**, *9*, 1215–1231. [CrossRef]
17. Pitt, J.M.; Marabelle, A.; Eggermont, A.; Soria, J.C.; Kroemer, G.; Zitvogel, L. Targeting the tumor microenvironment: Removing obstruction to anticancer immune responses and immunotherapy. *Ann. Oncol.* **2016**, *27*, 1482–1492. [CrossRef]
18. Khosravi, G.; Mostafavi, S.; Bastan, S.; Ebrahimi, N.; Gharibvand, R.S.; Eskandari, N. Immunologic tumor microenvironment modulators for turning cold tumors hot. *Cancer Commun.* **2024**, *44*, 521–553. [CrossRef]
19. Song, X.; Hao, C.; Li, Y.; Li, Y.; Dong, H.; Wei, Q.; Wei, M.; Li, H.; Zhao, L. Chiral inorganic nanomaterials in the tumor microenvironment: A new chapter in cancer therapy. *Pharmacol. Res.* **2024**, *208*, 107386. [CrossRef]
20. Ding, H.; Tan, P.; Fu, S.; Tian, X.; Zhang, H.; Ma, X.; Gu, Z.; Luo, K. Preparation and application of pH-responsive drug delivery systems. *J. Control. Release* **2022**, *348*, 206–238. [CrossRef]
21. Li, C.; Zhao, Z.; Luo, Y.; Ning, T.; Liu, P.; Chen, Q.; Chu, Y.; Guo, Q.; Zhang, Y.; Zhou, W.; et al. Macrophage-Disguised Manganese Dioxide Nanoparticles for Neuroprotection by Reducing Oxidative Stress and Modulating Inflammatory Microenvironment in Acute Ischemic Stroke. *Adv. Sci.* **2021**, *8*, 2101526. [CrossRef]
22. Yang, G.; Xu, L.; Chao, Y.; Xu, J.; Sun, X.; Wu, Y.; Peng, R.; Liu, Z. Hollow MnO_2 as a tumor-microenvironment-responsive biodegradable nano-platform for combination therapy favoring antitumor immune responses. *Nat. Commun.* **2017**, *8*, 902. [CrossRef] [PubMed]
23. Liang, X.; Wang, D.; Zhao, Y.; Wang, X.; Yao, S.; Huang, W.; Yang, Y.; Dong, X.; Zhang, L.; Yang, J. Tumor microenvironment-responsive manganese-based nano-modulator activate the cGAS-STING pathway to enhance innate immune system response. *J. Nanobiotechnol.* **2024**, *22*, 535. [CrossRef] [PubMed]
24. Anderson, N.M.; Simon, M.C. The tumor microenvironment. *Curr. Biol.* **2020**, *30*, R921–R925. [CrossRef]

25. Barkley, D.; Moncada, R.; Pour, M.; Liberman, D.A.; Dryg, I.; Werba, G.; Wang, W.; Baron, M.; Rao, A.; Xia, B.; et al. Cancer cell states recur across tumor types and form specific interactions with the tumor microenvironment. *Nat. Genet.* **2022**, *54*, 1192–1201. [CrossRef]
26. Zhao, Y.; Shen, M.; Wu, L.; Yang, H.; Yao, Y.; Yang, Q.; Du, J.; Liu, L.; Li, Y.; Bai, Y. Stromal cells in the tumor microenvironment: Accomplices of tumor progression? *Cell Death Dis.* **2023**, *14*, 587. [CrossRef]
27. Gabai, Y.; Assouline, B.; Ben-Porath, I. Senescent stromal cells: Roles in the tumor microenvironment. *Trends Cancer* **2023**, *9*, 28–41. [CrossRef]
28. Jin, H.R.; Wang, J.; Wang, Z.J.; Xi, M.J.; Xia, B.H.; Deng, K.; Yang, J.L. Lipid metabolic reprogramming in tumor microenvironment: From mechanisms to therapeutics. *J. Hematol. Oncol.* **2023**, *16*, 103. [CrossRef]
29. Gajewski, T.F.; Schreiber, H.; Fu, Y.-X. Innate and adaptive immune cells in the tumor microenvironment. *Nat. Immunol.* **2013**, *14*, 1014–1022. [CrossRef]
30. Mellman, I.; Chen, D.S.; Powles, T.; Turley, S.J. The cancer-immunity cycle: Indication, genotype, and immunotype. *Immunity* **2023**, *56*, 2188–2205. [CrossRef]
31. Pan, Y.; Yu, Y.; Wang, X.; Zhang, T. Tumor-Associated Macrophages in Tumor Immunity. *Front. Immunol.* **2020**, *11*, 583084. [CrossRef]
32. Mao, X.; Xu, J.; Wang, W.; Liang, C.; Hua, J.; Liu, J.; Zhang, B.; Meng, Q.; Yu, X.; Shi, S. Crosstalk between cancer-associated fibroblasts and immune cells in the tumor microenvironment: New findings and future perspectives. *Mol. Cancer* **2021**, *20*, 131. [CrossRef] [PubMed]
33. Yuan, Z.; Li, Y.; Zhang, S.; Wang, X.; Dou, H.; Yu, X.; Zhang, Z.; Yang, S.; Xiao, M. Extracellular matrix remodeling in tumor progression and immune escape: From mechanisms to treatments. *Mol. Cancer* **2023**, *22*, 48. [CrossRef] [PubMed]
34. Niland, S.; Riscanevo, A.X.; Eble, J.A. Matrix Metalloproteinases Shape the Tumor Microenvironment in Cancer Progression. *Int. J. Mol. Sci.* **2021**, *23*, 146. [CrossRef]
35. Sleeboom, J.J.F.; van Tienderen, G.S.; Schenke-Layland, K.; van der Laan, L.J.W.; Khalil, A.A.; Verstegen, M.M.A. The extracellular matrix as hallmark of cancer and metastasis: From biomechanics to therapeutic targets. *Sci. Transl. Med.* **2024**, *16*, eadg3840. [CrossRef] [PubMed]
36. Jiang, Y.; Zhang, H.; Wang, J.; Liu, Y.; Luo, T.; Hua, H. Targeting extracellular matrix stiffness and mechanotransducers to improve cancer therapy. *J. Hematol. Oncol.* **2022**, *15*, 34. [CrossRef]
37. Mukherjee, S.; Ray, S.K. Imitating Hypoxia and Tumor Microenvironment with Immune Evasion by Employing Three Dimensional In vitro Cellular Models: Impressive Tool in Drug Discovery. *Recent Pat. Anti-Cancer Drug Discov.* **2022**, *17*, 80–91. [CrossRef]
38. Bai, R.; Li, Y.; Jian, L.; Yang, Y.; Zhao, L.; Wei, M. The hypoxia-driven crosstalk between tumor and tumor-associated macrophages: Mechanisms and clinical treatment strategies. *Mol. Cancer* **2022**, *21*, 177. [CrossRef]
39. Dekker, Y.; Le Dévédec, S.E.; Danen, E.H.J.; Liu, Q. Crosstalk between Hypoxia and Extracellular Matrix in the Tumor Microenvironment in Breast Cancer. *Genes* **2022**, *13*, 1585. [CrossRef]
40. Huang, Y.; Lin, D.; Taniguchi, C.M. Hypoxia inducible factor (HIF) in the tumor microenvironment: Friend or foe? *Sci. China Life Sci.* **2017**, *60*, 1114–1124. [CrossRef]
41. Dharmaratne, N.U.; Kaplan, A.R.; Glazer, P.M. Targeting the Hypoxic and Acidic Tumor Microenvironment with pH-Sensitive Peptides. *Cells* **2021**, *10*, 541. [CrossRef]
42. Corn, K.C.; Windham, M.A.; Rafat, M. Lipids in the tumor microenvironment: From cancer progression to treatment. *Prog. Lipid Res.* **2020**, *80*, 101055. [CrossRef]
43. Chen, X.; Song, M.; Zhang, B.; Zhang, Y. Reactive Oxygen Species Regulate T Cell Immune Response in the Tumor Microenvironment. *Oxidative Med. Cell Longev.* **2016**, *2016*, 1580967. [CrossRef]
44. Jafari, M.; Sriram, V.; Premnauth, G.; Merino, E.; Lee, J.-Y. Modified peroxamide-based reactive oxygen species (ROS)-responsive doxorubicin prodrugs. *Bioorganic Chem.* **2022**, *127*, 105990. [CrossRef]
45. Saikolappan, S.; Kumar, B.; Shishodia, G.; Koul, S.; Koul, H.K. Reactive oxygen species and cancer: A complex interaction. *Cancer Lett.* **2019**, *452*, 132–143. [CrossRef] [PubMed]
46. Zhang, L.; Li, J.; Zong, L.; Chen, X.; Chen, K.; Jiang, Z.; Nan, L.; Li, X.; Li, W.; Shan, T.; et al. Reactive Oxygen Species and Targeted Therapy for Pancreatic Cancer. *Oxidative Med. Cell Longev.* **2016**, *2016*, 1616781. [CrossRef] [PubMed]
47. Zhang, J.; Li, H.; Wu, Q.; Chen, Y.; Deng, Y.; Yang, Z.; Zhang, L.; Liu, B. Tumoral NOX_4 recruits M_2 tumor-associated macrophages via ROS/PI3K signaling-dependent various cytokine production to promote NSCLC growth. *Redox Biol.* **2019**, *22*, 101116. [CrossRef] [PubMed]
48. Zhou, H.; Wang, L.; Lin, Z.; Jiang, C.; Chen, X.; Wang, K.; Liu, L.; Shao, L.; Pan, J.; Li, J.; et al. Methylglyoxal from gut microbes boosts radiosensitivity and radioimmunotherapy in rectal cancer by triggering endoplasmic reticulum stress and cGAS-STING activation. *J. ImmunoTher. Cancer* **2023**, *11*, e007840. [CrossRef]
49. Xu, Q.; Zhan, G.; Zhang, Z.; Yong, T.; Yang, X.; Gan, L. Manganese porphyrin-based metal-organic framework for synergistic sonodynamic therapy and ferroptosis in hypoxic tumors. *Theranostics* **2021**, *11*, 1937–1952. [CrossRef]

50. Wen, X.; Wang, C.; Bi, S.; Xu, Y.; Wu, Z.; Huang, H.; Liu, Z.; Zeng, S. Tumor Microenvironment Cascade Activated Biodegradable Nano-Enzymes for Glutathione-Depletion and Ultrasound-Enhanced Chemodynamic Therapy. *Small* **2024**, *20*, e2405457. [CrossRef]
51. Jiang, Y.; Zhao, J.; Zhang, D. Manganese Dioxide-Based Nanomaterials for Medical Applications. *ACS Biomater. Sci. Eng.* **2024**, *10*, 2680–2702. [CrossRef]
52. Huang, Y.; Ruan, Y.; Ma, Y.; Chen, D.; Zhang, T.; Fan, S.; Lin, W.; Huang, Y.; Lu, H.; Xu, J.-F.; et al. Immunomodulatory activity of manganese dioxide nanoparticles: Promising for novel vaccines and immunotherapeutics. *Front. Immunol.* **2023**, *14*, 1128840. [CrossRef] [PubMed]
53. Jia, C.; Guo, Y.; Wu, F. Chemodynamic Therapy via Fenton and Fenton-Like Nanomaterials: Strategies and Recent Advances. *Small* **2022**, *18*, 2103868. [CrossRef]
54. Zhang, Y.; Zhu, J.; Zhang, Z.; He, D.; Zhu, J.; Chen, Y.; Zhang, Y. Remodeling of tumor microenvironment for enhanced tumor chemodynamic/photothermal/chemo-therapy. *J. Nanobiotechnol.* **2022**, *20*, 388.
55. Tu, C.; Lu, H.; Zhou, T.; Zhang, W.; Deng, L.; Cao, W.; Yang, Z.; Wang, Z.; Wu, X.; Ding, J.; et al. Promoting the healing of infected diabetic wound by an anti-bacterial and nano-enzyme-containing hydrogel with inflammation-suppressing, ROS-scavenging, oxygen and nitric oxide-generating properties. *Biomaterials* **2022**, *286*, 121597. [CrossRef]
56. Guo, J.; Zhao, W.; Xiao, X.; Liu, S.; Liu, L.; Zhang, L.; Li, L.; Li, Z.; Li, Z.; Xu, M.; et al. Reprogramming exosomes for immunity-remodeled photodynamic therapy against non-small cell lung cancer. *Bioact. Mater.* **2024**, *39*, 206–223. [CrossRef]
57. Liang, L.; Jia, M.; Zhao, M.; Deng, Y.; Tang, J.; He, X.; Liu, Y.; Yan, K.; Yu, X.; Yang, H.; et al. Progress of Nanomaterials Based on Manganese Dioxide in the Field of Tumor Diagnosis and Therapy. *Int. J. Nanomed.* **2024**, *19*, 8883–8900. [CrossRef]
58. Yue, Z.; Zhao, Q.; Wang, S.; Yao, S.; Wan, X.; Hu, Q.; Wen, K.; Zhao, Y.; Li, L. Manganese Dioxide Coated Piezoelectric Nanosonosensitizer for Cancer Therapy with Tumor Microenvironment Remodeling and Multienzyme-Like Catalysis. *Small Methods* **2024**, *8*, e2400018. [CrossRef]
59. Wu, X.; Cheng, Y.; Zheng, R.; Xu, K.; Yan, J.; Song, P.; Wang, Y.; Rauf, A.; Pan, Y.; Zhang, H. Immunomodulation of Tumor Microenvironment by Arginine-Loaded Iron Oxide Nanoparticles for Gaseous Immunotherapy. *ACS Appl. Mater. Interfaces* **2021**, *13*, 19825–19835. [CrossRef]
60. Chen, X.; Cheng, D.; Ding, M.; Yu, N.; Liu, J.; Li, J.; Lin, L. Tumor-targeting biomimetic sonosensitizer-conjugated iron oxide nanocatalysts for combinational chemodynamic–sonodynamic therapy of colorectal cancer. *J. Mater. Chem. B* **2022**, *10*, 4595–4604. [CrossRef]
61. Zhang, W.; Li, L.; Wu, Y.; Li, C.; Xu, Z.; Zhang, N.; Wang, X.; Zhao, Y.; Zu, T.; He, Q.; et al. Biomimetic Iron-Based Nanoparticles Remodel Immunosuppressive Tumor Microenvironment for Metabolic Immunotherapy. *Int. J. Nanomed.* **2024**, *19*, 9333–9349. [CrossRef]
62. Li, Y.; Chen, J.; Xia, Q.; Shang, J.; He, Y.; Li, Z.; Chen, Y.; Gao, F.; Yu, X.; Yuan, Z.; et al. Photothermal Fe_3O_4 nanoparticles induced immunogenic ferroptosis for synergistic colorectal cancer therapy. *J. Nanobiotechnol.* **2024**, *22*, 630. [CrossRef] [PubMed]
63. Liu, X.; Chen, Q.; Yang, G.; Zhang, L.; Liu, Z.; Cheng, Z.; Zhu, X. Magnetic nanomaterials with near-infrared pH-activatable fluorescence via iron-catalyzed AGET ATRP for tumor acidic microenvironment imaging. *J. Mater. Chem. B* **2015**, *3*, 2786–2800. [CrossRef] [PubMed]
64. Cheng, F.; Wang, S.; Zheng, H.; Yang, S.; Zhou, L.; Liu, K.; Zhang, Q.; Zhang, H. Cu-doped cerium oxide-based nanomedicine for tumor microenvironment-stimulative chemo-chemodynamic therapy with minimal side effects. *Colloids Surf. B Biointerfaces* **2021**, *205*, 111878. [CrossRef] [PubMed]
65. Zeng, L.; Cheng, H.; Dai, Y.; Su, Z.; Wang, C.; Lei, L.; Lin, D.; Li, X.; Chen, H.; Fan, K.; et al. In Vivo Regenerable Cerium Oxide Nanozyme-Loaded pH/H_2O_2-Responsive Nanovesicle for Tumor-Targeted Photothermal and Photodynamic Therapies. *ACS Appl. Mater. Interfaces* **2021**, *13*, 233–244. [CrossRef]
66. Xia, Z.; Liu, N.; Wu, Q.; Chen, Z.; Wang, Y.; Fu, C.; Huang, Z.; Meng, X.; Qiao, B. Cerium oxide nanozymes enhance microwave immunotherapy by reshaping the tumor microenvironment. *Nanoscale* **2025**, *17*, 14614–14623. [CrossRef]
67. Yan, S.; Gao, Z.; Ding, J.; Chen, S.; Wang, Z.; Jin, W.; Qu, B.; Zhang, Y.; Yang, L.; Guo, D.; et al. Nanocomposites based on nanoceria regulate the immune microenvironment for the treatment of polycystic ovary syndrome. *J. Nanobiotechnol.* **2023**, *21*, 412. [CrossRef]
68. Li, X.; Yu, L.; Zhang, C.; Niu, X.; Sun, M.; Yan, Z.; Wang, W.; Yuan, Z. Tumor acid microenvironment-activated self-targeting & splitting gold nanoassembly for tumor chemo-radiotherapy. *Bioact. Mater.* **2021**, *7*, 377–388. [CrossRef]
69. Zhu, Y.; Chen, P.; Hu, B.; Zhong, S.; Yan, K.; Wu, Y.; Li, S.; Yang, Y.; Xu, Z.; Lu, Y.; et al. MDSC-targeting gold nanoparticles enhance PD-1 tumor immunotherapy by inhibiting NLRP3 inflammasomes. *Biomaterials* **2024**, *307*, 122533. [CrossRef]
70. Ding, Y.; Sun, Z.; Tong, Z.; Zhang, S.; Min, J.; Xu, Q.; Zhou, L.; Mao, Z.; Xia, H.; Wang, W. Tumor microenvironment-responsive multifunctional peptide coated ultrasmall gold nanoparticles and their application in cancer radiotherapy. *Theranostics* **2020**, *10*, 5195–5208. [CrossRef]

71. Jiang, W.; Wei, L.; Chen, B.; Luo, X.; Xu, P.; Cai, J.; Hu, Y. Platinum prodrug nanoparticles inhibiting tumor recurrence and metastasis by concurrent chemoradiotherapy. *J. Nanobiotechnol.* **2022**, *20*, 129. [CrossRef]
72. Chen, Q.; Liu, L.; Lu, Y.; Chen, X.; Zhang, Y.; Zhou, W.; Guo, Q.; Li, C.; Zhang, Y.; Zhang, Y.; et al. Tumor Microenvironment-Triggered Aggregated Magnetic Nanoparticles for Reinforced Image-Guided Immunogenic Chemotherapy. *Adv. Sci.* **2019**, *6*, 1802134. [CrossRef] [PubMed]
73. Wang, D.; Ma, W.; Zhang, Y.; Wang, Y.; Sun, L.; Jiang, J.; Jiao, L.; Li, R.; Zhang, Y.; Zhang, M.; et al. A versatile nanoplatform carrying cascade Pt nanozymes remodeling tumor microenvironment for amplified sonodynamic/chemo therapy of thyroid cancer. *Biomaterials* **2025**, *313*, 122778. [CrossRef] [PubMed]
74. Xu, S.; Luo, W.; Zhu, M.; Zhao, L.; Gao, L.; Liang, H.; Zhang, Z.; Yang, F. Human Serum Albumin-Platinum(II) Agent Nanoparticles Inhibit Tumor Growth Through Multimodal Action Against the Tumor Microenvironment. *Mol. Pharm.* **2023**, *21*, 346–357. [CrossRef]
75. Cai, C.; Zhang, Q.; Ye, J.; Yao, S.; Li, Q.; Fan, Z.; Ge, S.; Wang, Y.; Xu, D.; Zhou, J.; et al. Tumor microenvironment modulation innovates combinative cancer therapy via a versatile graphene oxide nanosystem. *Biomater. Sci.* **2025**, *13*, 3123–3148. [CrossRef]
76. Lin, B.; Chen, H.; Liang, D.; Lin, W.; Qi, X.; Liu, H.; Deng, X. Acidic pH and High-H_2O_2 Dual Tumor Microenvironment-Responsive Nanocatalytic Graphene Oxide for Cancer Selective Therapy and Recognition. *ACS Appl. Mater. Interfaces* **2019**, *11*, 11157–11166. [CrossRef]
77. Tao, Y.; Zhu, L.; Zhao, Y.; Yi, X.; Zhu, L.; Ge, F.; Mou, X.; Chen, L.; Sun, L.; Yang, K. Nano-graphene oxide-manganese dioxide nanocomposites for overcoming tumor hypoxia and enhancing cancer radioisotope therapy. *Nanoscale* **2018**, *10*, 5114–5123. [CrossRef]
78. Du, P.; Yan, J.; Long, S.; Xiong, H.; Wen, N.; Cai, S.; Wang, Y.; Peng, D.; Liu, Z.; Liu, Y. Tumor microenvironment and NIR laser dual-responsive release of berberine 9-*O*-pyrazole alkyl derivative loaded in graphene oxide nanosheets for chemo-photothermal synergetic cancer therapy. *J. Mater. Chem. B* **2020**, *8*, 4046–4055. [CrossRef]
79. Yao, L.; Zhao, M.-M.; Luo, Q.-W.; Zhang, Y.-C.; Liu, T.-T.; Yang, Z.; Liao, M.; Tu, P.; Zeng, K.-W. Carbon Quantum Dots-Based Nanozyme from Coffee Induces Cancer Cell Ferroptosis to Activate Antitumor Immunity. *ACS Nano* **2022**, *16*, 9228–9239. [CrossRef]
80. Bao, Y.; Li, G.; Li, S.; Zhang, H.; Wu, X.; Yan, R.; Wang, Z.; Guo, C.; Jin, Y. Multifunctional Tumor-Targeting Carbon Dots for Tumor Microenvironment Activated Ferroptosis and Immunotherapy in Cancer Treatment. *ACS Appl. Mater. Interfaces* **2023**, *15*, 56834–56845. [CrossRef]
81. Guo, L.; Ding, J.; Zhou, W. Converting bacteria into autologous tumor vaccine via surface biomineralization of calcium carbonate for enhanced immunotherapy. *Acta Pharm. Sin. B* **2023**, *13*, 5074–5090. [CrossRef]
82. Meng, X.; Liu, Z.; Deng, L.; Yang, Y.; Zhu, Y.; Sun, X.; Hao, Y.; He, Y.; Fu, J. Hydrogen Therapy Reverses Cancer-Associated Fibroblasts Phenotypes and Remodels Stromal Microenvironment to Stimulate Systematic Anti-Tumor Immunity. *Adv. Sci.* **2024**, *11*, e2401269. [CrossRef]
83. Gharehbaba, A.M.; Soltanmohammadi, F.; Eskandani, M.; Adibkia, K. From barriers to breakthroughs: Mesoporous silica nanoparticles in targeting the tumor microenvironment. *Int. J. Pharm.* **2025**, *682*, 125979. [CrossRef] [PubMed]
84. Liu, Y.; Zhang, Z.; Du, H.; Chen, X.; Hu, N.; Yu, T.; Hou, M.; Yu, X. An enzyme-responsive hydrogel functionalized with mesoporous silica nanoparticles for co-delivery of cisplatin and shRNA to overcome chemotherapy resistance in non-small cell lung cancer. *RSC Adv.* **2025**, *15*, 23966–23977. [CrossRef] [PubMed]
85. Yan, X.; Dong, H.; Gao, L.; Liu, M.; Wang, C. Mechanism of selenium-doped black phosphorus nanosheets wrapped with biomimetic tumor cell membrane for prostate cancer immunotherapy. *Mater. Sci. Eng. C* **2025**, *176*, 214339. [CrossRef] [PubMed]
86. Gao, Y.; Mu, M.; Wei, Y.; Yan, B.; Liu, H.; Guo, K.; Zhang, M.; Dai, X.; Sun, X.; Leong, D.T. Novel ultrathin ferrous sulfide nanosheets: Towards replacing black phosphorus in anticancer nanotheranostics. *Bioact. Mater.* **2025**, *43*, 564–578. [CrossRef]
87. Rahimkhoei, V.; Akbari, A.; Jassim, A.Y.; Hussein, U.A.-R.; Salavati-Niasari, M. Recent advances in targeting cancer stem cells by using nanomaterials. *Int. J. Pharm.* **2025**, *673*, 125381. [CrossRef]
88. Saini, S.; Kumar, K.; Saini, P.; Sethi, M.; Meena, P.; Dandia, A.; Weigand, W.; Parewa, V. Elucidating the Synergistic Promotional Mechanism of Water and Oxygen in the Aerobic C–C Homocoupling Reaction Catalyzed by Visible-Light-Derived Core–Shell Pd@A-CQDs Nanostructures. *ACS Appl. Mater. Interfaces* **2024**, *16*, 46200–46215. [CrossRef]
89. Mali, A.; Nayak, N.U.; van Doesburg, J.; Fokkink, R.; van Riessen, K.; de Kruijf, R.; Srinivas, M. Polymeric (Poly(lactic-co-glycolic acid)) Particles Entrapping Perfluorocarbons Are Stable for a Minimum of Six Years. *ACS Omega* **2025**, *10*, 6768–6779. [CrossRef]
90. Krafft, M.; Riess, J. Corrigendum to Therapeutic oxygen delivery by perfluorocarbon-based colloids. *Adv. Colloid Interface Sci.* **2021**, *295*, 102505. [CrossRef]
91. Srinivasan, M.K.; Aruchamy, M.; Namasivayam, N.; Muthupandian, S. Protective effect of pH-sensitive carvacrol zinc oxide quantum dots on Nrf-2, MAPK and NF-κB pathways and protein-bound carbohydrates in DMBA-induced mammary carcinogenesis. *Pathol. Res. Pract.* **2025**, *272*, 156117. [CrossRef]
92. Aydın, B.; Bozoğlu, S.; Karatepe, N.; Güner, F.S. Folic acid-conjugated magnetic carbon nanotube nanocarriers for targeted delivery of mitoxantrone. *Nanotechnology* **2025**, *36*, 305701. [CrossRef] [PubMed]

93. Zhu, M.; Zhou, Z.; Gu, Y.; Bing, L.; Xie, Y.; Shen, Z.; Chen, W. Powerful Tribocatalytic Degradation of Methyl Orange Solutions with Concentrations as High as 100 mg/L by $BaTiO_3$ Nanoparticles. *Nanomaterials* **2025**, *15*, 1135. [CrossRef] [PubMed]
94. Li, X.; Li, Q.; Chen, X.; Yan, B.; Li, S.; Deng, H.; Lu, H. Efficacy, Kinetics, and Mechanism of Tetracycline Degradation in Water by O_3/PMS/FeMoBC Process. *Nanomaterials* **2025**, *15*, 1108. [CrossRef] [PubMed]
95. Kalisz, A.; Kornaś, A.; Gil, J.; Sękara, A.; Húska, D.; Rudolphi-Szydło, E.; Sieprawska, A.; Gawrońska, K.; Kulak, M.; Fotopoulos, V.; et al. Exogenously applied putrescine and chitosan–putrescine nanocomposite alleviate the negative effects of chilling stress on iceberg lettuce seedlings. *Sci. Rep.* **2025**, *15*, 26963. [CrossRef]
96. Madbouly, N.A.; Ali, D.M.; Farid, A.A. Nanoparticles from grape seed extract inhibit inflammatory cytokines and ameliorate CCl4-induced hepatotoxicity. *BMC Complement. Med. Ther.* **2025**, *25*, 276. [CrossRef]
97. Chauhan, A.; Saini, A.; Sharma, D. The evolution of integrated magnetic hyperthermia and chemodynamic therapy for combating cancer: A comprehensive viewpoint. *Nanoscale Adv.* **2025**, *7*, 4820–4836. [CrossRef]
98. Li, Y.; Li, Y.; Chen, H.; Chen, Y.; Ni, Y.; Ji, J.; Wang, K.; Jin, Q. Self-assembled copper-amino acid nanoleaves for targeted treatment of deep-seated bacterial infections via chemodynamic therapy and cuproptosis-like death. *Biomaterials* **2025**, *325*, 123566. [CrossRef]
99. Kim, E.; Jang, D.; Kim, M.; Heo, J.; Shin, J.; Chung, J.-Y.; Choi, M.; Yang, W.-H.; Lee, H.; Cha, J.-H. High-throughput screening system for immunogenic cell death inducers using artificial intelligence-based real-time image analysis. *Comput. Biol. Med.* **2025**, *196*, 110727. [CrossRef]
100. Isaeva, A.S.; Yeriomenko, A.D.T.; Idota, E.; Volodina, S.I.; Porozova, N.O.; Bezsonov, E.E.; Malogolovkin, A.S. Shaping viral immunotherapy towards cancer-targeted immunological cell death. *Front. Oncol.* **2025**, *15*, 1540397. [CrossRef]
101. Yang, P.; Zhang, J.; Chang, Y.; Tang, L.; Ma, G.; Liu, X.; Gao, F.; Ma, X.; Guo, Y. Tumor microenvironment responsive smart nanoplatform for synergistic tumor therapy through co-enhancement of GSH depletion and hypoxia relief. *J. Inorg. Biochem.* **2025**, *272*, 113005. [CrossRef] [PubMed]
102. Du, H.; Tong, W.; Peng, B.; Gao, W.; Wang, Y.; Wang, H.; Liu, J.; Liu, Y.; Sun, B.; Ma, J.; et al. Mesoporous Nanoplatform That Efficiently Delivers DOX as an H_2O_2 Generator to Trigger Mutual Amplification of Cuproptosis and Chemodynamic Therapy. *Adv. Healthc. Mater.* **2025**, *14*, e2500933. [CrossRef] [PubMed]
103. Jiang, Y.; Lu, H.; Lei, L.; Scherman, D.; Liu, Y. Tumor microenvironment-responsive MIL-53(Fe)@MnO_2- induced glutathione depletion and sustained hydroxyl radical generation for enhanced chemodynamic cancer therapy. *J. Colloid Interface Sci.* **2025**, *700*, 138342. [CrossRef] [PubMed]
104. Chen, L.; Zhang, H.; Hou, Y.; Yunlong, W.; Feng, X. Immunomodulatory effects of photothermal therapy in breast cancer: Advances and challenges. *Front. Immunol.* **2025**, *16*, 1544693. [CrossRef]
105. Zhang, W.; Yue, W.; Shi, X.; Lv, L.; Fang, Z.; Li, J. A TME-activated photothermal agent with photodegradability for accurate breast tumor photothermal therapy. *RSC Adv.* **2025**, *15*, 25062–25066. [CrossRef]
106. Peng, W.; Zhou, T.; Hu, L.; Vankann, V.; Bohn, T.; Bopp, T.; Kuan, S.L.; Weil, T. Autonomous Activation of a Gated Chemiluminescent Photosensitizer Enables Targeted Photodynamic Therapy in Tumor Cells. *J. Am. Chem. Soc.* **2025**, *147*, 27822–27834. [CrossRef]
107. Yang, Z.; Deng, L.; Pan, Z.; Wang, X.; Lin, L.; Fang, Y.; Huang, Y.; Feng, X.; Chen, X. Oxygen-boosted nanodrug for amplified ferroptosis-photodynamic immunotherapy together with PD-1 checkpoint blockade against triple-negative breast cancer. *Colloids Surf. B Biointerfaces* **2025**, *255*, 114963. [CrossRef]
108. Wang, L.; Xu, D.; Hu, X.; Quan, R.; Lu, D.; Li, Z.; Yu, C.; Li, X.; Ma, S.; Li, X.; et al. Cascade reaction–driven biomimetic scintillant/metal–organic frameworks for X-ray triggered combinational therapy against glioma. *Mater. Today Bio* **2025**, *33*, 102069. [CrossRef]
109. Jiang, Y.; Zhang, W.; Liu, L.; Wu, Y.; Li, W.; Liang, J.; Shen, H.; Fang, S.; Huang, X.; Chu, Z.; et al. Gelatin Methacryloyl Xerogel Puncture Implants Loaded with $Cu_{0.5}Mn_{2.5}O_4$ Nanoparticles Synergizes Cuproptosis and STING Activation for Enhanced Breast Cancer Immunotherapy. *ACS Nano* **2025**, *19*, 27902–27918. [CrossRef]
110. Graván, P.; Peña-Martín, J.; de Andrés, J.L.; Pedrosa, M.; Villegas-Montoya, M.; Galisteo-González, F.; Marchal, J.A.; Sánchez-Moreno, P. Exploring the Impact of Nanoparticle Stealth Coatings in Cancer Models: From PEGylation to Cell Membrane-Coating Nanotechnology. *ACS Appl. Mater. Interfaces* **2024**, *16*, 2058–2074. [CrossRef]
111. Luo, Q.; Shi, W.; Wang, P.; Zhang, Y.; Meng, J.; Zhang, L. Tumor Microenvironment-Responsive Shell/Core Composite Nanoparticles for Enhanced Stability and Antitumor Efficiency Based on a pH-Triggered Charge-Reversal Mechanism. *Pharmaceutics* **2021**, *13*, 895. [CrossRef]
112. Liu, B.; Shen, J.; Wu, Y.; Wong, C.F.; Hei, T.Y.; Li, J.W.-T.; Hummel, S.N.; Jin, G.; Petrucci, N.R.; Wang, D.M.; et al. Novel strategy for tumor immunotherapy using FITC-folate bispecific adapter bridged CAR immune cell cocktails. *Bioact. Mater.* **2025**, *52*, 529–540. [CrossRef] [PubMed]
113. Xiao, D.; Dong, J.; Xie, F.; Feng, X.; Wang, J.; Xu, X.; Tang, B.; Sun, C.; Wang, Y.; Zhong, W.; et al. Polyvalent folate receptor-targeting chimeras for degradation of membrane proteins. *Nat. Chem. Biol.* **2025**, 1–11. [CrossRef] [PubMed]

114. Tian, Q.; Wang, M.; Mubeen, F.; Sundas, A.; Naz, R.; Hu, X.; Chen, X. CDKN2A and matrix metalloproteinases: Key regulators of cellular senescence in squamous cell carcinoma. *Am. J. Transl. Res.* **2025**, *17*, 4573–4589. [CrossRef]
115. Chen, Y.; Yan, Y.; Wei, W. Research advances of matrix metalloproteinases family in uveal melanoma. *Eur. J. Med. Res.* **2025**, *30*, 609. [CrossRef]
116. Sun, M.; Ni, C.; Li, A.; Liu, J.; Guo, H.; Xu, F.; Li, K.; Cao, X.; Shi, X.; Guo, R. A biomimetic nanoplatform mediates hypoxia-adenosine axis disruption and PD-L1 knockout for enhanced MRI-guided chemodynamic-immunotherapy. *Acta Biomater.* **2025**, *201*, 618–632. [CrossRef]
117. Tang, Z.; Luo, W.; Xu, M.; Liu, Y.; Yu, Q.; Wang, L. A TME-responsive oxygen-self-supplying hybridized polymersome for synergistic triple-modal therapy and precision theranostics in hypoxic tumors. *J. Mater. Chem. B* **2025**, *13*, 8136–8148. [CrossRef]
118. Zhou, M.; Wang, Y.; Cui, H.; Geng, H.; Ruan, L.; Zhao, Y.; Zhou, C.; Dai, W.; Chen, J.; Yu, J.; et al. EGCG and DOX dual-drug-loaded enzyme-responsive nanovesicles boost mitochondrial-mediated ICD for improved immunotherapy. *Front. Pharmacol.* **2025**, *16*, 1624109. [CrossRef]

MDPI AG
Grosspeteranlage 5
4052 Basel
Switzerland
Tel.: +41 61 683 77 34

Inorganics Editorial Office
E-mail: inorganics@mdpi.com
www.mdpi.com/journal/inorganics